T0296483

LONDON MATHEMATICAL SOCIETY STUDENT TEXTS

Managing Editor: Ian J. Leary,
Mathematical Sciences, University of Southampton, UK

London Mathematical Society Student Texts 103

A Course in Stochastic Game Theory

EILON SOLAN

Tel-Aviv University

CAMBRIDGE
UNIVERSITY PRESS

CAMBRIDGE
UNIVERSITY PRESS

University Printing House, Cambridge CB2 8BS, United Kingdom

One Liberty Plaza, 20th Floor, New York, NY 10006, USA

477 Williamstown Road, Port Melbourne, VIC 3207, Australia

314–321, 3rd Floor, Plot 3, Splendor Forum, Jasola District Centre,
New Delhi – 110025, India

103 Penang Road, #05–06/07, Visioncrest Commercial, Singapore 238467

Cambridge University Press is part of the University of Cambridge.

It furthers the University's mission by disseminating knowledge in the pursuit of
education, learning, and research at the highest international levels of excellence.

www.cambridge.org
Information on this title: www.cambridge.org/9781316516331
DOI: 10.1017/9781009029704

© Eilon Solan 2022

First published 2022

A catalogue record for this publication is available from the British Library.

Library of Congress Cataloging-in-Publication Data
Names: Solan, Eilon, author.
Title: A course in stochastic game theory / Eilon Solan,
School of Mathematical Sciences.
Description: First edition. | New York : Cambridge University Press, 2022.|
Series: LMST london mathematical society student texts |
Includes bibliographical references and index.
Identifiers: LCCN 2021055382 (print) | LCCN 2021055383 (ebook) |
ISBN 9781316516331 (hardback) | ISBN 9781009014793 (paperback) |
ISBN 9781009029704 (epub)
Subjects: LCSH: Game theory. | BISAC: MATHEMATICS / General
Classification: LCC QA269 .S65 2022 (print) | LCC QA269 (ebook) |
DDC 519.3–dc23/eng20220207
LC record available at https://lccn.loc.gov/2021055382
LC ebook record available at https://lccn.loc.gov/2021055383

ISBN 978-1-316-51633-1 Hardback
ISBN 978-1-009-01479-3 Paperback

The book is dedicated to the people who shared my life, my work, and my love of games in the last thirty years. To Abraham Neyman, who introduced me to game theory and to stochastic games; to Nicolas Vieille and Dinah Rosenberg, with whom I spent countless fun weeks of studying stochastic games; to Ehud Lehrer, who has been my colleague and partner at Tel Aviv University for the past 20 years; to my parents, Chaim and Zafrira; and to my two sons, Omri and Ron, who listened to game-theoretic problems since birth and eventually became coauthors.

Contents

Introduction

Stochastic games are a mathematical model that is used to study dynamic interactions among agents who influence the evolution of the environment. These games were first presented and studied by Lloyd Shapley (1953).[1,2] Since Shapley's seminal work, the literature on stochastic games expanded considerably, and the model was applied to numerous areas, such as arms race, fishery wars, and taxation.

A stochastic game is played in discrete time by a finite set I of players, and it consists of a finite number of states. In each state s, each player $i \in I$ has a given set of actions, denoted $A^i(s)$. In every stage $t \in \mathbb{N}$, the play is in one of the states, denoted s_t. Each player $i \in I$ chooses an action $a^i_t \in A^i(s_t)$ that is available to her at the current stage, receives a stage payoff, which depends on the current state s_t as well as on the actions $(a^j_t)_{j \in I}$ chosen by the players, and a new state s_{t+1} is chosen, according to a probability distribution that depends on the current state and on the actions of the players $(a^j_t)_{j \in I}$.

In a stochastic game, the players have two, seemingly contradicting, goals. First, they need to ensure that their future opportunities remain high. At the same time, they should make sure that their stage payoff is also high. This dichotomy makes the analysis of stochastic games intriguing and not trivial.

The study of stochastic games uses tools from many mathematical branches, such as probability, analysis, algebra, differential equations, and combinatorics. The goal of this book is to present the theory through the mathematical techniques that it employs. Thus, each chapter presents mathematical results

---

[1] Lloyd Stowell Shapley (Cambridge, Massachusetts, June 2, 1923 – Tucson, Arizona, March 12, 2016) was an American mathematician who made many influential contributions to Game Theory, like the Shapley value, stochastic games, and the defer-acceptance algorithm for stable marriages. Shapley shared the 2012 Nobel Prize in Economics together with game theorist Alvin Roth.

[2] All commentary is taken from Wikipedia.

from some branch of mathematics, and uses them to prove results on stochastic games. The goal is not to prove the most general theorems in stochastic games, but rather to present the beauty of the theory. Accordingly, we sometimes restrict the scope of the results that is proven, to allow for simpler proofs that bypass technical difficulties.

The material in this book is summarized by the following table:

Chapter	Tool + Result
1	Contracting mappings Stationary optimal strategies in Markov decision problems
2	Tauberian Theorem Uniform ϵ-optimality in hidden Markov decision problems
5	Contracting mappings Stationary discounted optimal strategies in zero-sum stochastic games
6	Semi-algebraic mappings Existence of the limit of the discounted value
7	B-graphs Continuity of the limit of the discounted value
8	Kakutani's fixed point theorem Stationary discounted equilibria in multiplayer stochastic games
9	Existence of the uniform value in zero-sum stochastic games
10	The vanishing discount factor approach Existence of uniform equilibrium in absorbing games
11	Ramsey's Theorem Existence of undiscounted equilibrium in two-player deterministic stopping games
12	Approximating infinite orbits Existence of undiscounted equilibrium in multiplayer quitting games
13	Linear complementarity problems Existence of undiscounted equilibrium in multiplayer quitting games

Each chapter contains exercises. Solutions are available as supplementary material on the book's page on the publisher's website. The book is based on a graduate level course that I taught at Tel Aviv University for more than a decade. I hope that the readers, as my students, will like the diversity of the topics and the elegance of the proofs. For the benefit of readers who would like to expand their knowledge in stochastic games, I added references to related results at the end of each chapter. Books and surveys that include material on different aspects of stochastic games include Raghavan et al. (1991), Raghavan and Filar (1991), Filar and Vrieze (1997), Başar and Olsder (1998), Mertens (2002), Vieille (2002), Neyman and Sorin (2003), Solan (2008), Chatterjee et al. (2009, 2013), Chatterjee and Henzinger (2012), Laraki and Sorin (2015), Mertens et al. (2015), Solan and Vieille (2015), Solan and Ziliotto (2016), Başar and Zaccour (2017), Jaśkiewicz and Nowak (2018a,b), and Renault (2019).

I end the introduction by thanking Ayala Mashiah-Yaakovi, who read the manuscript and the solution manual and made many comments that improved the text; Andrei Iacob, who copyedited the text; and John Yehuda Levy, Andrzej Nowak, Robert Simon, Bernhard von Stengel, Uri Zwick, and my students throughout the years for providing comments and spotting typos.

Notation

The set of positive integers is

$$\mathbb{N} := \{1, 2, 3, \ldots\}.$$

The number of elements in a finite set K is denoted by $|K|$. For every finite set K, the set of probability distributions over K is denoted by $\Delta(K)$. We identify each element $k \in K$ with the probability distribution in $\Delta(K)$ that assigns probability 1 to k. For a probability distribution $\mu \in \Delta(K)$, the *support* of μ, denoted supp(μ), is the set of all elements $k \in K$ that have positive probability under μ:

$$\text{supp}(\mu) := \{k \in K : \mu[k] > 0\}.$$

A probability distribution is *pure* if supp(μ) contains only one element: $|\text{supp}(\mu)| = 1$.

Let I be a finite set, and, for each $i \in I$, let A^i be a set. We denote by $A^I := \prod_{i \in I} A^i$ the cartesian product, and denote $A^{-i} := \prod_{j \in I \setminus \{i\}} A^j$. Similarly, if $a = (a^i)_{i \in I} \in A^I$, we denote by $a^{-i} := (a^j)_{j \in I \setminus \{i\}} \in A^{-i}$ the vector a with its i'th coordinate removed.

We will use two norms, the L_1-*norm* and the L_∞-*norm* (or the maximum norm). For a vector $x \in \mathbb{R}^n$, we define

$$\|x\|_1 := \sum_{i=1}^{n} |x_i|,$$

and

$$\|x\|_\infty := \max_{i=1,\ldots,n} |x_i|.$$

For a function $f: X \to \mathbb{R}$, $\text{argmax}_{x \in X} f(x)$ is the set of all points in X that maximize f:

$$\text{argmax}_{x \in X} f(x) := \left\{ y \in X : f(y) = \max_{x \in X} f(x) \right\}.$$

When the set X is compact and the function f is continuous, the set $\text{argmax}_{x \in X} f(x)$ is non-empty.

1

Markov Decision Problems

In this chapter, we introduce Markov decision problems, which are stochastic games with a single player. They serve as an appetizer. On the one hand, the basic concepts and basic proofs for zero-sum stochastic games are better understood in this simple model. On the other hand, some of the conclusions that we draw for Markov decision problems are different from those drawn for zero-sum stochastic games. This illustrates the inherent difference between single-player decision problems and multiplayer decision problems (=games). The interested reader is referred to, for example, Ross (1982) or Puterman (1994) for an exposition of Markov decision problems.

We will study both the T-stage evaluation and the discounted evaluation. We will introduce and study contracting mappings,[1] and will use such mappings to show that the decision maker has a stationary discounted optimal strategy. We will also define the concept of uniform optimality, and show that the decision maker has a stationary uniformly optimal strategy.

Definition 1.1 A *Markov decision problem*[2] is a vector $\Gamma = \langle S, (A(s))_{s \in S}, q, r \rangle$ where

- S is a finite set of states.
- For each $s \in S$, $A(s)$ is a finite set of actions available at state s. The set of pairs (state, action) is denoted by

$$SA := \{(s,a) \colon s \in S, a \in A(s)\}.$$

- $q \colon SA \to \Delta(S)$ is a transition rule.
- $r \colon SA \to \mathbf{R}$ is a payoff function.

---

[1] We adhere to the convention that a mapping is a function whose range is a general space or $\mathbb{R}^n$, while a function is always real-valued.

[2] Andrey Andreyevich Markov (Ryazan, Russia, June 14, 1856 – St. Petersburg, Russia, July 20, 1922) was a Russian mathematician. He is best known for his work on the theory of stochastic processes that now bear his name: Markov chains and Markov processes.

A Markov decision problem involves a decision maker, and it evolves as follows. The problem lasts for infinitely many stages. The initial state $s_1 \in S$ is given. At each stage $t \geq 1$, the following happens:

- The current state s_t is announced to the decision maker.
- The decision maker chooses an action $a_t \in A(s_t)$ and receives the stage payoff $r(s_t, a_t)$.
- A new state s_{t+1} is drawn according to $q(\cdot \mid s_t, a_t)$, and the game proceeds to stage $t + 1$.

Example 1.2   Consider the following situation. The technological level of a country can be High (H), Medium (M), or Low (L). The annual investment of the country in technological advances can also be high (2 billion dollars), medium (1 billion dollars), or low (0.5 billion dollars). The annual gain from technological level is increasing: the high, medium, and low technological level yield 10, 6, and 2 billion dollars, respectively. The technological level changes stochastically as a function of the investment in technological advancement, according to the following table:[3]

Technology level	High investment	Medium investment	Low investment
H	H	$\left[\frac{1}{2}(H), \frac{1}{2}(M)\right]$	$\left[\frac{1}{4}(H), \frac{3}{4}(M)\right]$
M	$\left[\frac{3}{5}(H), \frac{2}{5}(M)\right]$	M	$\left[\frac{2}{5}(M), \frac{3}{5}(L)\right]$
L	$\left[\frac{3}{5}(M), \frac{2}{5}(L)\right]$	$\left[\frac{2}{5}(M), \frac{3}{5}(L)\right]$	L

The situation can be presented as a Markov decision problem as follows:

- There are three states, which represent the three technological levels: $S = \{H, M, L\}$.
- There are three actions in each state, which represent the three investment levels: $A(s) = \{h, m, l\}$ for each $s \in S$.
- The transition rule is given by

---

[3] Here and in the sequel, a probability distribution is denoted by a list of probabilities and outcomes in square brackets, where the outcomes are written within round brackets. Thus, $\left[\frac{2}{3}(H), \frac{1}{3}(M)\right]$ means a probability distribution that assigns probability $\frac{2}{3}$ to H and probability $\frac{1}{3}$ to M.

$$q(H \mid H,h) = 1, \qquad q(M \mid H,h) = 0, \qquad q(L \mid H,h) = 0,$$

$$q(H \mid H,m) = \tfrac{1}{2}, \qquad q(M \mid H,m) = \tfrac{1}{2}, \qquad q(L \mid H,m) = 0,$$

$$q(H \mid H,l) = \tfrac{1}{4}, \qquad q(M \mid H,l) = \tfrac{3}{4}, \qquad q(L \mid H,l) = 0,$$

$$q(H \mid M,h) = \tfrac{3}{5}, \qquad q(M \mid M,h) = \tfrac{2}{5}, \qquad q(L \mid M,h) = 0,$$

$$q(H \mid M,m) = 0, \qquad q(M \mid M,m) = 1, \qquad q(L \mid M,m) = 0,$$

$$q(H \mid M,l) = 0, \qquad q(M \mid M,l) = \tfrac{2}{5}, \qquad q(L \mid M,l) = \tfrac{3}{5},$$

$$q(H \mid L,h) = 0, \qquad q(M \mid L,h) = \tfrac{3}{5}, \qquad q(L \mid L,h) = \tfrac{2}{5},$$

$$q(H \mid L,m) = 0, \qquad q(M \mid L,m) = \tfrac{2}{5}, \qquad q(L \mid L,m) = \tfrac{3}{5},$$

$$q(H \mid L,l) = 0, \qquad q(M \mid L,l) = 0, \qquad q(L \mid L,l) = 1.$$

- The payoff function (in billions of dollars) is given by

$$r(H,h) = 8, \qquad r(H,m) = 9, \qquad r(H,l) = 9\tfrac{1}{2},$$

$$r(M,h) = 4, \qquad r(M,m) = 5, \qquad r(M,l) = 5\tfrac{1}{2},$$

$$r(L,h) = 0, \qquad r(L,m) = 1, \qquad r(L,l) = 1\tfrac{1}{2}. \qquad \blacklozenge$$

Example 1.3 The Markov decision problem that is illustrated in Figure 1.1 is formally defined as follows:

- There are three states: $S = \{s(1), s(2), s(3)\}$.
- In state $s(1)$, there are two actions: $A(s(1)) = \{U, D\}$; in states $s(2)$ and $s(3)$, there is one action: $A(s(2)) = A(s(3)) = \{D\}$.
- Payoffs appear at the center of each entry and are given by:

$$r(s(1), U) = 10; \quad r(s(1), D) = 5; \quad r(s(2), D) = 10; \quad r(s(3), D) = -100.$$

- Transitions appear in parentheses next to the payoff and are given by:
  - If in state $s(1)$ the decision maker chooses U, the process moves to state $s(2)$, that is, $q(s(2) \mid s(1), U) = 1$.
  - If in state $s(1)$ the decision maker chooses D, the process remains in state $s(1)$, that is, $q(s(1) \mid s(1), D) = 1$.

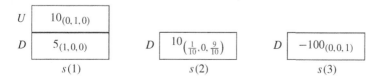

Figure 1.1 The Markov decision problem in Example 1.3.

- From state $s(2)$, the process moves to state $s(1)$ with probability $\frac{1}{10}$ and to state $s(3)$ with probability $\frac{9}{10}$, that is, $q(s(1) \mid s(2), D) = \frac{1}{10}$ and $q(s(3) \mid s(2), D) = \frac{9}{10}$.
- Once the process reaches state $s(3)$, it stays there, that is, $q(s(3) \mid s(3), D) = 1$.   ◆

1.1 On Histories

For $t \in \mathbb{N}$, the set of *histories of length t* is defined by

$$H_t := (SA)^{t-1} \times S,$$

where by convention $(SA)^0 = \emptyset$. This is the set of all histories that may occur until stage t. A typical element in H_t is denoted by h_t. The last state of history h_t is denoted by s_t. The set H_1 is identified with the state space S, and the history (s_1) is simply denoted by s_1.

We denote the set of all *histories* by

$$H := \bigcup_{t \in \mathbb{N}} H_t,$$

and the set of all *infinite histories* or *plays* by

$$H_\infty := (SA)^{\mathbb{N}}.$$

The set of plays H_∞ is a measurable space, with the sigma-algebra generated by the cylinder sets, which are defined as follows. For a history $\widetilde{h}_t = (\widetilde{s}_1, \widetilde{a}_1, \dots, \widetilde{s}_t) \in H_t$, the *cylinder set* $C(\widetilde{h}_t) \subset H_\infty$ is the collection of all plays that start with $\widetilde{h}_t$, that is,

$$C(\widetilde{h}_t) := \{h = (s_1, a_1, s_2, a_2, \dots) \in H_\infty : s_1 = \widetilde{s}_1, a_1 = \widetilde{a}_1, \dots, s_t = \widetilde{s}_t\}.$$

For every $t \in \mathbb{N}$, the collection of all cylinder sets $(C(\widetilde{h}_t))_{\widetilde{h}_t \in H_t}$ defines a finite partition, or an algebra, on H_∞. We denote by $\mathcal{H}_t$ this algebra and by $\mathcal{H}$ the sigma-algebra on H_∞ generated by the algebras $(\mathcal{H}_t)_{t \in \mathbb{N}}$.

1.2 On Strategies

A *mixed action* at state s is a probability distribution over the set of actions $A(s)$ available at state s. The set of mixed actions at state s is therefore $\Delta(A(s))$. A strategy of the decision maker specifies how the decision maker should play after each possible history.

Definition 1.4   A *strategy* is a mapping σ that assigns to each history $h = (s_1, a_1, \dots, a_{t-1}, s_t)$ a mixed action in $\Delta(A(s_t))$.

The set of all strategies is denoted by Σ.

A decision maker who follows a strategy σ behaves as follows: at each stage t, given the past history $(s_1, a_1, \ldots, s_t)$, the decision maker chooses an action a_t according to the mixed action $\sigma(\cdot \mid s_1, a_1, \ldots, s_t)$.

Comment 1.5   A strategy as defined in Definition 1.4 is termed in the literature *behavior strategy*.

Comment 1.6   The fact that the choice of the decision maker depends on past play implicitly assumes that the decision maker knows the past play; that is, the decision maker observes (and remembers) all past states that the process visited, and she remembers all her past choices. In Chapter 2, we will study the model of Markov decision problems when the decision maker does not observe the state.

Comment 1.7   A strategy contains a lot of irrelevant information. Indeed, when the initial state is $s_1 = s$, it is not important what the decision maker would play if the initial state were $s' \neq s$. Similarly, if in the first stage the decision maker played the action $a_1 = a$, it is irrelevant what she would play in the second stage if she played the action $a' \neq a$ in the first stage. We nevertheless regard a strategy as a mapping defined on the set of *all* histories, because of the simplicity of the definition; otherwise we would have to define for every strategy σ and every positive integer t the set of all histories of length t that can occur with positive probability when the decision maker follows strategy σ (which depend on the definition of σ up to stage $t - 1$), and define σ at stage t only for those histories.

Every strategy σ, together with the initial state s_1, defines a probability distribution $\mathbf{P}_{s_1, \sigma}$ on the space of measurable space $(H_\infty, \mathcal{H})$. To define this probability distribution formally, we define it on the collection of cylinder sets that generate $(H_\infty, \mathcal{H})$ by the rule

$$\mathbf{P}_{s_1, \sigma}(C(\widetilde{s}_1, \widetilde{a}_1, \ldots, \widetilde{s}_{t-1}, \widetilde{a}_{t-1}, \widetilde{s}_t)) \tag{1.1}$$

$$:= \mathbf{1}_{\{s_1 = \widetilde{s}_1\}} \cdot \prod_{k=1}^{t-1} \sigma(\widetilde{a}_k \mid \widetilde{s}_1, \widetilde{a}_1, \ldots, \widetilde{s}_1) \cdot \prod_{k=1}^{t-1} q(\widetilde{s}_{k+1} \mid \widetilde{s}_k, \widetilde{a}_k).$$

Let $\mathbf{P}_{s_1, \sigma}$ be the unique probability distribution on H_∞ that agrees with this definition on cylinder sets. The fact that, in this way, we indeed obtain a unique probability distribution is guaranteed by the Carathéodory[4] Extension Theorem (see, e.g., theorem 3.1 in Billingsley (1995)).

---

[4] Constantin Carathéodory (Berlin, Germany, September 13, 1873 – Munich, Germany, February 2, 1950) was a Greek mathematician who spent most of his career in Germany. He made significant contributions to the theory of functions of a real variable, the calculus of variations, and measure theory. His work also includes important results in conformal representations and in the theory of boundary correspondence.

Two simple classes of strategies are pure strategies that involve no random-ization, and stationary strategies that depend only on the current state and not on the whole past history.

Definition 1.8 A strategy σ is *pure* if $|\text{supp}(\sigma(h_t))| = 1$ for every history $h_t \in H$.

The set of pure strategies is denoted by Σ_P.

Definition 1.9 A strategy σ is *stationary* if, for every two histories $h_t = (s_1, a_1, s_2, \ldots, a_{t-1}, s_t)$ and $\widehat{h}_k = (\widehat{s}_1, \widehat{a}_1, \widehat{s}_2, \ldots, \widehat{a}_{k-1}, \widehat{s}_k)$ that satisfy $s_t = \widehat{s}_k$, we have $\sigma(h_t) = \sigma(\widehat{h}_k)$.

The set of stationary strategies is denoted by Σ_S.

A pure stationary strategy assigns to each state $s \in S$ an action in $A(s)$. Since the number of actions in $A(s)$ is $|A(s)|$, we can express the number of pure stationary strategies in terms of the data of the Markov decision problem.

Theorem 1.10 *The number of pure stationary strategies is* $\prod_{s \in S} |A(s)|$.

One can identify a stationary strategy σ with a vector $x \in \prod_{s \in S} \Delta(A(s))$. With this identification, $x(s)$ is the mixed action chosen when the current state is s. Thus, the set of stationary strategies Σ_S can be identified with the space $X := \prod_{s \in S} \Delta(A(s))$, which is convex and compact. For every element $x \in X$, the stationary strategy that corresponds to x is still denoted by x.

In Definition 1.4 we defined a strategy to be a mapping from histories to mixed actions. We now present another concept of a strategy that involves randomization – a mixed strategy.

Definition 1.11 A *mixed strategy* is a probability distribution over the set Σ_P of pure strategies.

Every strategy is equivalent to a mixed strategy. Indeed, a strategy σ is defined by $\aleph_0$ lotteries: to each history $h_t \in H$, it assigns a lottery $\sigma(h_t) \in \Delta(A(s_t))$. If the decision maker performs all the $\aleph_0$ lotteries before the play starts, then the realizations of the lotteries define a pure strategy. In particular, the strategy defines a probability distribution over the set of pure strategies.

Conversely, every mixed strategy is equivalent to a strategy. Indeed, given a mixed strategy τ, one can calculate for each history h_t the conditional probability $\sigma(a_t \mid h_t)$ that the action chosen after h_t is $a_t \in A(s_t)$. If the history h_t occurs with probability 0 under $\mathbf{P}_{s_1, \sigma}$, we set $\sigma(a_t \mid h_t)$ arbitrarily. One can show that the strategy σ is equivalent to the mixed strategy τ.

The equivalence just described is a special case of a more general result called *Kuhn's Theorem*;[5] see, for example, Maschler, Solan, and Zamir (2020, chapter 7).

1.3 The *T*-Stage Payoff

The decision maker receives the stage payoff $r(s_t, a_t)$ at every stage t. How does she compare sequences of stage payoffs? We will study two methods of evaluations. The first, which we consider in this section, is the T-stage evaluation. This evaluation is relevant when the process lasts T stages, and the goal of the decision maker is to maximize her expected average payoff during these stages. The second, which we will study in the next section, is the discounted evaluation, which is relevant when the play continues indefinitely, and the goal of the decision maker is to maximize the expected discounted sum of her stage payoffs.

The expectation operator for the probability distribution $\mathbf{P}_{s_1,\sigma}$ is denoted by $\mathbf{E}_{s_1,\sigma}[\cdot]$. In particular, $\mathbf{E}_{s_1,\sigma}[r(s_t,a_t)]$ is the expected payoff at stage t.

Definition 1.12 For every positive integer $T \in \mathbb{N}$, every initial state $s_1 \in S$, and every strategy $\sigma \in \Sigma$, define the *T-stage payoff* by:

$$\gamma_T(s_1;\sigma) := \mathbf{E}_{s_1,\sigma}\left[\frac{1}{T}\sum_{t=1}^{T} r(s_t,a_t)\right]. \qquad (1.2)$$

Example 1.13 The Markov decision problem in this example is given in Figure 1.2.

The initial state is $s(1)$. We will calculate the T-stage payoff of every pure strategy.

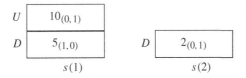

Figure 1.2 The Markov decision problem in Example 1.13.

[5] Harold William Kuhn (Santa Monica, California, July 29, 1925 – New York City, New York, July 2, 2014) was an American mathematician. He is known for the Karush–Kuhn–Tucker conditions, for Kuhn's theorem, and for developing Kuhn poker as well as the description of the Hungarian method for the assignment problem.

The strategy σ_D that always plays D yields a payoff 5 at every stage, and therefore its T-stage payoff is 5 as well:

$$\gamma_T(s(1); \sigma_D) = 5, \quad \forall T \in \mathbb{N}.$$

The strategy σ_U that plays U in the first stage yields 10 in the first stage and 2 in all subsequent stages. Therefore,

$$\gamma_T(s(1); \sigma_U) = 10 \cdot \frac{1}{T} + 2 \cdot \frac{T-1}{T} = 2 + \frac{8}{T}, \quad \forall T \in \mathbb{N}.$$

For every $0 \leq t < T$, the strategy $\sigma_{D_t U}$ that plays D in the first t stages and U in stage $t + 1$ yields 5 in the first t stages, 10 in stage $t + 1$, and 2 in all subsequent stages. Therefore,

$$\begin{aligned}
\gamma_T(s(1); \sigma_{D_t U}) &= 5 \cdot \frac{t}{T} + 10 \cdot \frac{1}{T} + 2 \cdot \frac{T-t-1}{T} \\
&= \frac{2T + 3t + 8}{T}, \quad \forall T \in \mathbb{N}, \forall 0 \leq t < T. \qquad \blacklozenge
\end{aligned}$$

Definition 1.14   Let $s \in S$ and let $T \in \mathbb{N}$. The real number $v_T(s)$ is the *T-stage value at the initial state s* if

$$v_T(s) := \sup_{\sigma \in \Sigma} \gamma_T(s; \sigma). \tag{1.3}$$

Any strategy in $\text{argmax}_{\sigma \in \Sigma} \gamma_T(s; \sigma)$ is *T-stage optimal at s*.

In other words, the T-stage value at s is the maximal amount that the decision maker can get when the initial state is s, and a strategy that guarantees this quantity is T-stage optimal.

Is the supremum in Eq. (1.3) attained? That is, is there a T-stage optimal strategy? As Theorem 1.15 states, the answer is positive.

Theorem 1.15   *For every $s \in S$ and every $T \geq 1$, there is a T-stage optimal strategy at the initial state s.*

Proof   In the T-stage game, the only relevant part of the strategy is its play up to stage T. In particular, for the purpose of studying the T-stage problem, we can define a strategy as a mapping $\sigma \colon \bigcup_{t=1}^{T} H_t \to \bigcup_{s \in S} \Delta(A(s))$, such that $\sigma(h_t) \in \Delta(A(s_t))$, for every history $h_t \in \bigcup_{t=1}^{T} H_t$. This set is a compact subset of a Euclidean space. The payoff function is continuous on this set. Since a continuous function defined on a compact set attains its maximum, the result follows. $\square$

Comment 1.16   We can strengthen Theorem 1.15 and prove that, for every $s \in S$ and every $T \geq 1$, there is a T-stage *pure* optimal strategy at the initial state s (see Theorem 1.18). To see this, consider the function that maps each mixed strategy σ into the T-stage payoff $\gamma_T(s; \sigma)$. This function is linear. Indeed, let σ_1 and σ_2 be two strategies, and let σ_3 be the following strategy: toss a fair coin; if the result is Head, follow σ_1, whereas if it is Tail, follow σ_2. Then

$$\gamma_T(s; \sigma_3) = \frac{1}{2}\gamma_T(s; \sigma_1) + \frac{1}{2}\gamma_T(s; \sigma_1).$$

By the Krein–Milman[6] Theorem, a linear function that is defined on a compact space attains its maximum at an extreme point. Since the pure strategies are the extreme points of the set of mixed strategies, it follows that the function $\sigma \mapsto \gamma_T(s; \sigma)$ attains its maximum at a pure strategy.

Example 1.3, continued   The quantity $\gamma_T(s(1); \sigma_{D_t U}) = \frac{2T+3t+8}{T}$ is maximized when $t = T - 1$: the decision maker plays $T - 1$ times D, and then she plays U once. The resulting average payoff is $5 + \frac{5}{T}$. The T-stage value at the initial state $s(1)$ is therefore $v_T(s(1)) = 5 + \frac{5}{T}$.   ◆

In general, the T-stage value, as well as the T-stage optimal strategies, can be found by *backward induction*, a method that is also known as the *dynamic programming principle*. We now formalize this method.

Theorem 1.17   *For every initial state $s_1 \in S$ and every $T \geq 2$, we have*

$$v_T(s_1) = \max_{a_1 \in A(s_1)} \left\{ \frac{1}{T}r(s_1, a_1) + \frac{T-1}{T}\sum_{s_2 \in S} q(s_2 \mid s_1, a_1)v_{T-1}(s_2) \right\}. \quad (1.4)$$

Eq. (1.4) states that, to calculate the T-stage value, we can break the problem into two parts: the first stage, and the last $T - 1$ stages. Since transitions and payoffs depend only on the current state and on the current action, the problem that starts at stage 2 is not affected by s_1 and a_1, the state and action at stage 1. This problem is a $(T - 1)$-stage Markov decision problem, whose value $v_{T-1}(s_2)$ depends on its initial state (and not on the initial state s_1). To calculate the T-stage value, we collapse the last $T - 1$ stages into a single number, the value of the $(T - 1)$-stage problem that starts

---

[6] Mark Grigorievich Krein (Kiev, Russia, April 3, 1907 – Odessa, Ukraine, October 17, 1989) was a Soviet mathematician who is best known for his work in operator theory.
David Pinhusovich Milman (Kiev, Russia, January 15, 1912 – Tel Aviv, Israel, July 12, 1982) was a Soviet and later Israeli mathematician specializing in functional analysis.

at stage 2, and we ask what is the optimal action in the first stage, assuming that if state s_2 is reached at stage 2, the continuation value is $v_{T-1}(s_2)$.

In Eq. (1.4) the weight of the payoff in the first stage, $r(s_1, a_1)$, is $\frac{1}{T}$, and the weight of the value of the $(T-1)$-stage problem that encapsulates the last $T-1$ stages is $\frac{T-1}{T}$. Why do we take these weights? The reason is that the quantity $r(s_1, a_1)$ represents the payoff in the first stage, while the quantity $v_{T-1}(s_2)$ captures the average payoff in $T-1$ stages: stages $2, 3, \ldots, T$. The weights of each of the two quantities reflect this point.

To prove Theorem 1.17, we will consider conditional expectation. Recall that $\mathbf{E}_{s_1, \sigma}[r(s_t, a_t)]$ is the expected payoff at stage t. For every $t' \leq t$ and every history $\widetilde{h}_{t'} = (\widetilde{s}_1, \widetilde{a}_1, \ldots, \widetilde{s}_{t'}) \in H_{t'}$ with $\widetilde{s}_1 = s_1$, the quantity $\mathbf{E}_{s_1, \sigma}[r(s_t, a_t) \mid \widetilde{h}_{t'}]$ is the expected payoff at stage t, conditional that the history $\widetilde{h}_{t'}$ has occurred, that is, conditional that the action in the initial state is $\widetilde{a}_1$, the state at stage 2 is $\widetilde{s}_2$, and so on. Formally, for every history $\widetilde{h}_{t'} = (\widetilde{s}_1, \widetilde{a}_1, \ldots, \widetilde{s}_{t'}) \in H_{t'}$, the probability distribution $\mathbf{P}_{s_1, \sigma}(\cdot \mid \widetilde{h}_{t'})$ is defined as follows:

- For histories that are not longer than $\widetilde{h}_{t'}$: For every $t \leq t'$ we have

$$\mathbf{P}_{s_1, \sigma}(C(s_1, a_1, \ldots, s_t) \mid \widetilde{h}_{t'}) := \mathbf{1}_{\{s_1 = \widetilde{s}_1, a_1 = \widetilde{a}_1, \ldots, s_t = \widetilde{s}_t\}}.$$

- For histories that are longer than $\widetilde{h}_{t'}$: For every $t > t'$, we have

$$\mathbf{P}_{s_1, \sigma}(C(s_1, a_1, \ldots, s_{t-1}, a_{t-1}, s_t) \mid h_{t'})$$

$$:= \mathbf{1}_{\{s_1 = \widetilde{s}_1, a_1 = \widetilde{a}_1, \ldots, s_{t'} = \widetilde{s}_{t'}\}} \cdot \prod_{k=t'}^{t-1} \sigma(a_k \mid s_1, a_1, \ldots, s_k)$$

$$\times \prod_{k=t'}^{t-1} q(s_{k+1} \mid s_k, a_k).$$

Denote by $\mathbf{E}_{s_1, \sigma}[\cdot \mid \widetilde{h}_{t'}]$ the expectation with respect to $\mathbf{P}_{s_1, \sigma}(\cdot \mid \widetilde{h}_{t'})$.

Proof of Theorem 1.17   For $T = 1$, the T-stage problem concerns the first stage only, and

$$v_1(s_1) = \max_{a_1 \in A(s_1)} r(s_1, a_1).$$

In particular, Eq. (1.4) holds. For $T \geq 2$, by definition and by the law of iterated expectations,

$$v_T(s_1)$$

$$= \max_{\sigma \in \Sigma} \mathbf{E}_{s_1,\sigma} \left[\frac{1}{T} \sum_{t=1}^{T} r(s_t, a_t) \right]$$

$$= \max_{\sigma \in \Sigma} \mathbf{E}_{s_1,\sigma} \left[\frac{1}{T} r(s_1, a_1) + \frac{T-1}{T} \cdot \frac{1}{T-1} \sum_{t=2}^{T} r(s_t, a_t) \right]$$

$$= \max_{\sigma \in \Sigma} \left(\mathbf{E}_{s_1,\sigma} \left[\frac{1}{T} r(s_1, a_1) \right] + \mathbf{E}_{s_1,\sigma} \left[\frac{T-1}{T} \cdot \frac{1}{T-1} \sum_{t=2}^{T} r(s_t, a_t) \mid h_2 \right] \right). \tag{1.5}$$

The term within the maximization in these equalities depends only on the part of the strategy σ that follows the initial state s_1. This part is composed of the mixed action $\sigma(s_1) \in \Delta(A(s_1))$ that is played in the first stage and the continuation strategies played from the second stage and on. We denote these continuation strategies by $(\sigma'_{s_1,a_1})_{a_1 \in A(s_1)}$. Formally, for every action $a_1 \in A(s_1)$, σ'_{s_1,a_1} is a strategy in the $T-1$ stage problem that is defined by

$$\sigma'_{s_1,a_1}(h_{t-1}) := \sigma(s_1, a_1, h_{t-1}), \quad \forall 2 \le t \le T, \forall h_{t-1} = (s_2, a_2, \ldots, s_t) \in H_{t-1}.$$

With this notation, the right-hand side in Eq. (1.5) is equal to

$$\max_{\alpha \in \Delta(A(s_1))} \max_{(\sigma'_{s_1,a_1})_{a_1 \in A_1(s_1)}} \mathbf{E}_{s_1,\alpha,(\sigma'_{s_1,a_1})_{a_1 \in A_1(s_1)}} \left[\frac{1}{T} r(s_1, a_1) \right.$$

$$\left. + \mathbf{E}_{s_1,\sigma'_{s_1,a_1}} \left[\frac{T-1}{T} \cdot \frac{1}{T-1} \sum_{t=2}^{T} r(s_t, a_t) \mid a_1, s_2 \right] \right], \tag{1.6}$$

where α captures the mixed action played in the first stage. The continuation strategies $(\sigma'_{s_1,a_1})_{a_1 \in A_1(s_1)}$ do not affect the payoff in the first stage $r(s_1, a_1)$. The action a_1 that is chosen in the first stage affects the continuation payoff in two ways. First, it determines the probability $q(s_2 \mid s_1, a_1)$ that the state in the first stage is s_2. Second, it determines the continuation strategy σ'_{s_1,a_1}. Since the probability distribution $\mathbf{P}_{s_1,\sigma}$ conditional on a_1 and s_2 is equal to the probability distribution $\mathbf{P}_{s_2,\sigma'_{s_1,a_1}}$, it follows that we can split the maximization problem in Eq. (1.6) into two parts, and obtain that

$$v_T(s_1) = \max_{\alpha \in \Delta(A(s_1))} \left(\frac{1}{T} r(s_1, \alpha) + \sum_{s_2 \in S} q(s_2 \mid s_1, a_1) \right.$$

$$\left. \times \max_{(\sigma'_{s_1, a_1})_{a_1 \in A_1(s_1)}} \mathbf{E}_{s_2, \sigma'_{s_1, a_1}} \left[\frac{T-1}{T} \cdot \frac{1}{T-1} \sum_{t=2}^{T} r(s_t, a_t) \right] \right). \tag{1.7}$$

Note that

$$v_{T-1}(s_2) = \max_{(\sigma'_{s_1, a_1})_{a_1 \in A_1(s_1)}} \mathbf{E}_{s_2, \sigma'_{s_1, a_1}} \left[\frac{1}{T-1} \sum_{t=2}^{T} r(s_t, a_t) \right];$$

hence, the right-hand side of Eq. (1.7) is equal to

$$\max_{\alpha \in \Delta(A(s_1))} \left(\frac{1}{T} r(s_1, \alpha) + \sum_{a_1 \in A_1(s_1)} \alpha(a_1) q(s_2 \mid s_1, a_1) \frac{T-1}{T} v_{T-1}(s_2) \right).$$

The function within the parentheses is linear in α, and $\Delta(A(s_1))$ is a compact set whose extreme points are the Dirac measures concentrated at the points a_1 with $a_1 \in A(s_1)$. A linear function that is defined on a compact set attains its maximum in an extreme point. The result follows. $\qquad\square$

The proof of Theorem 1.17 yields an algorithm that calculates the T-stage value and a T-stage optimal strategy σ^*. We will calculate by induction a k-stage optimal strategy σ_k^* for every $k = 1, 2, \ldots, T$. We start with $k = 1$, and calculate a one-stage optimal strategy for every initial state $s \in S$. Let $a_1^*(s) \in A(s)$ be an action that maximizes the quantity $r(s, a)$ over $a \in A(s)$, and set

$$\sigma_1^*(s) := a_1^*(s).$$

The value of the one-stage problem with initial state s is $v_1(s) = r(s_1, a_1^*(s))$. We continue recursively. Suppose that, for every initial state s, we already calculated $v_{k-1}(s)$ and already defined a $(k-1)$-stage optimal strategy σ_{k-1}^*. To calculate $v_k(s)$ and define a k-stage optimal strategy σ_k^*, we take

$$\max_{a \in A(s)} \left(\frac{1}{k} r(s, a) + \frac{k-1}{k} q(s' \mid s, a) v_{k-1}(s) \right), \tag{1.8}$$

and denote by $a_k^*(s) \in A(s)$ an action that achieves the maximum in Eq. (1.8). This is the quantity on the right-hand side of Eq. (1.4); hence, it is equal to $v_k(s)$. We can now define an optimal strategy σ^* for the decision maker as follows:

- At stage 1, play the action $a_k^*(s_1)$.
- From stage 2 on, follow the strategy σ_{k-1}^*; that is, at each stage t, when the current state is s_1 and $T - t + 1$ stages are left, play the action $a_{T-t+1}^*(s_t)$. Formally,

$$\sigma^*(h_t) := a_{T-t+1}^*(s_t), \quad \forall h_t = (s_1, a_1, \ldots, s_t) \in \bigcup_{t=1}^{T} H_t.$$

In Exercise 1.1, the reader is asked to prove that this strategy is indeed T-stage optimal.

The proof of Theorem 1.17 relies on the linearity of the payoff function: the goal of the decision maker is to maximize a linear function of the stage payoffs. If the sets of actions and states are not finite, the theorem still holds, provided that in Eq. (1.4) we replace maximum by supremum.

Theorem 1.17 admits the following corollary.

Theorem 1.18   *The T-stage value always exists. Moreover, there exists an optimal pure strategy $\sigma \in \Sigma$.*

One can show a stronger result concerning the structure of an optimal pure strategy: there exists an optimal pure strategy σ with the property that $\sigma(h_t)$ depends on the current state s_t and on the stage t, and is independent of the rest of the history $(s_1, a_1, \ldots, s_{t-1}, a_{t-1})$ (Exercise 1.3).

1.4 The Discounted Payoff

The discounted payoff depends on a parameter $\lambda \in (0, 1]$, called the *discount factor*, which measures how money grows with time: one dollar today is worth $\frac{1}{1-\lambda}$ dollars tomorrow, $\frac{1}{(1-\lambda)^2}$ dollars the day after tomorrow, and so on. In other words, the decision maker is indifferent between getting $1 - \lambda$ dollars today and one dollar tomorrow.

Definition 1.19   For every discount factor $\lambda \in (0, 1]$, every state $s \in S$, and every strategy $\sigma \in \Sigma$, the *λ-discounted payoff* under strategy profile σ at the initial state s is

$$\gamma_\lambda(s; \sigma) := \mathbf{E}_{s,\sigma} \left[\lambda \sum_{t=1}^{\infty} (1 - \lambda)^{t-1} r(s_t, a_t) \right]. \tag{1.9}$$

The λ in Eq. (1.9) serves as a normalization factor: a player who receives one dollar at every stage evaluates this stream of payoffs as one dollar. Since there are finitely many states and actions, the payoff function r is bounded, and therefore γ_λ obeys the same bound (which is independent of λ, thanks to the multiplication by λ).

The dominated convergence theorem (see, e.g., Shiryaev (1995), theorem 6.3) implies that

$$\gamma_\lambda(s;\sigma) := \lambda \sum_{t=1}^{\infty} (1-\lambda)^{t-1} \mathbf{E}_{s,\sigma} \left[r(s_t, a_t) \right].$$

Simple algebraic manipulations yield

$$\gamma_\lambda(s;\sigma) := \mathbf{E}_{s,\sigma} \left[\lambda r(s_1, a_1) + (1-\lambda) \left(\lambda \sum_{t=2}^{\infty} (1-\lambda)^{t-2} r(s_t, a_t) \right) \right]. \quad (1.10)$$

For every two states $s, s' \in S$ and every action $a \in A(s)$, set

$$\gamma_\lambda(s';\sigma_{s,a}) := \mathbf{E}_{s,\sigma} \left[\lambda \sum_{t=2}^{\infty} (1-\lambda)^{t-2} r(s_t, a_t) \mid s_1 = s,\, a_1 = a,\, s_2 = s' \right].$$

This is the expected discounted payoff from stage 2 on, when conditioning on the history at stage 2. Alternatively, this is the expected discounted payoff when the initial state is s', and the decision maker follows that part of her strategy that follows the history (s, a). If σ is a stationary strategy, then the way it plays after the first stage does not depend on the play in the first stage. Hence, in this case, for every two states $s, s' \in S$ and every action $a \in A(s)$ we have

$$\gamma_\lambda(s';\sigma_{s,a}) = \gamma_\lambda(s';\sigma).$$

From Eq. (1.10) we obtain:

$$\gamma_\lambda(s;\sigma) := \mathbf{E}_{s,\sigma} \left[\lambda r(s_1, a_1) + (1-\lambda) \gamma_\lambda(s_2; \sigma_{s_1, a_1}) \right]. \quad (1.11)$$

Thus, the expected payoff is a weighted average of the payoff $r(s_1, a_1)$ at the first stage and the expected payoff $\gamma_\lambda(s_2; \sigma_{s_1, a_1})$ in all subsequent stages. When the discount factor λ is high, the weight of the first stage is high; whereas when the discount factor λ is low, the weight of the first stage is low.

Eq. (1.11) illustrates that the decision maker's payoff consists of two parts: today's payoff and the future's payoff. The discount factor indicates the relative importance of each part. The lower the discount factor, the higher the importance of the future, and therefore the decision maker should put more weight on future opportunities. The higher the discount factor, the higher the importance of the present, and the decision maker should concentrate on short-term gains.

$$U \quad \boxed{10_{(0,1,0)}}$$

$$D \quad \boxed{5_{(1,0,0)}} \qquad\qquad D \quad \boxed{10_{\left(\frac{1}{10},0,\frac{9}{10}\right)}} \qquad\qquad D \quad \boxed{-100_{(0,0,1)}}$$

$$\quad s(1) \qquad\qquad\qquad\qquad s(2) \qquad\qquad\qquad\qquad s(3)$$

Figure 1.3 The Markov decision problem in Example 1.3.

Comment 1.20   In the proof of Theorem 1.17 we in fact showed that the T-stage payoff satisfies the following formula:

$$\gamma_T(s;\sigma) := \mathbf{E}_{s,\sigma}\left[\frac{1}{T}r(s_1,a_1) + \frac{T-1}{T}\gamma_{T-1}(s_2;\sigma_{s_1,a_1})\right]. \tag{1.12}$$

Thus, similar to the discounted payoff, the T-stage payoff is a weighted average of the payoff $r(s_1,a_1)$ at the first stage and the expected payoff $\gamma_{T-1}(s_2;\sigma_{s_1,a_1})$ in all subsequent stages, with weights $\frac{1}{T}$ and $\frac{T-1}{T}$.

Example 1.3, continued   The Markov decision problem in Example 1.3 is reproduced in Figure 1.3.

The initial state is $s(1)$. The strategy σ_D that always plays D at state $s(1)$ yields a payoff 5 at every stage, and therefore its λ-discounted payoff is 5 as well. Let us calculate the λ-discounted payoff of the strategy σ_U that always plays U at state $s(1)$. Since this strategy is stationary,

$$\gamma_\lambda(s(1);\sigma_U)$$
$$= 10\lambda + (1-\lambda)\left(10\lambda + (1-\lambda)\left(\frac{9}{10}(-100) + \frac{1}{10}\gamma_\lambda(s(1);\sigma_U)\right)\right). \tag{1.13}$$

The term $\gamma_\lambda(s(1);\sigma_U)$ on the right-hand side is the discounted payoff from the third stage and on, if at the second stage the play moves from $s(2)$ to $s(1)$. Eq. (1.13) solves to

$$\gamma_\lambda(s(1);\sigma_U) = \frac{10\lambda + 10\lambda(1-\lambda) - 100\frac{9}{10}(1-\lambda)^2}{1 - \frac{1}{10}(1-\lambda)^2}.$$

For $\lambda = 1$ (only the first day matters), we get

$$\gamma_1(s(1);\sigma_U) = 10,$$

while for λ close to 0 (the far future matters), we get

$$\lim_{\lambda\to 0}\gamma_\lambda(s(1);\sigma_U) = -100.$$

Since the function $\lambda \mapsto \gamma_\lambda(s(1); \sigma_U)$ is continuous, and since $\gamma_\lambda(s(1); \sigma_D) = 5$ for every $\lambda \in [0, 1)$, for a high discount factor the strategy σ_U is superior to the strategy σ_D, while for a low discount factor the strategy σ_D is superior to the strategy σ_U.   ♦

Definition 1.21   Let $s \in S$ and let $\lambda \in (0, 1]$ be a discount factor. The real number $v_\lambda(s)$ is the λ-*discounted value at the initial state s* if

$$v_\lambda(s) := \sup_{\sigma \in \Sigma} \gamma_\lambda(s; \sigma). \tag{1.14}$$

The strategies in $\text{argmax}_{\sigma \in \Sigma}\, \gamma_\lambda(s; \sigma)$ are said to be λ-*discounted optimal at the initial state s*.

Thus, the λ-discounted value at s is the maximal λ-discounted payoff that the decision maker can get when the initial state is s, and a strategy that guarantees this quantity is λ-discounted optimal.

In Theorem 1.17 we stated the dynamic programming principle for the T-stage decision problem. We now provide the analogous principle for the discounted problem. The proof of the result is left to the reader (Exercise 1.5).

Theorem 1.22   *For every state $s \in S$ and every discount factor $\lambda \in (0, 1]$, we have*

$$v_\lambda(s) = \max_{a \in A(s)} \left\{ \lambda r(s, a) + (1 - \lambda) \sum_{s' \in S} q(s' \mid s, a) v_\lambda(s') \right\}. \tag{1.15}$$

In Eq. (1.15), the weight of the payoff at the first stage is λ, while the weight of the value at the second stage is $1 - \lambda$. The reason for these weights comes from the definition of the λ-discounted payoff in Eq. (1.9). In that equation, the weight of the payoff at stage t is $\lambda(1 - \lambda)^{t-1}$. In particular, the weight of the payoff at the first stage is λ, which is similar to the weight of the payoff at the first stage in Eq. (1.15). Since the sum of the weights of the payoffs in Eq. (1.9) is 1, it follows that the total weight of the payoffs at stages $2, 3, \ldots$ is $1 - \lambda$, which is the weight of the second term on the right-hand side of Eq. (1.15).

In Section 1.7, we will prove that, for every discount factor λ, there is a pure stationary strategy that is λ-discounted optimal at *all* initial states. The proof uses contracting mappings, which will be defined in Section 1.5. Moreover, we will show that there is a pure stationary strategy that is optimal for every discount factor sufficiently close to 0.

Comment 1.23 Like we did for the T-stage problem, one can provide a direct argument for the existence of a discounted optimal strategy. Since the set of histories is countable, the set of strategies, which is $\prod_{h_t \in H} \Delta(A(s_t))$, is compact in the product topology. Moreover, the discounted payoff function is continuous in this topology. Hence, the supremum in Eq. (1.14) is attained.

1.5 Contracting Mappings

A *metric space* is a pair (X, d), where X is a set and $d: X \times X \to [0, \infty)$ is a *metric*, that is, d satisfies the following conditions:

- $d(x, y) = 0$ if and only if $x = y$.
- Symmetry: $d(x, y) = d(y, x)$ for all $x, y \in X$.
- Triangle inequality: $d(x, z) \leq d(x, y) + d(y, z)$ for all $x, y, z \in X$.

A sequence $(x_n)_{n \in \mathbb{N}}$ in a metric space is *Cauchy*[7] if, for every $\epsilon > 0$, there is an $n_0 \in \mathbb{N}$ such that $n_1, n_2 \geq n_0$ implies $d(x_{n_1} x_{n_2}) \leq \epsilon$. A metric space is *complete* if every Cauchy sequence has a limit. For every $m \in \mathbb{N}$, the Euclidean space $\mathbb{R}^m$ equipped with the distance induced by the Euclidean norm, the L_1-norm, or the L_∞-norm is complete. Readers who are not familiar with metric spaces can think of a metric space as $\mathbb{R}^m$ equipped with the Euclidean distance.

Definition 1.24 Let (X, d) be a metric space. A mapping $f: X \to X$ is *contracting* if there exists $\rho \in [0, 1)$ such that $d(f(x), f(y)) \leq \rho d(x, y)$ for all $x, y \in X$.

Example 1.25 Let $\rho \in [0, 1)$ and $a \in \mathbb{R}^n$. The mapping $f: \mathbb{R}^n \to \mathbb{R}^n$ that is defined by

$$f(x) := a + \rho x, \quad \forall x \in \mathbb{R}^n,$$

is contracting.

Theorem 1.26 *Let (X, d) be a complete metric space. Every contracting mapping $f: X \to X$ has a unique fixed point; that is, there exists a unique point $x \in X$ such that $x = f(x)$.*

Proof Let $f: X \to X$ be a contracting mapping.

---

[7] Augustin-Louis Cauchy (Paris, France, August 21, 1789 – Sceaux, France, May 23, 1857) was a French mathematician. He started the project of formulating and proving the theorems of calculus in a rigorous manner and was thus an early pioneer of analysis. He also developed several important theorems in complex analysis and initiated the study of permutation groups.

Step 1:　f has at most one fixed point.

If $x, y \in X$ are fixed points of f, then

$$d(x,y) = d(f(x), f(y)) \leq \rho d(x,y).$$

Since $\rho \in [0,1)$, this implies that $d(x,y) = 0$, and therefore $x = y$.

Step 2:　f has at least one fixed point.

Let $x_0 \in X$ be arbitrary, and define inductively $x_{n+1} = f(x_n)$ for every $n \geq 0$. Then for any $k, m > 0$,

$$d(x_k, x_{k+m}) \leq \sum_{l=0}^{m-1} d(x_{k+l}, x_{k+l+1})$$

$$\leq d(x_0, f(x_0))\rho^k \sum_{l=0}^{m-1} \rho^l < d(x_0, f(x_0))\frac{\rho^m}{1-\rho},$$

where the first inequality follows from the triangle inequality, and the second inequality holds since by induction: $d(x_l, x_{l+1}) \leq \rho^l d(x_0, x_1) = \rho^l d(x_0, f(x_0))$. Thus, $(x_k)_{k \in \mathbb{N}}$ is a Cauchy sequence, and therefore it converges to a limit x. By the triangle inequality,

$$d(x, f(x)) \leq d(x, x_k) + d(x_k, x_{k+1}) + d(x_{k+1}, f(x)), \tag{1.16}$$

for all $k \in \mathbb{N}$. Let us show that all three terms on the right-hand side of Eq. (1.16) converge to 0 as k goes to infinity; this will imply that $d(x, f(x)) = 0$, hence $x = f(x)$, that is, x is a fixed point of f. Indeed, $\lim_{k \to \infty} d(x, x_k) = 0$ because x is the limit of $(x_k)_{k \in \mathbb{N}}$; $\lim_{k \to \infty} d(x_k, x_{k+1})$ because $(x_k)_{k \in \mathbb{N}}$ is a Cauchy sequence; and finally, since f is contracting,

$$\lim_{k \to \infty} d(x_{k+1}, f(x)) = \lim_{k \to \infty} d(f(x_k), f(x)) \leq \lim_{k \to \infty} \rho d(x_k, x) = 0. \quad \square$$

1.6　Existence of an Optimal Stationary Strategy

In this section, we prove the following result, due to Blackwell (1965).[8]

Theorem 1.27　*For every $\lambda \in (0, 1]$, there exists a λ-discounted pure stationary optimal strategy.*

---

[8] David Harold Blackwell (Centralia, Illinois, April 24, 1919 – Berkeley, California, July 8, 2010) was an American statistician and mathematician who made significant contributions to game theory, probability theory, information theory, and Bayesian statistics.

The existence of a λ-discounted optimal strategy was discussed in Comment 1.23, while the existence of a λ-discounted pure optimal strategy is established by the same arguments as in Comment 1.16. We now explain the intuition behind the existence of a λ-discounted pure *stationary* optimal strategy. Let h_t and $\widehat{h_{\hat{t}}}$ be two histories that end at the same state s. Since the payoffs and transitions depend only on the current state, and not on past play, if the decision maker plays in the same way after h_t and after $\widehat{h_{\hat{t}}}$, the evolution of the Markov decision problem after h_t is the same as after $\widehat{h_{\hat{t}}}$. Suppose now that the optimal strategy σ prescribes to play differently after h_t and after $\widehat{h_{\hat{t}}}$, that is, $\sigma(h_t) \neq \sigma(\widehat{h_{\hat{t}}})$. Assume without loss of generality that the expected payoff after h_t is at least as high as the expected payoff after $\widehat{h_{\hat{t}}}$. Define a new strategy σ_1 as follows: σ_1 is similar to σ, except that after the history $\widehat{h_{\hat{t}}}$, it plays as σ plays after h_t. It is easy to see that $\gamma_\lambda(s_1; \sigma_1) \geq \gamma_\lambda(s_1; \sigma)$. Repeating this process over all histories shows that one can modify σ to be a stationary strategy, without lowering the λ-discounted payoff, thus establishing Theorem 1.27. The proof of Theorem 1.27 that we will provide will use a different idea – contracting mappings. This approach will be useful when we will later study stochastic games.

Before we can prove Theorem 1.27, we need a bit of preparation. Fix a function $w \colon S \to \mathbb{R}$. This function will capture the "discounted payoff from the next stage on," given the state at the next stage. Given the initial state s and the strategy σ, let $h_t \in H$ be a history with positive probability of realization, such that $\mathbf{P}_{s,\sigma}(C(h_t)) > 0$. Consider the situation in which, when the decision maker follows the strategy σ, once some history h_t is realized, the decision maker is told that after she chooses the action a_t and the new state s_{t+1} is announced, the process will terminate, and she will get a terminal payoff $w(s_{t+1})$. As in Eq. (1.15), the weights of the payoff at stage t is λ, and the weight of the terminal payoff[9] is $1 - \lambda$. The expected payoff from stage t and on is then given by

$$
\mathbf{E}_{s,\sigma}\left[\lambda r^i(s_t, a_t) + (1 - \lambda)\sum_{s' \in S} q(s' \mid s_t, a_t)w(s') \mid h_t\right] \qquad (1.17)
$$
$$
= \mathbf{E}_{s,\sigma}\left[\lambda r^i(s_t, a_t) + (1 - \lambda)w(s_{t+1}) \mid h_t\right].
$$

The first term in the expectation measures the expected stage payoff, while the second term measures the expected terminal payoff. Note that in Eq. (1.17)

---

[9] Setting the weight of the terminal payoff to $1 - \lambda$ is equivalent to considering a standard discounted payoff, assuming the payoff in *all* stages after stage t is $w(s_{t+1})$.

the expectation is a conditional expectation, given the history at stage t. The following result relates the expectation in Eq. (1.17) to the discounted payoff.

Lemma 1.28 *Let σ be a strategy, let $s \in S$, and let $w \colon S \to \mathbb{R}$ be a function. If for every $t \in \mathbb{N}$ and every $h_t \in H_t$,*

$$\mathbf{E}_{s,\sigma}\big[\lambda r(s_t, a_t) + (1 - \lambda)w(s_{t+1}) \mid h_t\big] \geq w(s_t) \qquad (1.18)$$

then

$$\gamma_\lambda(s;\sigma) \geq w(s). \qquad (1.19)$$

If the inequality in Eq. (1.18) *is reversed for every $t \in \mathbb{N}$ and every $h_t \in H_t$, so is the inequality in* Eq. (1.19). *If the inequality in* Eq. (1.18) *is an equality for every $t \in \mathbb{N}$ and every $h_t \in H_t$, then* Eq. (1.19) *becomes an equality as well.*

Proof   Recall the law of iterated expectation: for every function $f \colon S \to \mathbb{R}$, every $t \in \mathbb{N}$, and every history $h_t \in H_t$,

$$\mathbf{E}_{s,\sigma}[\mathbf{E}_{s,\sigma}[f(s_{t+1}) \mid h_t]] = \mathbf{E}_{s,\sigma}[f(s_{t+1})].$$

Taking expectations in Eq. (1.18), we deduce that

$$\mathbf{E}_{s,\sigma}[\lambda r(s_t,a_t)] \geq \mathbf{E}_{s,\sigma}[w(s_t)] - (1-\lambda)\mathbf{E}_{s,\sigma}[w(s_{t+1})], \quad \forall t \in \mathbb{N}. \qquad (1.20)$$

Multiplying both sides of Eq. (1.20) by $(1 - \lambda)^{t-1}$ and summing over $t \in \mathbb{N}$, we obtain Eq. (1.19):

$$
\begin{aligned}
\gamma_\lambda(s;\sigma) &= \sum_{t=1}^{\infty}(1-\lambda)^{t-1}\mathbf{E}_{s,\sigma}[\lambda r(s_t,a_t)] \\
&\geq \sum_{t=1}^{\infty}(1-\lambda)^{t-1}\big(\mathbf{E}_{s,\sigma}[w(s_t)] - (1-\lambda)\mathbf{E}_{s,\sigma}[w(s_{t+1})]\big) \qquad (1.21) \\
&= w(s),
\end{aligned}
$$

where the last equality holds because the sum involved is telescopic.

If the inequality in Eq. (1.18) is reversed for every $t \in \mathbb{N}$ and every $h_t \in H_t$, then the inequality in Eq. (1.20) is reversed as well, and therefore so is the equality in Eq. (1.21). The last conclusion follows from the first two statements. $\qquad \square$

We need the following technical result.

Lemma 1.29   *Let $x = (x_1, \ldots, x_n), y = (y_1, \ldots, y_n) \in \mathbb{R}^n$. Then*

$$\left| \max_{1 \leq i \leq n} x_i - \max_{1 \leq i \leq n} y_i \right| \leq \max_{1 \leq i \leq n} |x_i - y_i|.$$

Proof   Without loss of generality, we can assume that $\max_{1 \leq i \leq n} x_i \geq \max_{1 \leq i \leq n} y_i$. Suppose also that $x_{i_0} = \max_{1 \leq i \leq n} x_i$ and $y_{i_1} = \max_{1 \leq i \leq n} y_i$. Then

$$\left| \max_{1 \leq i \leq n} x_i - \max_{1 \leq i \leq n} y_i \right| = \max_{1 \leq i \leq n} x_i - \max_{1 \leq i \leq n} y_i$$

$$= x_{i_0} - y_{i_1}$$

$$\leq x_{i_0} - y_{i_0}$$

$$\leq \max_{1 \leq i \leq n} |x_i - y_i|. \qquad \square$$

Proof of Theorem 1.27   We define a mapping $T \colon \mathbb{R}^S \to \mathbb{R}^S$, prove that it is contracting, and conclude that it has a unique fixed point w. We then show that the decision maker has a pure stationary strategy x^* such that $\gamma_\lambda(s; x^*) = w(s)$ for every initial state $s \in S$, and that $\gamma_\lambda(s; \sigma) \leq w(s)$ for every initial state $s \in S$ and every strategy σ.

For every vector $w = (w(s))_{s \in S} \in \mathbb{R}^S$, define

$$(T(w))(s) := \max_{a \in A(s)} \left(\lambda r(s,a) + (1 - \lambda) \sum_{s' \in S} q(s' \mid s,a) w(s) \right).$$

Step 1:   The mapping T is contracting.

Let $w, u \in \mathbb{R}^S$. By Lemma 1.29,

$$|(T(w))(s) - (T(u))(s)| = \left| \max_{a \in A(s)} \left(\lambda r(s,a) + (1 - \lambda) \sum_{s' \in S} q(s' \mid s,a) w(s') \right) \right.$$

$$\left. - \max_{a \in A(s)} \left(\lambda r(s,a) + (1 - \lambda) \sum_{s' \in S} q(s' \mid s,a) u(s') \right) \right|$$

$$\leq \max_{a \in A(s)} \left| \left(\lambda r(s,a) + (1 - \lambda) \sum_{s' \in S} q(s' \mid s,a) w(s') \right) \right.$$

$$\left. - \left(\lambda r(s,a) + (1 - \lambda) \sum_{s' \in S} q(s' \mid s,a) u(s') \right) \right|$$

$$= \max_{a \in A(s)} (1 - \lambda) \sum_{s' \in S} q(s' \mid s,a) |w(s') - u(s')|$$

$$\leq (1 - \lambda) \|w - u\|_\infty.$$

It follows that $\|T(w) - T(u)\|_\infty \leq (1 - \lambda)\|w - u\|_\infty$, hence T is contracting. By Theorem 1.26, T has a unique fixed point w. For each $s \in S$, let $a_s \in A(s)$ be an action that maximizes the expression

$$\lambda r(s,a) + (1 - \lambda) \sum_{s' \in S} q(s' \mid s,a) w(s).$$

There might be more than one such action. Then,

$$(T(w))(s) = \lambda r(s,a_s) + (1 - \lambda) \sum_{s' \in S} q(s' \mid s,a_s) w(s). \qquad (1.22)$$

Let x_* be the pure stationary strategy that plays the action a_s at state s, for every $s \in S$. We prove that $w(s) = v_\lambda(s)$ for every $s \in S$, and that x^* is λ-discounted optimal.

Step 2: $\gamma_\lambda(s; x^*) = w(s)$ for every initial state $s \in S$.
This follows from Eq. (1.22) and Lemma 1.28.

Step 3: $\gamma_\lambda(s; \sigma) \le w(s)$ for every strategy σ and every initial state $s \in S$.
By the definition of $T(w)$,

$$(T(w))(s_t) = \max_{a \in A(s_t)} (\lambda r(s_t,a) + (1 - \lambda) w(s_{t+1})))$$

$$\ge \mathbf{E}_{s_t,\sigma} \left[\lambda r(s_t,a) + (1 - \lambda) w(s_{t+1}) \right],$$

for all $t \in \mathbb{N}$. The claim follows from Lemma 1.28. $\square$

We in fact proved the following characterization of the set of optimal strategies in Markov decision problems, whose proof is left for the reader (Exercise 1.16). In this characterization and later in the book, we will use the following notations:

$$r(s,x(s)) := \sum_{a \in A(s)} \left(\prod_{i \in I} x^i(s,a^i) \right) r(s,a), \quad \forall s \in S, x(s) \in \prod_{i \in I} \Delta(A^i(s)),$$

$$q(s' \mid s,x(s)) := \sum_{a \in A(s)} \left(\prod_{i \in I} x^i(s,a^i) \right) q(s' \mid s,a),$$

$$\forall s,s' \in S, x(s) \in \prod_{i \in I} \Delta(A^i(s)).$$

The quantity $\prod_{i \in I} x^i(s,a^i)$ is the probability that under the mixed action profile $x(s)$, the action profile a is chosen. Therefore, $r(s,x(s))$ is the expected stage payoff at stage s when the players play the stationary strategy profile x, and $q(s' \mid s,x(s))$ is the probability that the play moves from s to s' when the players play the stationary strategy profile x.

Theorem 1.30   *Let $\Gamma = \langle S, (A(s))_{s \in S}, q, r \rangle$ be a Markov decision problem, and let $\lambda \in (0,1]$ be a discount factor. A stationary strategy x is λ-discounted* *optimal at all initial states if and only if for every state $s \in S$ the mixed action $x(s)$ satisfies*

$$v_\lambda(s) = \lambda r(s, x(s)) + (1 - \lambda) \sum_{s' \in S} q(s' \mid s, x(s)) v_\lambda(s').$$

1.7 Uniform Optimality

For each $s \in S$ consider the function $\lambda \mapsto v_\lambda(s)$, which assigns to each discount factor its discounted value. How does this function depend on λ? Can it be equal to $\sin(\lambda)$ or e^λ? In this section we will answer this question, among others.

Recall that a function $f\colon \mathbb{R} \to \mathbb{R}$ is *rational* if it is the ratio of two polynomials.

Theorem 1.31 *Two rational functions $f, g\colon \mathbb{R} \to \mathbb{R}$ either coincide, or they (i.e., their graphs) have finitely many intersection points: the set $\{x \in \mathbb{R}\colon f(x) = g(x)\}$ is either $\mathbb{R}$ or finite.*

Proof Let $f = \frac{P_1}{Q_1}$ and $g = \frac{P_2}{Q_2}$, where P_1, Q_1, P_2, and Q_2 are polynomials. Then

$$\left\{x \in \mathbb{R}\colon f(x) = g(x)\right\} = \left\{x \in \mathbb{R}\colon \frac{P_1(x)}{Q_1(x)} = \frac{P_2(x)}{Q_2(x)}\right\}$$
$$= \left\{x \in \mathbb{R}\colon P_1(x)Q_2(x) - P_2(x)Q_1(x) = 0\right\}.$$

That is, $\{x \in \mathbb{R}\colon f(x) = g(x)\}$ is the set of all zeroes of a polynomial. Since a nonzero polynomial has finitely many zeros, the result follows. □

An $n \times n$ matrix $Q = (Q_{ij})_{i, j \in \{1, \dots, n\}}$ is *stochastic* if the sum of entries in every row is 1, that is, $\sum_{j=1}^n Q_{ij} = 1$ for all $i \in \{1, 2, \dots, n\}$. Let Id denote the identity matrix.

Theorem 1.32 *For every stochastic matrix Q and every $\lambda \in (0, 1]$, the matrix $Id - (1 - \lambda)Q$ is invertible, that is, the inverse matrix $(Id - (1 - \lambda)Q)^{-1}$ exists.*

Proof Setting $P := Id - (1 - \lambda)Q$ and $R := \sum_{k=0}^\infty (1 - \lambda)^k Q^k$, we note that $P \cdot R = Id$, and therefore P is invertible.

Alternatively, $P_{ii} > 0$ for every $i \in \{1, 2, \dots, n\}$ and $P_{ij} \leq 0$ for every $i, j \in \{1, 2, \dots, n\}$ such that $i \neq j$, which implies that P is invertible. □

Theorem 1.33   *For any fixed pure stationary strategy x and any fixed initial state $s \in S$, the function $\lambda \mapsto \gamma_\lambda(s;x)$ is rational.*

Our proof for Theorem 1.33 is valid for any stationary (and not necessarily pure) strategy (see Exercise 1.12).

Proof   Recall that a pure stationary strategy is a vector of actions, one for each state. Fix a pure stationary strategy $x = (a_s)_{s \in S}$. Denote by Q the transition matrix induced by x. This is a matrix with $|S|$ rows and $|S|$ columns, with entries (s, s') given by

$$Q_{s,s'} = q(s' \mid s, a_s).$$

Using the matrix Q, we can easily calculate the distribution of the state s_t at stage t. Suppose that one chooses an initial state according to a probability distribution $p \in \Delta(S)$ (which is expressed as a row vector), and then one plays the action a_s. What is the probability that the next state will be s'? This probability is $\sum_{s \in S} p_s q(s' \mid s, a_s)$, which is the s' coordinate of the vector pQ. Similarly, since $(pQ)_s$ is the probability that the state at stage 2 is s, the probability that the state at stage 3 is s' is given by $\sum_{s \in S}(pQ)_s q(s' \mid s, a_s)$, which is the s' coordinate of the vector pQ^2. By induction, it follows that the probability that the state at stage t is s' is the s' coordinate of the vector pQ^{t-1}.

For a state $s \in S$ denote by $\mathbf{1}(s) = (0, \ldots, 0, 1, 0, \ldots, 0)$ the row vector with the s coordinate equal to 1 and all the other coordinates equal to 0. Then $\mathbf{1}(s)Q^{t-1}$ represents the probability distribution of the state s_t at stage t, given that the initial state is s. Therefore, the λ-discounted payoff can be expressed as

$$\gamma_\lambda(s;x) = \sum_{t=1}^{\infty} \lambda(1 - \lambda)^{t-1} \mathbf{1}(s) Q^{t-1} R,$$

where R is the row vector $(r(s, a_s))_{s \in S}$. Therefore,

$$\gamma_\lambda(s;x) = \lambda \mathbf{1}(s) \left(\sum_{t=1}^{\infty} (1 - \lambda)^{t-1} Q^{t-1} \right) R$$

$$= \lambda \mathbf{1}(s)(I - (1 - \lambda)Q)^{-1} R.$$

By Theorem 1.32 the matrix $I - (1 - \lambda)Q$ is invertible, and by Cramer's rule,[10] the inverse matrix $(I - (1 - \lambda)Q)^{-1}$ can be represented as the ratio of

---

[10]   Gabriel Cramer (Geneva, Italy, July 31, 1704 – Bagolns-sur-Cèze, France, January 4, 1752) was a mathematician from the Republic of Geneva. In addition to presenting Cramer's rule for the calculation of the inverse of a matrix, Cramer worked on algebraic curves.

two polynomials in the entries of the matrix $I - (1 - \lambda)Q$. We conclude that for every fixed pure stationary strategy x, the function $\lambda \mapsto \gamma_\lambda(s; x)$ is rational. □

We can now prove a general structure theorem regarding the value function.

Corollary 1.34 *For any fixed state $s \in S$, the function $\lambda \mapsto v_\lambda(s)$ is continuous. Moreover, there exist $K \in \mathbb{N}$ and $0 = \lambda_0 < \lambda_1 < \cdots < \lambda_K = 1$ such that for every $k = 0, 1, \ldots, K - 1$, the following holds:*

- *The restriction of the function $\lambda \mapsto v_\lambda(s)$ to the interval $(\lambda_k, \lambda_{k+1})$ is rational.*
- *There is a pure stationary strategy $x_k \in \prod_{s\in S} A(s)$ that is λ-discounted optimal for all $\lambda \in (\lambda_k, \lambda_{k+1})$.*

Proof Let Σ_{SP} denote the finite set of all pure stationary strategies. For any fixed pure stationary strategy $x \in \Sigma_{SP}$ and any fixed state $s \in S$, consider the function $\lambda \mapsto \gamma_\lambda(s; x)$, which we denote by $\gamma_\bullet(s; x)$. By Theorem 1.33, $\gamma_\bullet(s; x)$ is a rational function; in particular, $\gamma_\bullet(s; x)$ is continuous. Since there exists a pure stationary optimal strategy, the λ-discounted value at the initial state s is given by

$$v_\lambda(s) = \max_{x \in \Sigma_{SP}} \gamma_\lambda(s; x).$$

Since the function $\lambda \mapsto v_\lambda(s)$ is the maximum of a finite family of rational functions, it is continuous.

By Theorem 1.31, two distinct rational functions intersect in finitely many points. Let Λ_s be the set of all intersection points of the rational functions $(\gamma_\bullet(s; x))_{x\in\Sigma_{SP}}$, and set $\Lambda := \bigcup_{s\in S} \Lambda_s$. Since the set Σ_{SP} is finite, the set Λ_s is finite for every state $s \in S$, and so the set Λ is finite as well. Add the points 0 and 1 to the set Λ, and denote $\Lambda = \{\lambda_0, \lambda_1, \ldots, \lambda_K\}$ where $0 = \lambda_0 < \lambda_1 < \cdots < \lambda_K = 1$.

Fix $k \in \{0, 1, \ldots, K-1\}$. By the choice of λ_k and λ_{k+1}, for every state $s \in S$ the functions $(\gamma_\bullet(s; x))_{x\in\Sigma_{SP}}$ have no common intersection point in the interval $(\lambda_k, \lambda_{k+1})$. Let $x_k \in \Sigma_{SP}$ be a pure stationary strategy that is λ-discounted optimal for some $\lambda \in (\lambda_k, \lambda_{k+1})$. We claim that x_k is λ'-discounted optimal at all initial states, for every $\lambda' \in (\lambda_k, \lambda_{k+1})$, as needed. Indeed, since x_k is λ-discounted optimal at all initial states, for every fixed pure stationary strategy $x \in \Sigma_{SP}$ and every fixed state $s \in S$, either $\gamma_\gamma(s; x_k) > \gamma_\lambda(s; x)$, or $\gamma_\lambda(s; x_k) = \gamma_\lambda(s; x)$. In the former case, since the set of intersection points of the functions $\gamma_\bullet(s; x_k)$ and $\gamma_\bullet(s; x)$ is disjoint from $(\lambda_k, \lambda_{k+1})$, it follows that $\gamma_{\lambda'}(s; x_k) > \gamma_{\lambda'}(s; x)$ for every $\lambda' \in (\lambda_k, \lambda_{k+1})$. In the latter case, for the same reason $\gamma_{\lambda'}(s;x_k) = \gamma_{\lambda'}(s;x)$ for every $\lambda' \in (\lambda_k, \lambda_{k+1})$. Hence, x_k is indeed λ'-discounted optimal for all $\lambda' \in (\lambda_k, \lambda_{k+1})$.   □

The significance of Corollary 1.34 is that the decision maker does not need to know precisely the discount factor for her to play optimally. If all the decision maker knows is that the discount factor is within an interval in which a specific pure stationary strategy x is optimal, by following x she ensures that she plays optimally, regardless of the exact value of the discount factor.

In particular, we get the following.

Corollary 1.35   *There is a pure stationary strategy that is optimal for every discount factor sufficiently close to 0.*

In many situations the decision maker is patient, that is, her discount factor is close to 0. For example, countries negotiating a peace treaty are often patient. Another example concerns an investor who may execute many transactions along the day, sometimes even selling a stock that she bought earlier in the day. For such an investor, one period of the game may last one hour or one minute, and subsequently her discount factor is quite close to 0. When the discount factor is close to 0, by Corollary 1.35, to play optimally the decision maker does not need to know the exact value of the discount factor.

Definition 1.36   A strategy σ is *uniformly optimal* at the initial state s if there is a $\lambda_0 > 0$ such that σ is λ-discounted optimal at the initial state s for every $\lambda \in (0, \lambda_0)$.

In the literature, uniformly optimal strategies are also called *Blackwell optimal*. By Corollary 1.35, we deduce the following result.

Theorem 1.37   *In every Markov decision problem, there is a pure stationary strategy that is uniformly optimal at all initial states.*

If $f : (0, 1] \to \mathbb{R}$ is a bounded rational function, then the limit $\lim_{\lambda \to 0} f(\lambda)$ exists. We therefore deduce that the discounted value is continuous at 0.

Corollary 1.38 $\lim_{\lambda \to 0} v_\lambda(s)$ *exists for every initial state* $s \in S$.

1.8 Comments and Extensions

Markov decision problems were first studied by Blackwell (1962). The model, as introduced by Definition 1.1, include finitely many states, and the set of actions available at each state is finite. Markov decision problems with general state and action sets were considered in the literature, and the existence of T-stage optimal strategies as well as of stationary λ-discounted optimal strategies was established under various topological conditions on the set SA of pairs (state, action) and continuity conditions on the payoff function and on the transition rule. For more details, the reader is referred to Puterman (1994).

By Theorem 1.34, for every state $s \in S$ the function $\lambda \mapsto v_\lambda(s)$ is piecewise rational. A natural goal is to characterize the set of all functions that can arise as the value function of some Markov decision problem. Such a characterization was provided by Lehrer et al. (2016).

Here we considered two types of evaluations for the decision maker: the T-stage evaluation and the discounted evaluations. Other evaluations have also been considered, see Puterman (1994, Section 5.4), where algorithms for approximating optimal strategies for various evaluations are described.

We proved that the limit $\lim_{\lambda \to 0} v_\lambda(s)$ of the discounted value exists for every initial state s. We did not touch upon the convergence of the T-stage value as T goes to infinity, namely, $\lim_{T \to \infty} v_T(s)$. For Markov decision problems with finitely many states and actions, the fact that $\lim_{T \to \infty} v_T(s)$ exists and is equal to $\lim_{\lambda \to 0} v_\lambda(s)$ follows from a result of Hardy and Littlewood, see Korevaar (2004, chapter I.7). We will not prove this result directly, as it will follow from a much more general result that we will obtain later in this book (see Theorem 9.13 on Page 139). A rich literature extends this result to Markov decision problems with general state and action sets, see, for example, Lehrer and Sorin (1992), Monderer and Sorin (1993), and Lehrer and Monderer (1994).

When the decision maker follows a uniformly optimal strategy, she guarantees that the discounted payoff is close to the value. This does not rule out the possibility that the payoff fluctuates along the play: during some long blocks of stages, the payoff is high; in other long blocks of stages, the payoff is low; and the blocks are arranged in such a way that the average payoff is close to the value. Sorin et al. (2010) proved that this is not the case: if the decision maker follows a uniformly optimal strategy, then for every sufficiently large positive integer m there is a $T \in \mathbb{N}$ such that for every $t \geq T$, the expected average payoff in stages $t, t+1, \ldots, t+T-1$ is close to $\lim_{\lambda \to 0} v_\lambda(s_1)$.

1.9 Exercises

Exercise 1.3 is used in the solution Exercise 5.1.

1. Prove that the strategy σ^* that is described after the proof of Theorem 1.17 is T-stage optimal.
2. In this exercise, we bound the variation of the sequence of the T-stage values $(v_T(s))_{T \in \mathbb{N}}$. Prove that for every $T, k \in \mathbb{N}$ and every state $s \in S$,

$$|(T + k)v_{T+k}(s) - T v_T(s)| \le k\|r\|_\infty.$$

3. Let Γ be a Markov decision problem. Prove that there is a pure T-stage optimal strategy with the following property: the action played in each stage depends only on the current stage and on the number of stages left. That is, for every $t \in \{1, 2, \ldots, T\}$ and every two histories $h_t = (s_1, a_1, \ldots, s_{t-1}, a_{t-1}, s_t)$ and $h'_t = (s'_1, a'_1, \ldots, s'_{t-1}, a'_{t-1}, s'_t)$, if $s_t = s'_t$, then $\sigma(h_t) = \sigma(h'_t)$.

4. For $\lambda \in (0, 1]$, calculate the λ-discounted value and the λ-discounted optimal strategy at the initial state $s(1)$ in Example 1.3.

5. Prove Theorem 1.22 about the dynamic programming principle for discounted decision problems: For every initial state $s \in S$ and every discount factor $\lambda \in (0, 1]$,

$$v_\lambda(s) = \max_{a \in A(s)} \left\{ \lambda r(s, a) + (1 - \lambda) \sum_{s' \in S} q(s' \mid s, a) v_\lambda(s') \right\}.$$

6. Find the discounted payoff of each pure stationary strategy in the following Markov decision problem and determine the discounted value for every discount factor.

U	$1_{\left(\frac{2}{3}, \frac{1}{3}\right)}$
D	$0_{(0, 1)}$

$s(1)$

U	$2_{\left(\frac{1}{2}, \frac{1}{2}\right)}$
D	$3_{(1, 0)}$

$s(2)$

7. Let σ_1 and σ_2 be two pure stationary strategies. Let σ_3 be a stationary strategy that at every state s chooses an action a that maximizes

$$\lambda r(s, a) + (1 - \lambda) \sum_{s' \in S} q(s' \mid s, a) \max\{\gamma_\lambda(s'; \sigma_1), \gamma_\lambda(s'; \sigma_2)\}.$$

Prove that

$$\gamma_\lambda(s; \sigma_3) \ge \max\{\gamma_\lambda(s; \sigma_1), \gamma_\lambda(s; \sigma_2)\}, \quad \forall s \in S.$$

8. Let Γ be a Markov decision problem and let s be a state. In view of Comment 1.20, is it true that for $\lambda = \frac{1}{T}$ we have $v_\lambda(s) = v_T(s)$? If so, prove it. If not, explain why an inequality does not necessarily hold.

9. Show that every contracting mapping is continuous.

10. Show that for every polynomial P there exist a Markov decision problem and an initial state s such that $v_\lambda(s) = P(\lambda)$ for all $\lambda \in (0, 1]$.

11. Let σ be a strategy in a Markov decision problem Γ, and let $\lambda \in (0, 1)$. Prove that σ is λ-discounted optimal at the initial state s if and only if the following condition holds: For every history $h_t \in H$ that satisfies $\mathbf{P}_{s,\sigma}(h_t) > 0$ and every action $a' \in A(s_t)$ that satisfies $\sigma(a' \mid h_t) > 0$,

$$a' \in \operatorname{argmax}_{a \in A(s_t)} \left\{ \lambda r(s_t, a) + (1 - \lambda) \sum_{s' \in S} q(s' \mid s_t, a) v_\lambda(s') \right\}.$$

12. Prove that for each fixed stationary (not necessarily pure) strategy x, the function $\lambda \mapsto \gamma_\lambda(s; x)$ is rational.

13. Find a Markov decision problem that satisfies the following two properties:

    - There is a strategy σ that is λ-discounted optimal for $\lambda = 1$ and for every λ sufficiently close to 0, but is not optimal for $\lambda = \frac{1}{2}$.
    - There is a strategy σ' that is not λ-discounted optimal for $\lambda = 1$ and for every λ sufficiently close to 0, but is optimal for $\lambda = \frac{1}{2}$.

14. Let $X \subseteq \mathbb{R}^n$ and $Y \subseteq \mathbb{R}^m$ be two closed sets. A *correspondence* $F \colon X \rightrightarrows Y$ is a mapping that assigns to each point $x \in X$ a subset $F(x) \subseteq Y$. We say that the correspondence F has non-empty values if $F(x) \neq \emptyset$ for every $x \in X$. The graph of a correspondence F is $\operatorname{Graph}(F) = \{(x, y) \in \mathbb{R}^{n+m} : y \in F(x)\}$.

    Let $X \subseteq \mathbb{R}^n$ be a compact set. Let $F \colon X \times X \rightrightarrows \mathbb{R}$ and $G \colon X \rightrightarrows X$ be two correspondences with non-empty values and compact graphs and let $\lambda \in (0, 1)$. Prove that there exists a unique function $f \colon X \rightrightarrows \mathbb{R}$ such that

    $$f(x) = \max_{y \in G(x)} (F(x, y) + \lambda f(y)).$$

15. Let $\Gamma = \langle S, (A(s))_{s \in S}, q, r \rangle$ be a Markov decision problem, and consider the following linear program in the variables $(v(s))_{s \in S}$:

    Minimize $\displaystyle \sum_{s \in S} v(s)$

    Subject to $v(s) \geq \lambda r(s, a) + (1 - \lambda) \displaystyle\sum_{s' \in S} q(s' \mid s, a) v(s'), \ \forall s \in S, a \in A.$

    Show that the solution $(v(s))_{s \in S}$ of this linear program has the property that $v(s)$ is the λ-discounted value at the initial state s.

16. Prove Theorem 1.30: Let $\Gamma = \langle S, (A(s))_{s \in S}, q, r \rangle$ be a Markov decision problem, let $\lambda \in (0, 1]$ be a discount factor, and let $v_\lambda(s)$ be the λ-discounted value at the initial state s, for every $s \in S$. A stationary

strategy x is λ-discounted optimal at all initial states if and only if, for every state $s \in S$, the mixed action $x(s)$ satisfies

$$v_\lambda(s) = \lambda r(s, x(s)) + (1 - \lambda) \sum_{s' \in S} q(s' \mid s, x(s)) v_\lambda(s').$$

17. Let $\Gamma = \langle S, (A(s))_{s \in S}, q, r \rangle$ be a Markov decision problem where S is countable, $A(s)$ is finite for every $s \in S$, and r is bounded. Prove that for every $\lambda \in (0, 1]$ the λ-discounted value exists at all initial states, and moreover the decision maker has a pure stationary λ-discounted optimal strategy.

18. Does $\lim_{\lambda \to 0} v_\lambda(s)$ exist in every Markov decision problem $\Gamma = \langle S, (A(s))_{s \in S}, q, r \rangle$ for every $s \in S$, where S is countable, $A(s)$ is finite for every $s \in S$, and r is bounded? Prove or provide a counterexample.

2

A Tauberian Theorem and Uniform ϵ-Optimality in Hidden Markov Decision Problems

In Chapter 1 we assumed that at every stage t the decision maker observes the state s_t before choosing an action. In practice, this is not necessarily the case. For example, suppose that a fishing company operates all fishing boats on a certain lake. The state variable is the number of fish in the lake, which is affected by the fishing intensity as well as by random factors. In each period (day, week, or month), the company has to decide what is the target quantity of fish. In this example, the state variable is not known to the company, and can only be estimated.

In this chapter, we will address the case where the decision maker does not observe the current state. All that the decision maker knows, besides the data of the problem (which includes the set of states, her sets of actions, the transition rule, and the payoff function) is the initial state (or the probability distribution according to which the initial state is chosen) and the actions she chose in the past. In particular, for every stage $t \in \mathbb{N}$ the decision maker can calculate for each state s the probability that $s_t = s$. This model is called a *hidden Markov decision problem*.

For our study, we will prove a Tauberian Theorem that relates the Abel limit and the Cesàro limit of a sequence of real numbers, and apply it to prove that a uniformly ϵ-optimal strategy exists in hidden Markov decision problems.

2.1 A Tauberian Theorem

Tauberian theorems[1] provide conditions that ensure that two ways of calculating limits give the same result. The following result relates the Abel limit[2] to the Cesàro limit.[3]

Theorem 2.1   *Let $(z_n)_{n=1}^{\infty}$ be a sequence of real numbers in the interval $[0,1]$. For each $k \in \mathbb{N}$, set $\bar{z}_k := \frac{1}{k}\sum_{n=1}^{k} z_n$, and for each $\lambda \in (0,1]$ set $\bar{z}_\lambda := \sum_{n=1}^{\infty} \lambda(1-\lambda)^{n-1}z_n$. Then*

$$\liminf_{n\to\infty} \bar{z}_n \leq \liminf_{\lambda\to 0} \bar{z}_\lambda \leq \limsup_{\lambda\to 0} \bar{z}_\lambda \leq \limsup_{n\to\infty} \bar{z}_n.$$

Proof   Since $\limsup_{\lambda\to 0} \bar{z}_\lambda = -\liminf_{\lambda\to 0}(-\bar{z}_\lambda)$ and $\limsup_{n\to\infty} \bar{z}_n = -\liminf_{n\to\infty}(-\bar{z}_n)$, it is sufficient to prove the leftmost inequality, namely $\liminf_{n\to\infty} \bar{z}_n \leq \liminf_{\lambda\to 0} \bar{z}_\lambda$.

Note that, for every $n \geq 0$,

$$\sum_{k=n}^{\infty} \lambda^2(1-\lambda)^k = \lambda^2(1-\lambda)^n \left(1 + (1-\lambda) + (1-\lambda)^2 + \cdots \right)$$

$$= \lambda(1-\lambda)^n. \tag{2.1}$$

Using Eq. (2.1), we now present the discounted sum $\bar{z}_\lambda$ as a weighted average of the arithmetic means $(\bar{z}_n)_{n\in\mathbb{N}}$:

$$\bar{z}_\lambda = \sum_{n=1}^{\infty} \lambda(1-\lambda)^{n-1}z_n \tag{2.2}$$

$$= \sum_{n=1}^{\infty} \sum_{k=n-1}^{\infty} \lambda^2(1-\lambda)^k z_n \tag{2.3}$$

$$= \sum_{n=1}^{\infty} \sum_{k=n}^{\infty} \lambda^2(1-\lambda)^{k-1} z_n \tag{2.4}$$

$$= \sum_{k=1}^{\infty} \sum_{n=1}^{k} \lambda^2(1-\lambda)^{k-1} z_n \tag{2.5}$$

---

[1] Alfred Tauber (Pressburg, Austrian Empire, November 5, 1866 – Theresienstadt Concentration Camp, Czechoslovakia, July 26, 1942) was a Jewish mathematician known for his contributions to mathematical analysis and to the theory of functions of a complex variable.

[2] Niels Henrik Abel (Nedstrand, Denmark–Norway, August 5, 1802 – Froland, Norway, April 6, 1829) was a Norwegian mathematician who demonstrated the impossibility of solving the general quintic equation in radicals. He was also an innovator in the field of elliptic functions and the discoverer of Abelian functions. The Abel Prize in mathematics, originally proposed in 1899 to complement the Nobel Prizes, is named in his honor.

[3] Ernesto Cesàro (Naples, Italy, March 12, 1859 – Torre Annunziata, Italy, September 12, 1906) was an Italian mathematician who worked in the field of differential geometry. Among his other contributions, he described fractals, space filling curves, and the averaging method called after him – Cesàro summation of divergent series.

$$= \sum_{k=1}^{\infty} \left(\lambda^2 (1 - \lambda)^{k-1} \sum_{n=1}^{k} z_n \right) \tag{2.6}$$

$$= \sum_{k=1}^{\infty} \left(k \lambda^2 (1 - \lambda)^{k-1} \cdot \frac{1}{k} \sum_{n=1}^{k} z_n \right) \tag{2.7}$$

$$= \sum_{k=1}^{\infty} k \lambda^2 (1 - \lambda)^{k-1} \bar{z}_k. \tag{2.8}$$

Eq. (2.2) is the definition of $\bar{z}_\lambda$; Eq. (2.3) holds by Eq. (2.1); Eq. (2.5) is a change of the order of summation; to obtain Eq. (2.7), we multiplied and divided by k; and Eq. (2.8) follows by the definition of $\bar{z}_k$.

Denote $a := \liminf_{n \to \infty} \bar{z}_n$, fix $\epsilon > 0$, and choose $N_0 \in \mathbb{N}$ sufficiently large such that $\bar{z}_k \geq a - \epsilon$ for all $k \geq N_0$. Let λ_0 be sufficiently small such that

$$\sum_{k=1}^{N_0} k \cdot (\lambda_0)^2 \cdot (1 - \lambda_0)^{k-1} < \epsilon, \quad \forall \lambda \in (0, \lambda_0]; \tag{2.9}$$

for example, take $\lambda_0 \leq \sqrt{\epsilon}/N_0$. Since the payoffs are between 0 and 1, and since $\bar{z}_k \geq a - \epsilon$ for every $k \geq N_0$, we obtain that for every $\lambda \in (0, \lambda_0)$,

$$\bar{z}_\lambda = \sum_{k=1}^{\infty} k \lambda^2 (1 - \lambda)^{k-1} \bar{z}_k \tag{2.10}$$

$$= \sum_{k=1}^{N_0 - 1} k \lambda^2 (1 - \lambda)^{k-1} \bar{z}_k + \sum_{k=N_0}^{\infty} k \lambda^2 (1 - \lambda)^{k-1} \bar{z}_k \tag{2.11}$$

$$\geq \sum_{k=N_0}^{\infty} k \lambda^2 (1 - \lambda)^{k-1} \bar{z}_k \tag{2.12}$$

$$\geq (a - \epsilon) \sum_{k=N_0}^{\infty} k \lambda^2 (1 - \lambda)^{k-1} \tag{2.13}$$

$$= (a - \epsilon) \left(\sum_{k=1}^{\infty} k \lambda^2 (1 - \lambda)^{k-1} - \sum_{k=1}^{N_0 - 1} k \lambda^2 (1 - \lambda)^{k-1} \right) \tag{2.14}$$

$$= (a - \epsilon) \left(1 - \sum_{k=1}^{N_0 - 1} k \lambda^2 (1 - \lambda)^{k-1} \right) \tag{2.15}$$

$$\geq (a - \epsilon)(1 - \epsilon) > a - 2\epsilon, \tag{2.16}$$

where Eq. (2.10) follows from Eqs. (2.2)–(2.8); Eq. (2.12) holds since payoffs are nonnegative; Eq. (2.13) holds by the choice of N_0; and Eq. (2.16) follows from Eq. (2.9). To see that Eq. (2.15) holds as well, substitute $z_n = 1$ for all $n \in \mathbb{N}$ in Eqs. (2.2)–(2.8), to obtain that the left-hand side is 1 and the right-hand side is $\sum_{k=1}^{\infty} k \lambda^2 (1 - \lambda)^{k-1}$. Thus, $\bar{z}_\lambda \geq a - 2\epsilon$ for every $\lambda \in (0, \lambda_0)$. Consequently, $\liminf_{\lambda \to 0} \bar{z}_\lambda \geq a - 2\epsilon$. Since $\epsilon > 0$ is arbitrary, we conclude that $\liminf_{\lambda \to 0} \bar{z}_\lambda \geq a$. □

2.2 Hidden Markov Decision Problems

Similar to a Markov decision problem, a *hidden Markov decision problem* is given by a vector $\Gamma = \langle S, (A(s))_{s \in S}, q, r \rangle$, where S is a finite set of states; $A(s)$ is a finite set of actions available to the decision maker at state s for each $s \in S$; q is the transition rule; and r is the payoff function.

The decision maker does not observe the state, yet has to choose an action at each stage. To ensure that the chosen action is possible at the current unobserved state, we will assume in this chapter that the same actions are available in all states.

Assumption 2.1 $A(s) = A(s')$ *for any two states* $s, s' \in S$.

Let A denote the common set of actions. To simplify the calculations, we will also assume without loss of generality that payoffs are nonnegative and bounded by 1.

Assumption 2.2   *For every state* $s \in S$ *and every action* $a \in A$, *we have* $r(s, a) \in [0, 1]$.

Since along the process the decision maker does not know the state, it is more convenient to assume that the initial state s is not necessarily chosen deterministically. Rather, the choice is made according to a probability distribution $p_1 \in \Delta(S)$, which is known to the decision maker.

The probability that the state at stage t is equal to s depends on the actions that the decision maker played up to stage $t - 1$, and can be calculated by the law of total probability. This probability, which is denoted by $y_t(s)$, is given by

$$y_1(s) = p_1(s),$$

$$y_t(s) = \sum_{s' \in S} y_{t-1}(s') q(s \mid s', a_{t-1}), \quad \forall t \geq 2.$$

Note that y_t is a random variable that depends on the actions played by the decision maker up to stage $t - 1$. Because the number of actions is finite, the set of all possible conditional probabilities that can arise at stage t is finite.

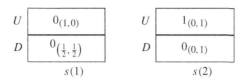

Figure 2.1 The Markov decision problem in Example 2.2.

Example 2.2 Consider the Markov decision problem with two states and two actions in each state that is depicted in Figure 2.1.

The quantity $y_t(s(2))$ is the probability that the state in stage t is $s(2)$. Since once the play visits state $s(2)$, it remains there forever, the sequence $(y_t(s(2)))_{t \in \mathbb{N}}$ is nondecreasing.

Suppose that the initial state is $s(1)$, that is, p_1 is the Dirac measure concentrated at $s(1)$. When the decision maker plays U, the state does not change, and therefore $y_t = y_{t-1}$. When the decision maker plays D, the probability that the new state is $s(1)$ is divided by 2 and we have

$$y_t(s(1)) = \frac{y_{t-1}(s(1))}{2}, \quad y_t(s(2)) = y_{t-1}(s(2)) + \frac{1}{2}y_{t-1}(s(1)).$$

In particular, if the number of times that the decision maker played the action D in the past is k, then the probability that the current state is $s(1)$ is equal to $\frac{1}{2^{k+1}}$, and the probability that the current state is $s(2)$ is equal to $1 - \frac{1}{2^{k+1}}$. ◆

The information available to the decision maker at stage t consists of her own past actions. Consequently, a history of length t in a hidden Markov decision problem includes the actions that were chosen by the decision maker in the first $t-1$ stages, and therefore the set of histories is $H := \bigcup_{t \in \mathbb{N}} A^t$. The set of *plays* is A^∞.

Definition 2.3 A *strategy* in a hidden Markov decision problem $\Gamma = \langle S, (A(s))_{s \in S}, q, r \rangle$ is a mapinng $\sigma : H \to \Delta(A)$.

Denote by $\Sigma := (\Delta(A))^H = \{\sigma : H \to \Delta(A)\}$ the set of all strategies. The T-stage value and the λ-discounted value in a hidden Markov decision problem at the initial distribution p_1 are defined analogously to Definitions 1.14 and 1.21, and are denoted by $v_T(p_1)$ and $v_\lambda(p_1)$, respectively.

Definition 2.4 Let $p_1 \in \Delta(S)$. The real number $v(p_1)$ is the *value* of the problem with initial distribution p_1 if $v(p_1) = \lim_{T \to \infty} v_T(p_1) = \lim_{\lambda \to 0} v_\lambda(p_1)$. Let $\epsilon \geq 0$. A strategy σ is *uniformly ϵ-optimal* at the initial distribution p_1 if there are $T_0 \in \mathbb{N}$ and $\lambda_0 \in (0, 1]$ such that

$$\gamma_T(p_1;\sigma) \geq v_T(p_1) - \epsilon, \quad \forall T \geq T_0, \tag{2.17}$$

$$\gamma_\lambda(p_1;\sigma) \geq v_\lambda(p_1) - \epsilon, \quad \forall \lambda \in (0,\lambda_0). \tag{2.18}$$

This definition differs from Definition 1.36 of a uniformly optimal strategy in two aspects.

- Whereas a uniformly optimal strategy is optimal for every discount factor sufficiently close to 0, a strategy is uniformly ϵ-optimal if the decision maker may gain by switching to another strategy, but by doing so she can gain at most ϵ.
- Whereas a uniformly optimal strategy is required to be optimal only in the discounted evaluation, a uniformly ϵ-optimal strategy is required to be ϵ-optimal both in the discounted evaluation and in the T-stage evaluation.

In Example 1.13, the strategy σ_D that always plays the action D is uniformly ϵ-optimal for every $\epsilon > 0$, but not for $\epsilon = 0$. In Example 2.2, the strategy that plays T times the action D and thereafter plays the action U is uniformly ϵ-optimal, provided $2^T < \epsilon$ (Exercise 2.2).

Theorem 2.5   *The value exists for every initial distribution $p_1 \in \Delta(S)$. Moreover, there exists a pure uniformly ϵ-optimal strategy for every $\epsilon > 0$.*

The rest of this section is devoted to the proof of Theorem 2.5. We first define the expected average payoff between two stages. For two positive integers $t_1, t_2 \in \mathbb{N}$ such that $t_1 < t_2$, let

$$\gamma_{t_1,t_2}(p_1;\sigma) := \mathbf{E}_{p_1,\sigma}\left[\frac{1}{t_2 - t_1 + 1}\sum_{t=t_1}^{t_2} r(s_t,a_t)\right].$$

Note that $\gamma_T(p_1;\sigma) = \gamma_{1,T}(p_1;\sigma)$.

As the following lemma states, the function $p_1 \mapsto \gamma_{t_1,t_2}(p_1;\sigma)$ is 1-Lipshitz.

Lemma 2.6   *For any strategy σ, any two initial distributions $p_1, p_1' \in \Delta(S)$, and any two integers $t_1 < t_2$,*

$$|\gamma_{t_1,t_2}(p_1;\sigma) - \gamma_{t_1,t_2}(p_1';\sigma)| \leq \|p_1 - p_1'\|_1.$$

Proof   By the law of total expectation,

$$\mathbf{E}_{p_1,\sigma}[r(s_t,a_t)] = \sum_{s\in S} p_1(s) \cdot \mathbf{E}_{s,\sigma}[r(s_t,a_t)].$$

This implies that

$$\gamma_{t_1,t_2}(p_1;\sigma) = \sum_{s\in S} p_1(s) \cdot \gamma_{t_1,t_2}(s;\sigma).$$

Similarly,

$$\gamma_{t_1,t_2}(p_1';\sigma) = \sum_{s\in S} p_1'(s) \cdot \gamma_{t_1,t_2}(s;\sigma).$$

Subtracting the second equality from the first and using the fact that payoffs are nonnegative and bounded by 1, we obtain

$$|\gamma_{t_1,t_2}(p_1;\sigma) - \gamma_{t_1,t_2}(p_1';\sigma)| \le \sum_{s\in S} |p_1(s) - p_1'(s)| \cdot \gamma_{t_1,t_2}(s;\sigma)$$

$$\le \sum_{s\in S} |p_1(s) - p_1'(s)| = \|p_1 - p_1'\|_1. \quad \square$$

Since the decision maker does not observe the state, a pure strategy defines uniquely the sequence of actions that the decision maker plays. Because a pure strategy is defined for *every* history, it also indicates how to play after histories that were not played, but this part of the pure strategy does not affect the decision maker's payoff.

In what follows, each infinite sequence of actions $(a_t)_{t\in\mathbb{N}}$ represents a pure strategy, namely, a pure strategy σ that satisfies

$$\sigma(a_t \mid a_1,a_2,\ldots,a_{t-1}) = 1, \quad \forall t \in \mathbb{N}.$$

To show that there is a pure uniformly ϵ-optimal strategy, it suffices to define the sequence of actions that the decision maker plays and yields high payoff in all discounted games (for discount factors close to 0) and in all T-stage games (for all T sufficiently large).

For a strategy σ, an $\epsilon \in (0,1)$, and a $T \in \mathbb{N}$, define

$$T_0 = T_0(\sigma,\epsilon,T) := 1 + \max\{t \le T : \gamma_t(p_1;\sigma) < \gamma_T(p_1;\sigma) - \epsilon\}, \quad (2.19)$$

where the maximum of an empty set is 0. The maximum in Eq. (2.19) is strictly smaller than T, hence $T_0(\sigma,\epsilon,T) \le T$. Note that if $T_0 > 1$, then, in particular,

$$\gamma_{T_0-1}(p_1;\sigma) < \gamma_T(p_1;\sigma) - \epsilon, \quad (2.20)$$

and for every $t = \{T_0, T_0+1, \ldots, T\}$, we have

$$\gamma_T(p_1;\sigma) - \epsilon \le \gamma_t(p_1;\sigma). \quad (2.21)$$

The next result asserts that T_0 cannot be too close to T, and that for every $t \in \{T_0, T_0+1, \ldots, T\}$, the average payoff between stage T_0 and t is high.

Lemma 2.7   *We have $T_0 \leq 1 + (1 - \epsilon)T$. Moreover, for every t such that $T_0 \leq t \leq T$,*

$$\gamma_{T_0,t}(p_1;\sigma) \geq \gamma_T(p_1;\sigma) - \epsilon.$$

Proof   If $T_0 = 1$, then clearly $T_0 \leq 1 + (1 - \epsilon)T$, and the definition of T_0 implies that $\gamma_k(p_1;\sigma) \geq \gamma_T(p_1;\sigma) - \epsilon$ for every k such that $1 \leq k \leq T$, which is the desired result.

Suppose then that $T_0 > 1$, and assume to the contrary that $T_0 > 1+(1-\epsilon)T$, that is,

$$\frac{T - T_0 + 1}{T} < \epsilon. \tag{2.22}$$

Since $T_0 > 1$, it follows that $\gamma_T(p_1;\sigma) > \epsilon$. Indeed, if $\gamma_T(p_1;\sigma) \leq \epsilon$ then, since payoffs are nonnegative, we would have $T_0 = 1$. Since the payoffs are bounded by 1, Eqs. (2.20) and (2.22) imply that

$$\begin{aligned}
\gamma_T(p_1;\sigma) &\leq \frac{T_0 - 1}{T} \cdot \gamma_{T_0-1}(p_1;\sigma) + \frac{T - T_0 + 1}{T} \\
&< \frac{T_0 - 1}{T} \cdot (\gamma_T(p_1;\sigma) - \epsilon) + \epsilon \\
&< 1 \cdot (\gamma_T(p_1;\sigma) - \epsilon) + \epsilon = \gamma_T(p_1;\sigma),
\end{aligned}$$

a contradiction; the last inequality above holds because $\gamma_T(p_1;\sigma) > \epsilon$. The first claim of the lemma is established. Let us prove the second claim. Take t such that $T_0 \leq t \leq T$. By the definition of T_0,

$$\begin{aligned}
\gamma_T(p_1;\sigma) - \epsilon &\leq \gamma_t(p_1;\sigma) \\
&= \frac{T_0 - 1}{T} \cdot \gamma_{T_0-1}(p_1;\sigma) + \frac{T - T_0 + 1}{T} \cdot \gamma_{T_0,t}(p_1;\sigma).
\end{aligned}$$

Thus, $\gamma_T(p_1;\sigma) - \epsilon$ is smaller than the weighted average of two terms. By Eq. (2.20), the first term is smaller than $\gamma_T(p_1;\sigma) - \epsilon$. It follows that the second term is larger than $\gamma_T(p_1;\sigma) - \epsilon$, which is what we wanted to prove.   □

The first conclusion of Lemma 2.7 can be recast as follows.

Corollary 2.8   *For every strategy σ, every $T \in \mathbb{N}$, and every $\epsilon > 0$,*

$$T - T_0(\sigma,\epsilon,T) \geq \epsilon T - 1.$$

We are now ready to prove Theorem 2.5.

Proof of Theorem 2.5   Denote $v_*(p_1) := \limsup_{T \to \infty} v_T(p_1)$. We will prove that for every $\epsilon > 0$, there is a strategy σ_* such that $\gamma_T(p_1;\sigma_*) \geq v_*(p_1) - \epsilon$, for all sufficiently large $T \in \mathbb{N}$.

For $T \in \mathbb{N}$, let σ_T be a pure optimal strategy in the T-stage problem. Such a strategy exists by Theorem 1.17. Can this theorem be applied in our setup? Theorem 1.17 holds for T-stage problems with finite state space. In our setup, the state at stage t is the conditional distribution y_t. As we mentioned earlier, the number of possible conditional distributions y_t in each stage is finite, and since the number of stages in the T-stage problem is finite (it is precisely T), it follows that the total number of conditional probabilities up to stage T is finite.

Fix $\epsilon > 0$ and set $z_T := y_{T_0(\sigma_T, \epsilon, T)}$. This is the conditional distribution over S under the strategy σ_T at stage $T_0(\sigma_T, \epsilon, T)$. Since the strategy σ_T is pure, it plays a deterministic sequence of actions up to stage T. Denote the part of the sequence of actions between stages $T_0(\sigma_T, \epsilon, T)$ and T by

$$\vec{a}^T = \left(a_1^T, a_2^T, \ldots, a_{T-T_0(\sigma_T, \epsilon, T)+1}^T\right). \tag{2.23}$$

By Lemma 2.7, for every $T \in \mathbb{N}$ and every t such that $T_0(\sigma_T, \epsilon, T) \leq t \leq T$ we have

$$v_T(p_1) - \epsilon \leq \gamma_{T_0(\sigma_T, \epsilon, T), t}(p_1; \sigma_T) = \gamma_{t-T_0(\sigma_T, \epsilon, T)+1}(z_T, \vec{a}^T). \tag{2.24}$$

Since payoffs are bounded, there is an increasing sequence of positive integers $(T_k)_{k \in \mathbb{N}}$ such that

$$\lim_{k \to \infty} v_{T_k}(p_1) = v_*(p_1).$$

For each $T \in \mathbb{N}$, we have $z_T \in \Delta(S)$. Since the set of states is finite, the set $\Delta(S)$ is compact, and therefore by taking a subsequence we can assume without loss of generality that the limit

$$z_* := \lim_{k \to \infty} z_{T_k}$$

exists.

The set of actions A is finite, and therefore the set $A^{\mathbb{N}}$ is compact in the product topology. An element in $A^{\mathbb{N}}$ is denoted by $\vec{a} = (\vec{a}_t)_{t \in \mathbb{N}}$, and a sequence of elements $(\vec{a}^{(k)})_{k \in \mathbb{N}}$ in $A^{\mathbb{N}}$ converges to a limit $\vec{a}$ if $\lim_{k \to \infty} \vec{a}_t^{(k)} = \vec{a}_t$ for all $t \in \mathbb{N}$. Again thanks to the finiteness of the set of actions A, this condition means that for every t there is a $K_t \in \mathbb{N}$ such that $\vec{a}_t^{(k)} = \vec{a}_t$ for all $k \geq K_t$; or, equivalently, for every $t \in \mathbb{N}$ there is a $K_t \in \mathbb{N}$ such that the prefix of length t of $\vec{a}^{(k)}$ coincides with the prefix of length t of $\vec{a}$, for every $k \geq K_t$.

We return to the sequence $(\vec{a}^T)_{T \in \mathbb{N}}$, see Eq. (2.23). Since the set $A^{\mathbb{N}}$ is compact, $(\vec{a}^T)_{T \in \mathbb{N}}$ has a convergent subsequence in the product topology. That is, without loss of generality we can assume that there is a vector $\vec{a}^* = (\vec{a}_t^*)_{t \in \mathbb{N}}$ such that for every $t \in \mathbb{N}$ there is a $K_0 \in \mathbb{N}$ such that the prefix of length t of $\vec{a}^{T_k}$ coincides with the prefix of length t of $\vec{a}^*$, for every $k \geq K_0$.

Letting $k \to \infty$ in Eq. (2.24) and using Lemma 2.6, we obtain that

$$\gamma_t(z_*; \vec{a}^*) \geq v_*(p_1) - 2\epsilon, \quad \forall t \in \mathbb{N}. \tag{2.25}$$

Thus, we found a pure strategy whose t-stage payoff is at least $v_*(p_1) - 2\epsilon$ for every $t \in \mathbb{N}$, when the initial distribution is z_*. Unfortunately, the initial distribution is p_1 and not z_*. All that is left to do is to add to the pure strategy $\vec{a}^*$ a prefix, which ensures that the distribution becomes close to z_* when the initial distribution is p_1.

Let then $k \in \mathbb{N}$ be sufficiently large such that $\|z_{T_k} - z_*\|_1 \leq \epsilon$. Consider the pure strategy σ_*, which follows σ_{T_k} in the first $T_0(\sigma_{T_k}, \epsilon, T_k)$ stages and after that follows $\vec{a}^*$:

$$\left(a_1^{T_k}, a_2^{T_k}, \ldots, a_{T_0(\sigma_{T_k}, \epsilon, T_k)}^{T^k}, a_1^*, a_2^*, \ldots\right).$$

Under σ_*, the distribution over states at stage $T_0(\sigma_{T_k}, \epsilon, T_k)$ is z_{T_k}. Therefore,

$$\gamma_{T_0(\sigma_{T_k}, \epsilon, T_k), t}(p_1; \sigma_*) = \gamma_{t - T_0(\sigma_{T_k}, \epsilon, T_k)}(z_{T_k}; \vec{a}^*), \tag{2.26}$$

for all $t \geq T_0(\sigma_{T_k}, \epsilon, T_k)$. By Lemma 2.6,

$$|\gamma_{t - T_0(\sigma_{T_k}, \epsilon, T_k)}(z_{T_k}; \vec{a}^*) - \gamma_{t - T_0(\sigma_{T_k}, \epsilon, T_k)}(z_*; \vec{a}^*)| \leq \|z_{T_k} - z_*\|_1 \leq \epsilon. \tag{2.27}$$

By Eq. (2.25),

$$\lim_{t \to \infty} \gamma_{t - T_0(\sigma_{T_k}, \epsilon, T_k)}(z_*; \vec{a}^*) \geq v_*(p_1) - 2\epsilon. \tag{2.28}$$

Combining Eqs. (2.26)–(2.28) we obtain

$$\gamma_{T_0(\sigma_{T_k}, \epsilon, T_k), t}(p_1; \sigma_*) \geq v_*(p_1) - 3\epsilon,$$

provided t is large enough. For every $t > T_0(\sigma_{T_k}, \epsilon, T_k)$,

$$\gamma_t(p_1; \sigma_{T_k}) = \frac{T_0 - 1}{t} \cdot \gamma_{T_0 - 1}(p_1; \sigma_{T_k}) + \frac{t - T_0 + 1}{t} \cdot \gamma_{T_0, t}(p_1; \sigma_{T_k})$$

$$\geq \frac{t - T_0 + 1}{t} \cdot \gamma_{T_0, t}(p_1; \sigma_{T_k}).$$

Hence, for t sufficiently large,

$$\gamma_t(p_1; \sigma_*) \geq v_*(p_1) - 4\epsilon, \tag{2.29}$$

as claimed. Eq. (2.29) implies that $v_t(p_1) \geq v_*(p_1) - 4\epsilon$ for t sufficiently large. Since $\epsilon > 0$ is arbitrary, and since $v_*(p_1) = \limsup_{t \to \infty} v_t(p_1)$, we conclude that the limit $\lim_{t \to \infty} v_t(p_1)$ exists and is equal to $v_*(p_1)$. From this and from Eq. (2.29), we deduce that Eq. (2.17) holds with 4ϵ for σ_*.

We will finally show that Eq. (2.18) holds with 6ϵ for σ_*. Applying Theorem 2.1 to the sequence $(z_n)_{n\in\mathbb{N}}$ defined by $z_n = \mathbf{E}_{p_1,\sigma_*}[r(s_n,a_n)]$, we obtain that

$$\liminf_{\lambda\to 0} \gamma_\lambda(p_1;\sigma_*) \geq v_*(p_1) - 4\epsilon.$$

Since $v_t(p_1)$ is the value of the T-stage problem at the initial distribution p_1,

$$\gamma_t(p_1;\sigma) \leq v_*(p_1) + \epsilon, \quad \forall t \geq T, \forall \sigma \in \Sigma.$$

Since payoffs are bounded by 1, the proof of Theorem 2.1 implies that for every $\lambda \in (0, \sqrt{\epsilon}/T)$ we have

$$\gamma_\lambda(p_1;\sigma) \leq v_*(p_1) + 2\epsilon, \quad \forall t \geq T, \forall \sigma \in \Sigma.$$

We deduce that for every $\lambda > 0$ sufficiently small,

$$v_\lambda(p_1) \leq v_*(p_1) + 2\epsilon \leq \gamma_\lambda(p_1;\sigma_*) + 6\epsilon,$$

as desired. $\square$

2.3 Comments and Extensions

In Chapter 1, we studied the case in which the decision maker observed the current state s_t, and in this chapter, we studied the case in which the only information available to the decision maker is the initial distribution and her own past actions. There are intermediate cases in which the decision maker does not know the state for sure, yet at every stage she does receive some information on the state. Theorem 2.5 can be generalized to this setup, see Rosenberg et al. (2002), from where the proof of Theorem 2.5 is taken. Further extensions of Theorem 2.5 can be found in Renault (2011, 2014).

We studied the cases where the players are interested in the discounted sum of their stage payoff or in the average payoff in the first T stages. These two evaluations are special cases of a general family of evaluations. Let $\theta = (\theta_t)_{t=1}^\infty$ be a sequence of nonnegative reals that sum to 1. The θ-payoff of a strategy σ at the initial state s is the quantity

$$\gamma_\theta(s;\sigma) := \mathbf{E}_{s,\sigma}\left[\sum_{t=1}^\infty \theta_t r(s_t,a_t)\right],$$

and the θ-value at the initial state s is the quantity

$$v_\theta(s) := \sup_{\sigma\in\Sigma} \gamma_\theta(s;\sigma).$$

For every $\lambda \in (0, 1]$, the λ-discounted evaluation is a θ-evaluation, where $\theta_t := \lambda(1 - \lambda)^{t-1}$ for every $t \in \mathbb{N}$. For every $T \in \mathbb{N}$, the T-stage evaluation is a θ-evaluation, where $\theta_t = \frac{1}{T}$ for $t = \{1, 2, \ldots, T\}$ and $\theta_t = 0$ for $t > T$. One can extend the definition of the value to include θ-evaluations: the quantity $v(s)$ is the *value* at the initial state s if for every $\epsilon > 0$ there exists $\delta > 0$ such that $|v_\theta(s) - v(s)| < \epsilon$ for every sequence $\theta = (\theta_t)_{t=1}^\infty$ of nonnegative reals that sum to 1 and is such that $\max_{t \in \mathbb{N}} \theta_t \leq \delta$. θ-evaluations and the existence of the value were studied, by, for example, Renault (2011, 2014), Venel and Ziliotto (2016), Ziliotto (2016a), and Renault and Venel (2017).

2.4 Exercises

Exercise 2.5 is used in the solution of Exercise 10.8.

1. Let $\Gamma = \langle S, (A(s))_{s \in S}, q, r \rangle$ be a Markov decision problem, and let σ be a strategy that is uniformly optimal at all initial states according to Definition 1.36. Prove that there is a $\lambda_0 > 0$ such that $\gamma_\lambda(p_1; \sigma) \geq v_\lambda(p_1)$ for all $\lambda \in (0, \lambda_0)$ and all $p_1 \in \Delta(S)$.

2. Show that in Example 2.2, the strategy that plays T times D and thereafter plays U is uniformly ϵ-optimal, provided that $2^T < \epsilon$.

3. For the following hidden Markov decision problem with two states at all initial distributions p_1, find the value and a pure strategy that is uniformly ϵ-optimal for every $\epsilon > 0$.

U	$0_{(0,1)}$
D	$1_{(1,0)}$

$s(1)$

U	$1_{(1,0)}$
D	$0_{(0,1)}$

$s(2)$

4. Find the value of the following hidden Markov decision problem with two states at all initial distributions p_1.

U	$0_{(\frac{1}{2}, \frac{1}{2})}$
D	$0_{(0,1)}$

$s(1)$

U	$1_{(1,0)}$
D	$0_{(0,1)}$

$s(2)$

5. Alice has M unfair coins; for each $i \in \{1, 2, \ldots, M\}$, the probability that the outcome of coin i is Head is $p_i \in (0, 1]$. Bob has one unfair coin; the

probability that the outcome of Bob's coin is Head is $\delta \in (0,1]$. Alice and Bob would like to choose a number in $\{1, 2, \ldots, M\}$ such that the probability that each number $j \in \{1, 2, \ldots, M\}$ is chosen is close to x_j, where $x = (x_j)_{j=1}^{M} \in \Delta(\{1, 2, \ldots, M\})$ is some given probability distribution. To this end, they do the following:

- Alice chooses a sequence $\vec{i} = (i_n)_{n \in \mathbb{N}}$ of numbers in $\{1, 2, \ldots, M\}$. The sequence must satisfy the following property: for every $j \in \{1, 2, \ldots, M\}$, the frequency of stages in which Alice chose the number j should be close to x_j: there are $\epsilon > 0$ and $N_0 \in \mathbb{N}$ such that

$$\left| \frac{1}{n} \#\{k \leq n : i_k = j\} - x_j \right| < \epsilon, \quad \forall n \geq N_0.$$

- At each period $n \in \mathbb{N}$ until a number is selected, Alice tosses the coin i_n and Bob tosses his coin. If the outcomes of both tosses is Head, the number i_n is selected. Otherwise, Alice and Bob continue to the next period.

Do the following.

(a) Prove that for every $n \geq N_0$, we have

$$\left| \frac{1}{n} \sum_{k=1}^{n} p_{i_k} - \sum_{i=1}^{M} p_i x_i \right| \leq \epsilon.$$

(b) For every $j \in \{1, 2, \ldots, M\}$, calculate the probability $A_j(\delta, \vec{i})$ that the number j is selected as a function of Alice's choices.

(c) Denote $C_n := \prod_{k=1}^{n-1} (1 - \delta p_{i_k})$ and $d_n := C_n - C_{n+1}$ for every $n \in \mathbb{N}$. Show that for every $\delta \in (0,1]$,

$$\sum_{n=1}^{\infty} \delta p_{i_n} C_n = 1.$$

(d) Show that there exists $\delta_0 \in (0,1]$ such that for every $\delta \in (0, \delta_0)$, we have

$$1 - 2\epsilon \leq \left(\sum_{i=1}^{M} p_i x_i \right) \cdot \delta \sum_{k=1}^{\infty} k d_k \leq 1 + 2\epsilon.$$

(e) Show that there is $\delta_0 \in (0,1]$ such that, provided $\delta \in (0, \delta_0)$, we have

$$\left| A_j(\delta, \vec{i}) - \frac{p_j x_j}{\sum_{i=1}^{M} p_i x_i} \right| \leq 2\epsilon.$$

6. Prove that in every Markov decision problem as studied in Chapter 1 and every initial state $s \in S$, the two limits $lim_{T \to \infty} v_T(s)$ and $lim_{\lambda \to 0} v_\lambda(s)$ exist and coincide.

   Hint: Adapt the proof of Theorem 2.5.

3

Strategic-Form Games: A Review

In the present chapter, we review material on strategic-form games that will be needed in the sequel. Readers who are interested in expanding their knowledge of strategic-form games are referred to Maschler, Solan, and Zamir (2020, chapters 4 and 5).

Definition 3.1   A *strategic-form game* is a triplet $G = \langle I, (A^i)_{i \in I}, (u^i)_{i \in I} \rangle$ where

- $I = \{1, 2, \ldots, n\}$ is a finite set of *players*.
- For each $i \in I$, A^i is a finite set of *actions* of player i. The set of all *action profiles* is denoted by $A := \prod_{i \in I} A^i$.
- $u^i : A \to \mathbb{R}$ is a *payoff function*.

Definition 3.2   A *mixed action* of player i is a probability distribution over her action set A^i.

In other words, a mixed action of player i is an element in $\Delta(A^i)$. An action is sometimes called *pure strategy*, and a mixed action is sometimes called *mixed strategy*.

Definition 3.3   A *mixed action profile* is a vector $\alpha = (\alpha^1, \alpha^2, \ldots, \alpha^n) \in \prod_{i \in I} \Delta(A^i)$ of mixed actions, one for each player.

Definition 3.4   The *expected payoff* associated with a mixed action profile $\alpha \in \prod_{i \in I} \Delta(A^i)$ is

$$\gamma^i(\alpha) := \sum_{a \in A} \left(\prod_{i \in I} \alpha^i[a^i] \right) u^i(a).$$

This is the expectation of u^i when each player i uses the mixed action α^i.

The basic solution concept that we use is equilibrium.

Definition 3.5   Let $\epsilon \geq 0$. A mixed action profile α_* is an ϵ-*equilibrium* if for every player $i \in I$,

$$\gamma^i(\alpha) \geq \sup_{\alpha^i \in \Delta(A^i)} \gamma^i(\alpha^i, \alpha_*^{-i}) - \epsilon.$$

The corresponding payoff vector $(\gamma^i(\alpha_*))_{i \in I}$ is an ϵ-*equilibrium payoff*.

A 0-equilibrium is called an *equilibrium*, and a 0-equilibrium payoff is called an *equilibrium payoff*. Thus, an equilibrium is a mixed action profile such that no player can gain by a unilateral deviation, and an ϵ-equilibrium is a mixed action profile such that no player can gain more than ϵ by a unilateral deviation.

As the following theorem, due to Nash (1950),[1] asserts, every strategic-form game (with finitely many players and actions) has an equilibrium.

Theorem 3.6   *Let $G = \langle I, (A^i)_{i \in I}, (u^i)_{i \in I} \rangle$ be a strategic-form game. If the action sets $(A^i)_{i \in I}$ are finite, then the game has at least one equilibrium.*

Denote by $E(G)$ the set of equilibrium payoffs of the strategic-form game G. The set $E(G)$ is subset of $\mathbb{R}^n$. Simple continuity arguments show the following.

Theorem 3.7   *The set $E(G)$ is compact.*

Definition 3.8   A game $G = \langle I, (A^i)_{i \in I}, (u^i)_{i \in I} \rangle$ is *zero-sum* if $|I| = n = 2$ and $u^1(a^1, a^2) + u^2(a^1, a^2) = 0$ for every $a^1 \in A^1$ and $a^2 \in A^2$.

In a zero-sum game, the payoff function of Player 2 is determined by the payoff function of Player 1. We denote the payoff function of Player 1 by u, and then the payoff function of Player 2 is $-u$. The expected payoff of Player 1 that corresponds to the mixed action profile (α^1, α^2) is denoted by $\gamma(\alpha^1, \alpha^2)$, and the expected payoff of Player 2 is $-\gamma(\alpha^1, \alpha^2)$.

For zero-sum games, the concept of equilibrium payoff is reduced to the following definition of the value.

Definition 3.9   Let $G = \langle \{1, 2\}, A^1, A^2, u \rangle$ be a zero-sum game. The real number $v \in \mathbb{R}$ is the *value* of G if

$$v = \sup_{\alpha^1 \in \Delta(A^1)} \left\{ \inf_{\alpha^2 \in \Delta(A^2)} \gamma(\alpha^1, \alpha^2) \right\} = \inf_{\alpha^2 \in \Delta(A^2)} \left\{ \sup_{\alpha^1 \in \Delta(A^1)} \gamma(\alpha^1, \alpha^2) \right\}. \quad (3.1)$$

---

[1] John Forbes Nash Jr. (Bluefield, West Virginia, June 13, 1928 – Monroe Township, New Jersey, May 23, 2015) was an American mathematician who worked in game theory and differential geometry. He shared the 1994 Nobel Prize in Economics with two other game theorists, Reinhard Selten and John Harsanyi, and the 2015 Abel Prize in Mathematics together with mathematician Louis Nirenberg.

We denote the value of G, if it exists, by val(G). A mixed action α^1 that attains the supremum in the middle term in Eq. (3.1) is called an *optimal strategy* of Player 1. Similarly, a mixed action α^2 that attains the infimum in the right-hand side term in Eq. (3.1) is called an *optimal strategy* of Player 2.

The following theorem, due to von Neumann (1928),[2] is a special case of Theorem 3.6 and preceded it by 22 years; it states that the value exists.

Theorem 3.10 *Let $G = \langle \{1,2\}, A^1, A^2, \rangle$ be a two-player zero-sum strategic-form game. If the action sets A^1 and A^2 are finite, then the game has a value.*

The analogue of the concept of ϵ-equilibrium in two-player zero-sum games is the concept of optimal strategies.

Definition 3.11 Let $\epsilon \geq 0$. A mixed action $\alpha^1 \in \Delta(A^1)$ is *ϵ-optimal* for Player 1 if

$$\gamma(\alpha^1, \alpha^2) \geq v - \epsilon, \quad \forall \alpha^2 \in \Delta(A^2).$$

Thus, a mixed action is ϵ-optimal if it guarantees that Player 1 receives at least the value minus ϵ, whatever her opponent plays. The analogous definition for Player 2 is the following.

Definition 3.12 Let $\epsilon \geq 0$. A mixed action $\alpha^2 \in \Delta(A^2)$ is *ϵ-optimal* for Player 2 if

$$\gamma(\alpha^1, \alpha^2) \leq v + \epsilon, \quad \forall \alpha^1 \in \Delta(A^1).$$

A 0-optimal strategy is called *optimal*.

Example 3.13 Consider the two-player zero-sum strategic-form game displayed in Figure 3.1. Here and in the sequel, Player 1 chooses a row and Player 2 chooses a column. Since the game is zero-sum, only the payoffs of Player 1 are listed. Formally,

- The set of players is $I = \{1, 2\}$.
- The sets of actions are $A^1 = \{T, B\}$ and $A^2 = \{L, R\}$.
- The payoff function is

$$u(T, L) = u(B, R) = 1, \quad u(T, R) = u(B, L) = 0.$$

---

[2] John von Neumann (Budapest, Austria–Hungary, December 28, 1903 – Washington, D.C., United States, February 8, 1957) was a Hungarian–American mathematician who made important contributions in quantum physics, functional analysis, set theory, computer science, economics, and many other mathematical fields. Most notably, von Neumann was a pioneer of the modern digital computer and the application of operator theory to quantum mechanics, member of the Manhattan Project Team, and creator of game theory and the concept of cellular automata.

		Player 2	
		L	R
Player 1	T	1	0
	B	0	1

Figure 3.1 The strategic-form game in Example 3.13.

If Player 1 plays $\left[\frac{1}{2}(T), \frac{1}{2}(B)\right]$, then she guarantees an expected payoff $\frac{1}{2}$:

$$\gamma\left(\left[\frac{1}{2}(T), \frac{1}{2}(B)\right], \alpha^2\right) = \frac{1}{2}, \quad \forall \alpha^2 \in \Delta(A^2).$$

Similarly, if Player 2 plays $\left[\frac{1}{2}(L), \frac{1}{2}(R)\right]$, then she guarantees $\frac{1}{2}$:

$$\gamma\left(\alpha^1, \left[\frac{1}{2}(L), \frac{1}{2}(R)\right]\right) = \frac{1}{2}, \quad \forall \alpha^1 \in \Delta(A^1).$$

Hence the value of the game is $\frac{1}{2}$, and the optimal strategies of the two players are $\left[\frac{1}{2}(T), \frac{1}{2}(B)\right]$ and $\left[\frac{1}{2}(L), \frac{1}{2}(R)\right]$, respectively.   ♦

Since the expected payoff is at most the maximal payoff in absolute values, the value is at most this quantity as well. Formally,

Theorem 3.14   *Let* $G = \langle\{1,2\}, A^1, A^2, u\rangle$ *be a two-player zero-sum strategic-form game. Then*

$$|\mathrm{val}(G)| \leq \|u\|_\infty = \max_{(a^1, a^2) \in A^1 \times A^2} |u(a^1, a^2)|.$$

The proof of Theorem 3.14 is left to the reader (Exercise 3.1). The value operator maps each payoff function to the value of the corresponding zero-sum game. As the following theorem states, it has norm 1.

Theorem 3.15   *Let* $G = \langle\{1,2\}, A^1, A^2, u\rangle$ *and* $\widehat{G} = \langle\{1,2\}, A^1, A^2, \widehat{u}\rangle$ *be two zero-sum strategic-form games with the same action sets. Then*

$$|\mathrm{val}(G) - \mathrm{val}(\widehat{G})| \leq \|u - \widehat{u}\|_\infty.$$

The proof is left to the reader (Exercise 3.2).

Comment 3.16   Our definition of strategic-form games requires that the sets of actions of the players are finite. Strategic-form games with general measurable action sets have been studied in the literature; see, for example, Fan (1952), who proved the existence of the value in two-player zero-sum strategic-form games where the action sets of the players are non-empty, compact, and convex subsets of locally convex topological linear spaces, and Glicksberg (1952), who proved the existence of equilibria when the strategy sets of the players are non-empty and compact Hausdorff spaces.

3.1 Exercises

Exercise 3.3 is used in the proof of Lemma 7.15. Exercise 3.5 is used in Section 9.5 and in the proof of Theorem 10.4.

1. Prove Theorem 3.14.
2. Prove Theorem 3.15.
3. Let $G = \langle \{1,2\}, A^1, A^2, u \rangle$ be a two-player zero-sum strategic-form game and let $\rho \in \mathbb{R}$. Let $G' = \langle \{1,2\}, A^1, A^2, u' \rangle$ be the two-player zero-sum strategic-form game defined by $u'(a_1, a_2) = u(a_1, a_2) + \rho$ for every $(a_1, a_2) \in A_1 \times A_2$. Prove that $\mathrm{val}(G') = \mathrm{val}(G) + \rho$.
4. Fix the set of players I, and for each $i \in I$ fix a finite action set A^i. For every vector of payoff functions $u = (u^i)_{i \in I}$, denote by $E(u)$ the set of Nash equilibrium payoffs in the multiplayer strategic-form game $G = \langle I, (A^i)_{i \in I}, u \rangle$. In this exercise we study the correspondences (set-valued functions) that maps to each vector of payoff functions u the set $E(u)$.

   (a) Prove that this function has a closed graph. Are its values convex?
   (b) Show that this function is not non-expansive: there exist two vectors of payoff functions $u, \widetilde{u}$ and an equilibrium payoff $w \in E(u)$ such that for every equilibrium payoff $\widehat{w} \in G(\widehat{u})$ one has

   $$\|w - \widehat{w}\|_\infty > \|u - \widehat{u}\|_\infty.$$

5. The *min-max value* of player i in a strategic-form game $G = \langle I, (A^i)_{i \in I}, (u^i)_{i \in I} \rangle$ is the quantity

   $$\overline{v}^i := \inf_{\alpha^{-i} \in \prod_{j \neq i} \Delta(A^j)} \sup_{\alpha^i \in \Delta(A^i)} u^i(\alpha^i, \alpha^{-i}). \tag{3.2}$$

   Prove that for every equilibrium α of G we have $u^i(x) \geq \overline{v}^i$.
6. In the following three-player strategic-form game each player has two actions, Player 1 chooses a row, Player 2 chooses a column, and Player 3 chooses a matrix.

(a) Prove that

$\min_{\alpha^2 \in \Delta(\{L,R\}), \alpha^3 \in \Delta(\{E,W\})} \max_{\alpha^1 \in \Delta(\{T,B\})} u^1(\alpha^1, \alpha^2, \alpha^3) = \frac{3}{4}$.

(b) Prove that

$\max_{\alpha^1 \in \Delta(\{T,B\})} \min_{\alpha^2 \in \Delta(\{L,R\}), \alpha^3 \in \Delta(\{E,W\})} u^1(\alpha^1, \alpha^2, \alpha^3) = \frac{1}{2}$.

	E — L	E — R		W — L	W — R
T	0	1	T	1	1
B	1	1	B	1	0

4

Stochastic Games: The Model

In this chapter, we define the model of stochastic games, which are the main subject of the book. These games involve several decision makers, called *players*; the stage payoff of each player, as well as the transitions, is determined by the current state and the actions chosen by all the players. Thus, each player influences the evolution of the state process and the payoffs of all players.

Definition 4.1   A *stochastic game* is a vector $\Gamma = \langle I, S, (A^i(s))_{s \in S}^{i \in I}, q, r \rangle$, where

- $I = \{1, 2, \ldots, n\}$ is a finite set of *players*.
- S is a finite set of *states*.
- For every $i \in I$ and every state $s \in S$, $A^i(s)$ is a finite set of *actions* available to player i at s. We denote the set of all *action profiles* at state s by $A(s) = \prod_{i \in I} A^i(s)$, and the set of all action profiles at all states by

$$SA = \{(s, a) \colon s \in S, a \in A(s)\}.$$

- $q \colon SA \to \Delta(S)$ is a *transition rule*.
- For every player $i \in I$, $r^i \colon SA \to \mathbf{R}$ is a *payoff function*.

   Play starts at an initial state $s_1 \in S$ and is played in stages. At each stage $t \geq 1$, the following takes place:

- The current state s_t is announced to the players.
- Each player $i \in I$ chooses an action $a_t^i \in A^i(s_t)$. The players' choices are made simultaneously and independently.
- The action profile $a_t = (a_t^i)_{i \in I}$ is publicly announced to all players.
- Each player $i \in I$ receives a stage payoff $r^i(s_t, a_t)$.
- A new state $s_{t+1} \in S$ is drawn according to the probability distribution $q(\cdot \mid s_t, a_t)$, and the game proceeds to stage $t + 1$.

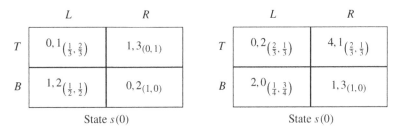

Figure 4.1 The game in Example 4.3.

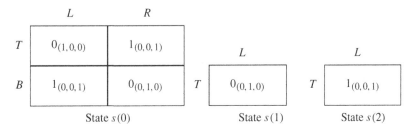

Figure 4.2 The game in Example 4.4.

We will deal with stochastic games with finitely many players, states, and actions for each player.

Example 4.2   Every repeated game is a stochastic game with a single state.   ◆

Example 4.3   Consider the two-player stochastic game with two states that is depicted in Figure 4.1. Formally, the data of the game are given as follows:

- There are two players: $I = \{1, 2\}$.
- There are two states: $S = \{s(0), s(1)\}$.
- In each state, both players have two actions.
- Each entry specifies the payoffs to players, as well as the transitions (in parentheses). For example, if in state $s(0)$ the players select the action pair (T, L), then the payoff of Player 1 is 0, the payoff of Player 2 is 1, with probability $\frac{1}{3}$ the play remains in $s(0)$, and with probability $\frac{2}{3}$ the play moves to $s(1)$.   ◆

Example 4.4   Consider the two-player zero-sum stochastic game with three states that is depicted in Figure 4.2. Formally, the data of the game are given as follows:

- There are two players: $I = \{1,2\}$.
- There are three states: $S = \{s(0), s(1), s(2)\}$.
- In state $s(0)$ both players have two available actions, whereas in states $s(1)$ and $s(2)$ both players have a single available action.
- Each entry specifies the payoffs to Player 1, as well as the transitions (in parentheses). ♦

4.1 On Histories and Strategies

The definitions of histories and strategies are analogous to the corresponding definitions in Markov decision problems. For $t \in \mathbb{N}$, the set of *histories of length t* is

$$H_t := (S \times A)^{t-1} \times S.$$

We let

$$H := \bigcup_{t \in \mathbb{N}} H_t$$

denote the set of all *histories*, and let

$$H_\infty := (S \times A)^{\mathbb{N}}$$

denote the space of all *infinite histories* or *plays*. As for Markov decision problems, we denote by $\mathcal{H}_t$ the algebra over the set H_∞ that is spanned by the cylinder sets of length t.

A strategy of player $i \in I$ is a mapping σ^i that assigns to each history $h_t = (s_1, a_1, \ldots, a_{t-1}, s_t) \in H$ a mixed action in $\Delta(A^i(s_t))$. The set of all strategies of player i is denoted by Σ^i. A strategy σ^i of player i is *pure* if it involves no randomization: $|\mathrm{supp}(\sigma^i(h_t))| = 1$ for all histories $h_t \in H$.

A strategy σ^i of player i is *stationary* if the mixed action that is played after a given history depends only on the current state: $\sigma^i(h_t)$ is a function of s_t (and is independent of $(s_1, a_1, \ldots, s_{t-1}, a_{t-1})$). We identify a stationary strategy σ^i of player i with a vector $x^i \in \prod_{s \in S} \Delta(A^i(s))$. Under this identification, $x^i(s)$ is the mixed action that player i implements whenever the current state is s. Thus, the space of stationary strategies of player i is identified with the space $X^i := \prod_{s \in S} \Delta(A^i(s))$.

A *strategy profile* is a vector $\sigma = (\sigma^i)_{i \in I}$ of strategies. We define $X := \prod_{i \in I} X^i$. This space is (identified with) the space of stationary strategy profiles.

Every pair (initial state s, strategy profile σ) induces a probability distribution on the space of plays H_∞, equipped with the sigma-algebra generated by the finite cylinders. This probability distribution is denoted by $\mathbf{P}_{s,\sigma}$, and the corresponding expectation operator is denoted by $\mathbf{E}_{s,\sigma}$.

4.2 Absorbing States and Absorbing Games

In this section, we define absorbing states and the class of absorbing games, which is a special class of stochastic games. A state is *absorbing* if once the play reaches that state, it never leaves the state, regardless of what the players play.

Definition 4.5   A state $s \in S$ is *absorbing* if for every action profile $a \in A(s)$ we have $q(s \mid s,a) = 1$.

Once the game reaches an absorbing state, it reduces to a repeated game. Since every repeated game admits an equilibrium (e.g., the players repeatedly play an equilibrium of the base game), if we are interested in the existence of an equilibrium, we can assume without loss of generality that once the play reaches an absorbing state, the stream of payoffs is constant (and equals one of the equilibrium payoffs of this repeated game). In other words, to every absorbing state there corresponds an *absorbing payoff*, which is the payoff the players receive at every stage once this state is reached.

A stochastic game is *absorbing* if all its states except one are absorbing.

Definition 4.6   An *absorbing game* is a stochastic game Γ in which all states except one are absorbing.

For such games, we will assume that the initial state is the nonabsorbing state, and that in each absorbing state the payoff is independent of the actions of the players. Absorbing games form a simple class of stochastic games in which the state can change at most once along the play.

Since in absorbing games the play terminates once it leaves the initial state, the only part of the strategy of the players that is relevant for calculating payoffs is how they play as long as the play remains in the nonabsorbing state. Denoting by $s(0)$ the unique nonabsorbing state, a stationary strategy of player $i \in I$ reduces to a probability distribution in $\Delta(A^i(s(0)))$.

Example 4.4, continued   The game in Example 4.4 is an absorbing game: the states $s(1)$ and $s(2)$ are absorbing, and $s(0)$ is the unique nonabsorbing

	L	R
T	0	1 *
B	1 *	0 *

State $s(0)$

Figure 4.3 The absorbing game in Example 4.4 in a simplified form.

state. To simplify the presentation of absorbing games, we will write down only the nonabsorbing state, as shown in Figure 4.3. An asterisk indicates an entry where the transition leads to an absorbing state with probability 1, whose payoff is equal to the payoff displayed by that entry. In this representation, we implicitly assume that the stage payoff in the entry (T, R) is equal to the absorbing payoff this entry leads to. This will be the case in all the examples of absorbing games we will study. ♦

4.3 Comments and Extensions

As for Markov decision problems, the model of stochastic games can be extended to the case when the state and action spaces are Borel spaces. In such a case, the transition rule and the payoff functions of the players are required to be measurable, see, for example, Başar and Olsder (1998), Nowak (2003a, 2003b), or Jaśkiewicz and Nowak (2018a, 2018b). Some authors studied stochastic games with a continuum of players, which led to the development of mean field games, see, for example, Jovanovic and Rosenthal (1988), Khan and Sun (2002), or Chakrabarti (2003). Others studied stochastic games when the time that elapses between consecutive stages is small or the limit case, that is, continuous-time games, see, for example, Zachrisson (1964), Başar and Olsder (1998), Guo and Hernández-Lerma (2005a, 2005b), Jasso-Fuentes (2005), Levy (2013a), Neyman (2013, 2017), Wei and Chen (2016), or Zhang (2018).

We assumed that players observe (and never forget) the current state and the actions chosen by the other players. In the general model of stochastic games, at every stage each player observes a private signal, which depends on the current state and on the action profile that the players just chose. For the general model, see, for example, Mertens et al. (2015, chapter IV).

Exercise 4.5 is adapted from Maschler (1967).

4.4 Exercises

1. Write down the parameters of the following three-player absorbing game (sets of players, states, actions, transition, and payoffs). Here Player 1 chooses a row, Player 2 chooses a column, and Player 3 chooses a matrix.

<table>
<tr><td></td><td colspan="2" align="center">C^3</td><td></td><td colspan="2" align="center">Q^3</td></tr>
<tr><td></td><td align="center">C^2</td><td align="center">Q^2</td><td></td><td align="center">C^2</td><td align="center">Q^2</td></tr>
<tr><td>C^1</td><td align="center">0, 0, 0</td><td align="center">0, 1, 3 *</td><td></td><td align="center">3, 0, 1 *</td><td align="center">1, 1, 0 *</td></tr>
<tr><td>Q^1</td><td align="center">1, 3, 0 *</td><td align="center">1, 0, 1 *</td><td></td><td align="center">0, 1, 1 *</td><td align="center">0, 0, 0 *</td></tr>
</table>

2. Write down the parameters (sets of players, states, actions, transition, and payoffs) of the following zero-sum two-player game.

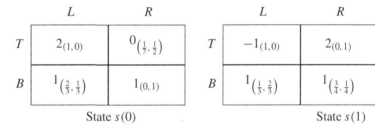

<table>
<tr><td></td><td align="center">L</td><td align="center">R</td><td></td><td align="center">L</td><td align="center">R</td></tr>
<tr><td>T</td><td align="center">$2_{(1,0)}$</td><td align="center">$0_{\left(\frac{1}{2},\frac{1}{2}\right)}$</td><td>$T$</td><td align="center">$-1_{(1,0)}$</td><td align="center">$2_{(0,1)}$</td></tr>
<tr><td>B</td><td align="center">$1_{\left(\frac{2}{3},\frac{1}{3}\right)}$</td><td align="center">$1_{(0,1)}$</td><td>$B$</td><td align="center">$1_{\left(\frac{1}{3},\frac{2}{3}\right)}$</td><td align="center">$1_{\left(\frac{3}{4},\frac{1}{4}\right)}$</td></tr>
</table>

State $s(0)$ State $s(1)$

3. Present the following situation as a stochastic game. The United States and Russia are competing for military superiority. Each of the two countries can lead in military power, but they can also be equal in military power. The "utility" of leading in military power is 10, while the "utility" of being inferior is -10. The "utility" of equal power is 0 for both countries. Each year both countries decide what their military budget is independently and simultaneously: a high budget of 7 or a low budget of 4. The total utility of a country in a given year is its utility from the mutual position minus the military budget of that year. In a year in which both countries make equal investments in armaments, the relative position of strength is maintained till the following year. In a year in which one country invests more than its rival, that country improves its position the following year with probability 0.6 and remains in the same position with probability 0.4 (that is, if it was inferior that year, then with probability 0.6 it will tie the following year and with probability 0.4 it will remain inferior; if it was a tie that year, then with probability 0.6 it will lead the following year and with probability 0.4 it will remain a tie; finally, if it was leading that year, then with probability 1 it will lead also the following year).

4. Fishermen of the UK and Iceland share a certain region of the Atlantic Ocean that lies between the two countries. At the beginning of the fishing season, each of the two countries sets a fishing quota for its fishermen, who fish the maximum amount they are allowed. For simplicity, assume that the quota is the number of fish that the fishermen are allowed to fish, and it cannot exceed half the number of fish in the region. The decision concerning this year's quota is determined according to the average number of fish in a square kilometer, a quantity that is measured at the beginning of the fishing season. For further simplicity, assume that the fish are uniformly spread in the region. The natural annual growth of the fish population is $1 - \exp(-cx)$, where x is the current number of fish per square kilometer and c is a fixed parameter. That is, if the number of fish per square kilometer at the end of the fishing season is x, then the number of fish at the beginning of the next fishing season is $x \cdot (1 - \exp(-cx))$. The gain for a country when its fishermen fish x fish, and the fishermen of the other country fish y fish, is given by the function $R(x, y)$.

   Present the situation as a stochastic game, where the state variable captures the average number of fish per square kilometer.

5. Back in the 1960s, the United States and USSR engaged in talks to abolish nuclear tests. To ensure that the other country does not carry a test, it was suggested that each country can perform each year K surprise inspections anywhere in the other country if the seismographs showed abnormal activity. The seismographs could not distinguish between seismographic movements that originate from real earthquakes and those that originate from nuclear tests.

   (a) What is the proper state variable to describe the situation as a stochastic game?

   (b) Present the situation as a stochastic game. View the situation as a zero-sum game in which a country wins 1 for every undetected test carried out, and loses 2 for every detected test carried out. Assume that the measurement of the seismographs can be either high or low; a nuclear test causes the measurement to be high with probability 1; if no nuclear test takes place, then the probability of a high measurement is p. Assume that a nuclear test is detected with probability 1 if an inspection is made within two days of the test.

5

Two-Player Zero-Sum Discounted Games

In this chapter, we extend the notion of discounted payoff to the model of stochastic games, and we define the concept of discounted equilibrium. We then prove that every two-player zero-sum stochastic game admits a discounted value, and that each player has a stationary discounted optimal strategy. The proof uses the same tools we employed in Chapter 2 to prove that in Markov decision problems the decision maker has a stationary discounted optimal strategy. We finally prove that the discounted value is continuous in the parameters of the game, namely, the payoff function, the transition rule, and the discount factor.

5.1 The Discounted Payoff

In this section, we define the discounted payoff in stochastic games, and study some of its properties.

Definition 5.1 For a discount factor $\lambda \in (0, 1]$, an initial state s, and a strategy profile $\sigma \in \Sigma$, the λ-*discounted payoff* associated with s and σ is

$$\gamma_\lambda^i(s; \sigma) := \mathbf{E}_{s,\sigma} \left[\lambda \sum_{t=1}^{\infty} (1 - \lambda)^{t-1} r^i(s_t, a_t) \right]$$

$$:= \lambda \sum_{t=1}^{\infty} (1 - \lambda)^{t-1} \mathbf{E}_{s,\sigma} \left[r^i(s_t, a_t) \right].$$

Example 4.4: continued The two-player zero-sum absorbing game in this example is reproduced in Figure 5.1.

Recall that in absorbing games (with nonabsorbing state $s(0)$), a stationary strategy of player i is equivalent to a probability distribution in $\Delta(A^i(s(0)))$. Consider the stationary strategy profile x_1 in which Player 1 always plays T

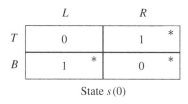

Figure 5.1 The game in Example 4.4.

and Player 2 always plays L. We denote this stationary strategy profile by $x_1 = (T, L)$. The stage payoff is 0 at every stage, and therefore $\gamma_\lambda(s(0); x_1) = 0$ for every discount factor $\lambda \in (0, 1]$.

Consider now the stationary strategy profile $x_2 = (B, R)$. As in the previous case, the stage payoff is 0 at every stage, hence $\gamma_\lambda(s(0); x_2) = 0$ for every discount factor $\lambda \in (0, 1]$.

In general, a stationary strategy profile in this game is characterized by two real numbers in $[0, 1]$: the per-stage probability according to which Player 1 chooses the action T, and the per-stage probability according to which Player 2 chooses the action L. Hence, a stationary strategy profile is a vector

$$x = ([c(T), (1 - c)(B)], [d(L), (1 - d)(R)]),$$

with $c, d \in [0, 1]$. Let us calculate the λ-discounted payoff $\gamma_\lambda(s(0); x)$ for every discount factor $\lambda \in (0, 1]$. With probability cd the entry (T, L) is chosen, the stage payoff is 0, and the payoff from the second stage and on is $\gamma_\lambda(s(0); x)$ due to the fact that x is a stationary strategy profile. With probability $(1 - c)(1 - d)$ the entry (B, R) is chosen, and the stage payoff as well as all future payoffs are 0. With probability $(1 - c)d + c(1 - d)$ one of the entries (T, R) or (B, L) is chosen, and the stage payoff as well as all future payoffs are 1. Consequently,

$$\gamma_\lambda(s(0); x) = cd\,(\lambda \cdot 0 + (1 - \lambda)\gamma_\lambda(s(0); x)) + ((1 - c)d + c(1 - d)).$$

The solution of this equation is

$$\gamma_\lambda(s(0); x) = \frac{c + d - 2cd}{1 - cd(1 - \lambda)}. \qquad \blacklozenge$$

This example leads us to a general method to calculate the discounted payoff of a stationary profile: this quantity is the solution of a certain system of linear equations.

Theorem 5.2 Let $\Gamma = \langle I, S, (A^i(s))_{s \in S}^{i \in I}, q, (r^i)_{i \in I} \rangle$ be a stochastic game, and let $x = (x^1, x^2, \ldots, x^n)$ be a stationary strategy profile. Then $(\gamma_\lambda^i(s; x))_{s \in S}$

is the unique solution of the following system of $|S|$ linear equations (one equation for each state $s \in S$):

$$\gamma_\lambda^i(s; x) = \sum_{a \in A(s)} \left[\left(\prod_{i \in I} x^i(s, a^i) \right) \left(\lambda r^i(s, a) + (1 - \lambda) \sum_{s' \in S} q(s' \mid s, a) \gamma_\lambda^i(s'; x) \right) \right].$$

(5.1)

Note that, by Theorem 1.32, this system of equations has a unique solution for every stationary strategy profile x and every discount factor $\lambda \in (0, 1]$. Since this system is linear in λ, its solution is a rational function of λ. In particular, we deduce that the function $\lambda \mapsto \gamma_\lambda^i(s; x)$ is a rational function. It is interesting to note that the same conclusion holds for Markov decision problems, see Theorem 1.33.

Corollary 5.3 *Let $\Gamma = \langle I, S, (A^i(s))_{s \in S}^{i \in I}, q, (r^i)_{i \in I} \rangle$ be a stochastic game. For every stationary strategy profile $x \in X$, every player $i \in I$, and every state $s \in S$, the function $\lambda \mapsto \gamma_\lambda^i(s; x)$ is a rational function.*

Proof of Theorem 5.2 Fix a player $i \in I$. Consider an auxiliary Markov decision problem where the set of state is S, the set of actions at each state $s \in S$ is $A(s)$, the payoff function is r^i, and the transition is q. In other words, we turn the stochastic game into a Markov decision process by assuming that there is one entity, a mediator, who chooses actions for all the players.

Next, in the Markov decision problem consider the stationary strategy y defined by

$$y(s, a) = \prod_{i \in I} x^i(s, a^i), \quad \forall s \in S, \forall a = (a^1, a^2, \ldots, a^n) \in A(s).$$

This is the stationary strategy of the mediator in which she effectively follows x. The probability of moving from state s to state s' under the stationary strategy y in the auxiliary Markov decision problem is

$$q(s' \mid s, y(s)) = q(s' \mid s, x(s)).$$

This quantity is also the probability to move from state s to state s' under the stationary strategy profile x in the stochastic game. Consequently, for every $s \in S$, the λ-discounted payoff of player i under the strategy profile x, which is $\gamma_\lambda^i(s_1; x)$, coincides with the λ-discounted payoff of player i under the stationary strategy y in the auxiliary Markov decision problem at the initial state s, which is $\gamma_\lambda(s_1; y)$. The result now follows from Theorem 1.33. $\square$

The coefficients in Eq. (5.1) are determined by the strategies of the players, the payoff function, the transition rule, and the discount factor. Moreover, these coefficients are continuous in these parameters. By Cramer's rule, it follows that the solution of this system, $(\gamma_\lambda^i(s;x))_{s\in S}$, is continuous in these parameters as well. Formally, denote by $\gamma_\lambda^i(s;x;r,q)$ the λ-discounted payoff at the initial state s when the payoff function is r and the transition rule is q.

Theorem 5.4 *For every player $i \in I$ and every initial state s, the function*

$$(\lambda, x, r, q) \mapsto \gamma_\lambda^i(s;x;r,q)$$

is continuous.

This theorem tells us that small perturbations in the data of the game or in the strategies of the players do not have a large effect on the discounted payoff.

Recall that the *support* of a mixed action is the set of actions that are played with positive probability under this mixed action. A strategy $\sigma^i \in \Sigma^i$ is a *best response* of player i against the strategy profile $\sigma^{-i} \in \Sigma^{-i}$ of the other players if it attains to player i the maximal payoff when the other players follow σ^{-i} at all initial states, that is, if

$$\gamma_\lambda^i(s;\sigma^i,\sigma^{-i}) = \max_{\sigma'^i \in \Sigma^i} \gamma_\lambda^i(s;\sigma'^i,\sigma^{-i}), \quad \forall s \in S.$$

The following lemma implies that for every stationary strategy profile x^{-i} of players $I \setminus \{i\}$, the set of stationary best responses of player i in the λ-discounted game is a polytope, the extreme points of which are pure stationary strategies.

Theorem 5.5 *Let Γ be a stochastic game, let $\lambda \in (0,1]$, let $i \in I$ be a player, and let $x^{-i} \in X^{-i}$ be a stationary strategy profile. Let $x^i \in X^i$ be a stationary strategy of player i that is a best response in the λ-discounted game against x^{-i}:*

$$\gamma_\lambda^i(s;x^i,x^{-i}) = \max_{z^i \in \Delta(A^i)} \gamma_\lambda^i(s;z^i,x^{-i}), \quad \forall s \in S.$$

Let $x'^i \in X^i$ be a stationary strategy of player i such that $\mathrm{supp}(x'^i(s)) \subseteq \mathrm{supp}(x^i(s))$ *for every state $s \in S$. Then x'^i is also a best response in the λ-discounted game against x^{-i}:*

$$\gamma_\lambda^i(s;x'^i,x^{-i}) = \gamma_\lambda^i(s;x^i,x^{-i}), \quad \forall s \in S.$$

Proof We will use the general observation that when the other players use a stationary strategy profile, the decision problem of player i reduces to a Markov decision problem. Indeed, given the stationary strategy profile x^{-i} of the other players, we introduce an auxiliary Markov decision problem $\widehat{\Gamma} = \langle S, (A^i(s))_{s\in S}, \widehat{q}, \widehat{r}\rangle$ as the Markov decision problem induced by the stochastic game Γ when the players $I \setminus \{i\}$ follow the stationary strategy profile x^{-i}:

$$\widehat{q}(\cdot \mid s, a^i) := \sum_{a^{-i} \in A^{-i}(s)} \left(\prod_{j \neq i} x^j(a^j) \right) q(\cdot \mid s, a^i, a^{-i}),$$

$$\widehat{r}(s, a^i) := \sum_{a^{-i} \in A^{-i}(s)} \left(\prod_{j \neq i} x^j(a^j) \right) r(s, a^i, a^{-i}).$$

Since x^i is a best response against x^{-i}, the highest payoff for player i from the initial state s when the other players follow x^{-i} is $\gamma_\lambda^i(s; x^i, x^{-i})$, hence $\gamma_\lambda^i(s; x^i, x^{-i})$ is the λ-discounted value of the auxiliary decision problem $\widehat{\Gamma}$.

Theorem 1.30 implies that the set of stationary strategies of player i that are best responses against x^{-i} is the set of all stationary strategies $y^i = (y^i(s))_{s \in S} \in X^i$ such that $y^i(s)$ maximizes the following quantity, for every initial state $s \in S$:

$$\lambda r^i(y^i(s), x^{-i}(s)) + (1 - \lambda) \sum_{s' \in S} q(s' \mid s, y^i(s), x^{-i}(s)) \gamma_\lambda^i(s'; x^i, x^{-i}). \quad (5.2)$$

For each fixed state $s \in S$, the function in Eq. (5.2) is linear in y^i. Since x^i is a best response against x^{-i}, every action $a^i \in \text{supp}(x^i(s))$ maximizes the quantity in Eq. (5.2). This in turn implies that every mixed action $x'^i(s) \in \Delta(A^i(s))$ such that $\text{supp}(x'^i(s)) \subseteq \text{supp}(x^i(s))$ maximizes the quantity in Eq. (5.2). By Theorem 1.30 once again, every stationary strategy x'^i such that $\text{supp}(x'^i(s)) \subseteq \text{supp}(x^i(s))$ for every state $s \in S$ is a best response against x^{-i}, as claimed. $\qquad\qquad\qquad\qquad\qquad\qquad\qquad\qquad\qquad\qquad\qquad\qquad\qquad\square$

5.2 The Discounted Value

In this section, we will study two-player zero-sum stochastic games. As for strategic-form games, a two-player stochastic game is zero-sum if the sum of the payoffs of the two players is always 0.

Definition 5.6   A two-player stochastic game is *zero-sum* if $r^1(s, a) + r^2(s, a) = 0$ for every state $s \in S$ and every action profile $a \in A(s)$.

In this case, we denote by r the payoff function of Player 1, and the payoff function of Player 2 is $-r$. We also denote

$$\gamma_\lambda(s; \sigma^1, \sigma^2) := \gamma_\lambda^1(s; \sigma^1, \sigma^2) = -\gamma_\lambda^2(s; \sigma^1, \sigma^2).$$

Definition 5.7   Let $s \in S$ be a state and let $\lambda \in (0, 1]$ be a discount factor. The real number $v_\lambda(s)$ is the λ-*discounted value* at the initial state s if

$$v_\lambda(s) = \sup_{\sigma^1 \in \Sigma^1} \inf_{\sigma^2 \in \Sigma^2} \gamma_\lambda(s; \sigma^1, \sigma^2) = \inf_{\sigma^2 \in \Sigma^2} \sup_{\sigma^1 \in \Sigma^1} \gamma_\lambda(s; \sigma^1, \sigma^2).$$

Thus, the quantity $v_\lambda(s)$ is the λ-discounted value at the initial state s if Player 1 can guarantee that the discounted payoff does not fall below $v_\lambda(s)$, and Player 2 can guarantee that the discounted payoff does not exceed this quantity.

Definition 5.8 Let $s \in S$ be a state and let $\lambda \in (0, 1]$ be a discount factor. Suppose that the value at the initial state s exists. A strategy $\sigma \in \Sigma^1$ of Player 1 is λ-*discounted optimal at the initial state s* if

$$v_\lambda(s) = \inf_{\sigma^2 \in \Sigma^2} \gamma_\lambda(s; \sigma^1, \sigma^2).$$

Similarly, a strategy $\sigma^2 \in \Sigma^2$ of Player 2 is λ-*optimal* at the initial state $s \in S$ if

$$v_\lambda(s) = \sup_{\sigma^1 \in \Sigma^1} \gamma_\lambda(s; \sigma^1, \sigma^2).$$

A strategy of a player is λ-discounted optimal if it is λ-discounted optimal at all initial states.

An optimal strategy of Player 1 guarantees that the discounted payoff does not fall below $v_\lambda(s)$ at every initial state s. If both players use a λ-discounted optimal strategy, then the discounted payoff is $v_\lambda(s)$, and no player can gain by deviating.

The main result presented in this section is that in two-player zero-sum stochastic games the λ-discounted value always exists, and, moreover, both players have stationary λ-discounted optimal strategies. As Example 3.13 shows, the optimal strategy need not be pure.

Example 4.4, continued The two-player zero-sum absorbing game in this example is reproduced again in Figure 5.2. Recall that the λ-discounted payoff when Player 1 plays the stationary strategy $x^1 = [c(T), (1 - c)(B)]$ and Player 2 plays the stationary strategy $x^2 = [d(L), (1 - d)(R)]$ is

$$\gamma_\lambda(x^1, x^2) = \frac{c + d - 2cd}{1 - cd(1 - \lambda)}.$$

We claim that Player 1 does not have a pure optimal strategy. Indeed, suppose Player 1 plays a pure strategy. Then Player 2 can ensure that the payoff is 0: Whenever Player 1 plays T, Player 2 will play L, and whenever Player 1 plays B, Player 2 will play R. However, by playing a mixed strategy, say $c = \frac{1}{2}$, Player 1 ensures that the λ-discounted payoff is positive. In particular, Player 1 does not have a pure optimal strategy. ♦

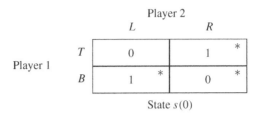

Figure 5.2 The absorbing game in Example 4.4.

5.3 Existence of the Discounted Value

In this section, we prove the following theorem, due to Shapley (1953).

Theorem 5.9   *Every two-player zero-sum stochastic game admits a λ-discounted value, for every $\lambda \in (0,1]$. Moreover, both players have stationary λ-discounted optimal strategies.*

Our proof is analogous to the proof of Theorem 1.27, which states that in every Markov decision problem the decision maker has a stationary optimal strategy: We define a suitable mapping $T \colon \mathbb{R}^S \to \mathbb{R}^S$, prove that it is contracting, and prove that its unique fixed point is the λ-discounted value of the game.

Let Γ be a stochastic game, and let $w \in \mathbb{R}^S$ be an arbitrary mapping. For each state $s \in S$, consider the following two-player zero-sum strategic-form game $G_{s,\lambda}(w)$:

- The set of players is $\{1,2\}$.
- The sets of actions of the two players are $A^1(s)$ and $A^2(s)$, respectively.
- The payoff function is

$$u_{s,w}(a^1, a^2; w) = \lambda r(s, a^1, a^2) + (1 - \lambda) \sum_{s' \in S} q(s' \mid s, a^1, a^2) w(s').$$

The game $G_{s,\lambda}(w)$ is the strategic-form game the players play at state s if the continuation payoff from the second stage and on is captured by the vector w, and λ measures the weight of the stage payoff at state s. By von Neumann's theorem (Theorem 3.10), this game has a value $\mathrm{val}(G_{s,\lambda}(w))$.

Example 4.4, continued   The game in this example is illustrated again in Figure 5.3, but here every state is described separately. Let $w \in \mathbb{R}^S$ be arbitrary.

The strategic-form games $(G_{s,\lambda}(w))_{s \in S}$ are described in Figure 5.4.   ◆

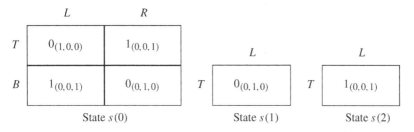

Figure 5.3 The absorbing game in Example 4.4.

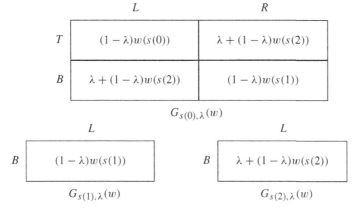

Figure 5.4 The strategic-form games $(G_{s,\lambda}(w))_{s \in S}$ in Example 4.4.

Proof of Theorem 5.9   We consider $\mathbb{R}^S$ as a metric space with the metric

$$d(w, v) := \|w - v\|_\infty = \max_{s \in S} |w(s) - v(s)|.$$

Define a mapping $T : \mathbb{R}^S \to \mathbb{R}^S$ by the formula

$(T(w))(s)$

$= \text{val}(G_{s,\lambda}(w))$

$= \displaystyle\max_{x^1 \in \Delta(A^1(s))} \min_{x^2 \in \Delta(A^2(s))} \left\{ \lambda r(s, x^1, x^2) + (1 - \lambda) \sum_{s' \in S} q(s' \mid s, x^1, x^2) w(s') \right\}.$

The mapping T is called the *Shapley operator*.

Step 1:   The Shapley operator is contracting.

By Theorem 3.15,

$$|(T(w))(s) - (T(v))(s)| = |\text{val}(G_{s,\lambda}(w)) - \text{val}(G_{s,\lambda}(v))|$$
$$\leq \|u_{s,w}(a^1, a^2) - u_{s,v}(a^1, a^2)\|_\infty$$

$$= (1 - \lambda) \sum_{s' \in S} q(s' \mid s, x^1, x^2) |w(s') - v(s')|$$

$$\leq (1 - \lambda) \max_{s' \in S} |w(s') - v(s')|$$

$$= (1 - \lambda) d(w, v),$$

whence

$$d(T(w), T(v)) = \|T(w) - T(v)\|_\infty \leq (1 - \lambda) d(w, v) < d(w, v).$$

By Theorem 1.26, there is a unique $v^* \in \mathbb{R}^S$ such that $v^* = T(v^*)$, that is, $v^*(s) = (T(v^*))(s)$ for every state $s \in S$. By the definition of T, we obtain

$$v^*(s)$$

$$= (T(v^*))(s)$$

$$= \mathrm{val}(G_{s,\lambda}(v^*))$$

$$= \max_{x^1 \in \Delta(A^1(s))} \min_{x^2 \in \Delta(A^2(s))} \left\{ \lambda r(s, x^1, x^2) + (1 - \lambda) \sum_{s' \in S} q(s' \mid s, x^1, x^2) v^*(s') \right\}'$$

$$\tag{5.3}$$

$$= \min_{x^2 \in \Delta(A^2(s))} \max_{x^1 \in \Delta(A^1(s))} \left\{ \lambda r(s, x^1, x^2) + (1 - \lambda) \sum_{s' \in S} q(s' \mid s, x^1, x^2) v^*(s') \right\}.$$

$$\tag{5.4}$$

For every state $s \in S$ let $x^1(s) \in \Delta(A^1(s))$ be an element that maximizes the quantity in the right-hand side of Eq. (5.3). Then $x^1 := (x^1(s))_{s \in S}$ is a stationary strategy of Player 1. We show that $\gamma_\lambda(s; x^1, \sigma^2) \geq v^*(s)$ for every initial state $s \in S$.

Step 2: $\sup_{\sigma^1 \in \Sigma^1} \inf_{\sigma^2 \in \Sigma^2} \gamma_\lambda(s; \sigma^1, \sigma^2) \geq v^*(s)$ for every initial state $s \in S$.

Fix a strategy $\sigma^2 \in \Sigma^2$. Since x^1 maximizes the quantity in the right-hand side of Eq. (5.3), for every $t \in \mathbb{N}$ and every history $h_t \in H$ we have

$$\lambda r(s_t, x^1_{s_t}, \sigma^2(h_t)) + (1 - \lambda) \sum_{s' \in S} q(s' \mid s_t, x^1_{s_t}, \sigma^2(h_t)) v^*(s_{t+1}) \geq v(s_t).$$

This inequality is equivalent to

$$\mathbf{E}_{s, x^1, \sigma^2} \left[\lambda r(s_t, a^1_t, a^2_t) + (1 - \lambda) v^*(s_{t+1}) \mid h_t \right] \geq v(s_t).$$

By Lemma 1.28,

$$\gamma_\lambda(s; x^1, \sigma^2) \geq v^*(s), \quad \forall s \in S. \tag{5.5}$$

Since the strategy σ^2 is arbitrary, it follows that

$$\sup_{\sigma^1 \in \Sigma^1} \inf_{\sigma^2 \in \Sigma^2} \gamma_\lambda(s; \sigma^1, \sigma^2) \geq \inf_{\sigma^2 \in \Sigma^2} \gamma_\lambda(s; x^1, \sigma^2) \geq v^*(s), \quad \forall s \in S.$$

Step 3: $\inf_{\sigma^2 \in \Sigma^2} \sup_{\sigma^1 \in \Sigma^1} \gamma_\lambda(s; \sigma^1, \sigma^2) \leq v^*(s)$ for every initial state $s \in S$.

For every state $s \in S$ let $x^2(s) \in \Delta(A^2(s))$ be an element that minimizes the quantity in the right-hand side of Eq. (5.4). Then $x^2 := (x^2(s))_{s \in S}$ is a stationary strategy of Player 2. An analogous argument to that in Step 2 shows that for every strategy σ^1 of Player 1,

$$\sup_{\sigma^1 \in \Sigma^1} \gamma_\lambda(s; \sigma^1, x^2)) \leq v_s^*, \quad \forall s \in S,$$

and therefore

$$\inf_{\sigma^2 \in \Sigma^2} \sup_{\sigma^1 \in \Sigma^1} \gamma_\lambda(s; \sigma^1, \sigma^2)) \leq v_s^*, \quad \forall s \in S.$$

Since

$$\inf_{\sigma^2 \in \Sigma^2} \sup_{\sigma^1 \in \Sigma^1} \gamma_\lambda(s; \sigma^1, \sigma^2)) \geq \sup_{\sigma^1 \in \Sigma^1} \inf_{\sigma^2 \in \Sigma^2} \gamma_\lambda(s; \sigma^1, \sigma^2),$$

we deduce from Steps 2 and 3 that

$$v_s^* \geq \inf_{\sigma^2 \in \Sigma^2} \sup_{\sigma^1 \in \Sigma^1} \gamma_\lambda(s; \sigma^1, \sigma^2))$$

$$\geq \sup_{\sigma^1 \in \Sigma^1} \inf_{\sigma^2 \in \Sigma^2} \gamma_\lambda(s; \sigma^1, \sigma^2)$$

$$\geq v_s^*,$$

and therefore $v^*(s)$ is the λ-discounted value of the game at the initial state s. Eq. (5.5) implies that the stationary strategy x^1 that was constructed in Step 2 is a λ-discounted optimal strategy of Player 1 at all initial states. An analogous argument shows that the stationary strategy x^2 that was constructed in Step 3 is a λ-discounted optimal strategy of Player 2. $\qquad\square$

The proof of Theorem 5.9 shows that the value function is a fixed point of the Shapley operator. This yields the following result, which, as we will see below, allows us to calculate the discounted value of games.

Theorem 5.10 *The value $v_\lambda = (v_\lambda(s))_{s \in S}$ is the unique vector in $\mathbb{R}^S$ that satisfies*

$$v_\lambda(s) = \mathrm{val}(G_{s,\lambda}(v_\lambda)), \quad \forall s \in S.$$

Example 4.4, continued   The strategic-form games $(G_{s,\lambda}(w))_{s \in S}$ that correspond to the game in this example are shown in Figure 5.4.

What is the fixed point of the Shapley operator T? This fixed point should satisfy

$$w(s(0)) = (T(w))(s(0)) = \mathrm{val}(G_{s(0),\lambda}(w)),$$
$$w(s(1)) = (T(w))(s(1)) = \mathrm{val}(G_{s(1),\lambda}(w)),$$
$$w(s(2)) = (T(w))(s(2)) = \mathrm{val}(G_{s(2),\lambda}(w)).$$

Since $G_{s(1),\lambda}(w)$ and $G_{s(2),\lambda}(w)$ are 1×1 games, their value is equal to the unique number that appears in the payoff matrix:

$$w(s(1)) = (1 - \lambda)w(s(1)) \implies w(s(1)) = 0.$$
$$w(s(2)) = \lambda + (1 - \lambda)w(s(2)) \implies w(s(2)) = 1.$$

Substituting $w(s(1)) = 0$ and $w(s(2)) = 1$ in $G_{s(1),\lambda}(w)$, we see that $w(s(0))$ should be the value of the 2×2 two-player strategic-form game as shown in Figure 5.5.

By Exercise 3.1, $w(s(0)) \leq 1$, hence in the game $G_{s(0),\lambda}(w)$ Player 1 does not have a pure optimal strategy. Therefore, when Player 2 plays her optimal strategy, Player 1 should be indifferent between T and B. Let $[y(L), (1 - y)(R)]$ denote the optimal strategy of Player 2, and write $w = w(s(0))$ for short. Then

$$w = y = (1 - \lambda)wy + 1 - y.$$

Therefore,

$$0 = 1 - 2w + (1 - \lambda)w^2,$$

which solves to

$$w = \frac{1 \pm \sqrt{\lambda}}{1 - \lambda}.$$

	L	R
T	$(1 - \lambda)w(s(0))$	1
B	1	0

Figure 5.5 The strategic-form game $G_{s(0),\lambda}(w)$ in Example 4.4.

Since $\frac{1+\sqrt{\lambda}}{1-\lambda} > 1$, while $w = w(s(0))$ must be between 0 and 1, we deduce that the λ-discounted value at the initial state $s(0)$ is given by

$$v_\lambda(s(0)) = \frac{1 - \sqrt{\lambda}}{1 - \lambda}.$$

Unlike the case of Markov decision problems (see Corollary 1.34), the discounted value is not a piecewise–rational function in λ, but rather a rational function in $\sqrt{\lambda}$. As we will see in Chapter 6, in every stochastic game and for every initial state s, the function $\lambda \mapsto v_\lambda(s)$ is a piecewise–rational function in $\lambda^{1/k}$, for some natural number k. ◆

5.4 Continuity of the Value Function

So far we restricted attention to a specific stochastic game Γ. In this section, we fix the set of states S and the sets of actions $(A^1(s))_{s\in S}$ and $(A^2(s))_{s\in S}$ available to the players in the various states. We will allow the payoff function, the transition rule, and the discount factor to vary, and we will prove that the discounted value, regarded as a function of the payoff function, the transition rule, and the discount factor, is continuous.

Continuity of the value function is important, since often the data of the game are not known precisely, and can only be estimated via sampling. In this case, the precision of the data depends on the precision of the measurement, as well as on the number of samples we have. If the value function is continuous, then the value of the true game is close to the value of the approximating game that is defined using the estimated parameters, and optimal strategies in the approximating game are almost optimal in the true game.

Once the set of states S and the sets of actions $(A^1(s))_{s\in S}$ and $(A^2(s))_{s\in S}$ are fixed, the set of all payoff functions is $\mathbb{R}^{SA}$, and the set of all transition rules is $(\Delta(S))^{SA}$. For a payoff function $r \colon SA \to \mathbb{R}$ and a transition rule $q \colon SA \to \Delta(S)$, we let $v_\lambda(s;r,q)$ denote the value of the two-player zero-sum stochastic game $\langle \{1,2\}, S, (A^1(s), A^2(s))_{s\in S}, q, r \rangle$. Our main result in this section is the following.

Theorem 5.11 *For every $s \in S$, the function $(\lambda, r, q) \mapsto v_\lambda(s;r,q)$ is continuous.*

We need the following technical result.

Lemma 5.12 *Let X be a metric space and let Y be a compact space. Let $f \colon X \times Y \to \mathbb{R}$ be a continuous function. Define two functions $g, h \colon X \to \mathbb{R}$ by*

$$g(x) := \min_{y \in Y} f(x, y), \quad \forall x \in X,$$

$$h(x) := \max_{y \in Y} f(x, y), \quad \forall y \in Y.$$

Then g and h are continuous.

Observe that since Y is compact and f is continuous, g and h are well defined.

Proof   We prove that g is continuous; the proof for h is analogous. Let $(x_n)_{n \in \mathbb{N}}$ be a sequence in X that converges to a limit x. To prove that $g(x) = \lim_{n \to \infty} g(x_n)$, we will show that $\limsup_{n \to \infty} g(x_n) \leq g(x)$ and $\liminf_{n \to \infty} g(x_n) \geq g(x)$.

We first prove that $\limsup_{n \to \infty} g(x_n) \leq g(x)$. Let y^* be such that $f(x, y^*) = \min_{y \in Y} f(x, y) = g(x)$. Then $g(x_n) = \min_{y \in Y} f(x_n, y) \leq f(x_n, y^*)$ for all $n \in \mathbb{N}$, and therefore

$$\limsup_{n \to \infty} g(x_n) \leq \lim_{n \to \infty} f(x_n, y^*) = f(x, y^*) = g(x).$$

Now let us prove that $\liminf_{n \to \infty} g(x_n) \geq g(x)$. For each $n \in \mathbb{N}$, let y_n^* be such that $f(x_n, y_n^*) = \min_{y \in Y} f(x_n, y) = g(x_n)$. Since Y is compact, there is a subsequence $(n_k)_{k \in \mathbb{N}}$ such that $\lim_{k \to \infty} g(x_{n_k})$ exists and is equal to $\liminf_{n \to \infty} g(x_n)$, and that $y^* := \lim_{k \to \infty} y_{n_k}^*$ exists. Then

$$\liminf_{n \to \infty} g(x_n) = \lim_{k \to \infty} g(x_{n_k})$$
$$= \lim_{k \to \infty} f(x_{n_k}, y_{n_k}^*)$$
$$= f(x, y^*) \geq \min_{y \in Y} f(x, y) = g(x). \qquad \square$$

Proof of Theorem 5.11   By Theorem 5.4 for each fixed initial state $s \in S$, the function

$$(\lambda, x^1, x^2, r, q) \mapsto \gamma_\lambda(s; x; r, q)$$

is continuous.

Since $\Delta(A^2)$ is compact, Lemma 5.12 implies that the function

$$(\lambda, x^1, r, q) \mapsto \min_{x^2 \in \Delta(A^2)} \gamma_\lambda(s; x^1, x^2; r, q)$$

is continuous. Since $\Delta(A^1)$ is compact, using again Lemma 5.12 we deduce that the function

$$(\lambda, r, q) \mapsto \max_{x^1 \in \Delta(A^1)} \min_{x^2 \in \Delta(A^2)} \gamma_\lambda(s; x^1, x^2; r, q)$$

is continuous. Since both players have stationary optimal strategies,

$$\max_{x^1 \in \Delta(A^1)} \min_{x^2 \in \Delta(A^2)} \gamma_\lambda(s; x^1, x^2; r, q) = v_\lambda(s; r, q).$$

In particular, the function

$$(\lambda, r, q) \mapsto v_\lambda(s; r, q)$$

is continuous, as desired. □

5.5 Comments and Extensions

In this chapter, we studied two-player zero-sum stochastic games with finitely many states and actions. Extensions of Shapley's result (Theorem 5.9) to stochastic games with more general sets of states and actions can be found in, for example, Maitra and Parthasarathy (1970), Couwenbourgh (1980), Kumar and Shiau (1981), Nowak (1985b, 1986), and Jaśkiewicz and Nowak (2011). A recent survey on these extensions is Jaśkiewicz and Nowak (2018a).

We were interested in the existence of the discounted value, the structure of discounted optimal strategies, and the continuity of the value. Algorithms to find the discounted value and optimal strategies of two-player zero-sum stochastic games can be found in, for example, Filar and Tolwinski (1991), Breton (1991) and the references therein, Filar and Vrieze (1997), Raghavan and Syed (2003), Hansen et al. (2011), and Bourque and Raghavan (2014).

In Example 4.4, we have seen that even when the payoffs, the transitions, and the discount factor are all rational numbers, the discounted value is not necessarily a rational number. A natural question that arises concerns the identification of classes of games where the discounted value lies in the smallest field that contains the payoffs, the transitions, and the discount factor. Results in this direction can be found in Raghavan, Ferguson, and Parthasarathy (1991) and the references therein and Raghavan (2003).

Exercise 5.2 is taken from Parthasarathy and Raghavan (1981). Exercise 5.5 is taken from Bewley and Kohlberg (1978). Exercises 5.7, 5.8, 5.12, 5.15–5.17, and 5.16 are taken from Maschler et al. (2020).

5.6 Exercises

Exercise 5.14 is used in the proof of Theorem 10.4. Exercise 5.11 is used in the solution of Exercise 5.12. Exercise 5.9 is used in the solution of Exercise 5.15. Exercise 5.15 is used in the solution of Exercises 5.16, 5.17, and Exercise 9.18.

1. In this exercise, we study the T-stage payoff. For $T \in \mathbb{N}$, a state $s \in S$, and a strategy profile $\sigma \in \Sigma$, the *T-stage payoff* under the strategy profile σ at the initial state s is

$$\gamma_T^i(s;\sigma) := \mathbf{E}_{s,\sigma}\left[\frac{1}{T}\sum_{t=1}^{T} r^i(s_t,a_t)\right].$$

   Do the following:

   (a) Prove that for every T, the stochastic game admits a *T-stage equilibrium*, that is, a strategy profile σ_* that satisfies

   $$\gamma_T^i(s;\sigma_*) \geq \gamma_T^i(s;\sigma^i,\sigma_*^{-i}), \quad \forall i \in I,\ s \in S,\ \sigma^i \in \Sigma^i.$$

   (b) Prove that for every T, the stochastic game admits a T-stage equilibrium σ_* with the property that the mixed action played by a player at stage t depends on t and s_t; that is, for every $t \in \{1,2,\dots,T\}$ and every two histories $h_t = (s_1,a_1,\dots,a_{t-1},s_t)$ and $h_t' = (s_1',a_1',\dots,a_{t-1}',s_t')$, if $s_t = s_t'$, then $\sigma_*^i(h_t) = \sigma_*^i(h_t')$ for every $i \in I$.

   (c) For the following game, find all equilibrium payoffs in the one-stage and two-stage games.

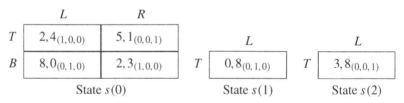

	L	R
T	$2,4_{(1,0,0)}$	$5,1_{(0,0,1)}$
B	$8,0_{(0,1,0)}$	$2,3_{(1,0,0)}$

State $s(0)$

	L
T	$0,8_{(0,1,0)}$

State $s(1)$

	L
T	$3,8_{(0,0,1)}$

State $s(2)$

2. Calculate the λ-discounted value at the initial state $s(0)$ of the following two-player zero-sum absorbing game for $\lambda = \frac{2}{3}$. Since the $\frac{2}{3}$-discounted value at $s(0)$ is irrational, it follows that even when the data that defines the stochastic game are rational, the discounted value is not necessarily rational.

	L	R
T	$1_{(1,0)}$	$0_{(0,1)}$
B	$0_{(0,1)}$	$3_{(1,0)}$

State $s(0)$

	L
T	$2_{(0,1)}$

State $s(1)$

3. Calculate the discounted value of the following two-player zero-sum absorbing game.

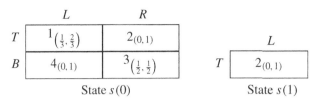

State $s(0)$ and State $s(1)$

4. Calculate the discounted value of the following two-player zero-sum absorbing game.

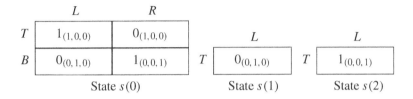

State $s(0)$, State $s(1)$, and State $s(2)$

5. Calculate the discounted value of the following two-player zero-sum game where $s(2)$ and $s(3)$ are absorbing states.

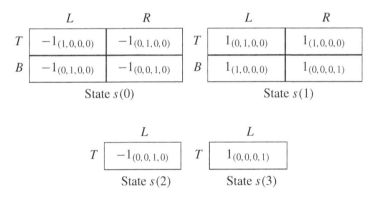

State $s(0)$, State $s(1)$, State $s(2)$, and State $s(3)$

6. Consider the following game with two nonabsorbing states. Using the symmetry between the two states, prove that the discounted value at state $s(1)$ is $v_\lambda(s(1)) = -\frac{\lambda}{4-\lambda}$. What is the optimal discounted strategy of each player?

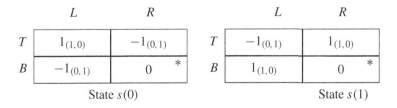

State $s(0)$ and State $s(1)$

7. In the stochastic game described below there are four states, denoted by $s(0)$, $s(1)$, $s(2)$, and $s(3)$. The states $s(1)$, $s(2)$, and $s(3)$ are absorbing states.

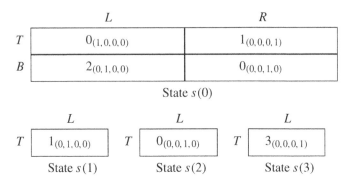

State $s(0)$

State $s(1)$ State $s(2)$ State $s(3)$

(a) Prove that for every $\lambda \in (0, 1]$, the λ-discounted value at the initial state $s(0)$ is positive.

(b) Prove that Player 1 has no stationary optimal strategy that chooses a pure action in state $s(0)$.

(c) For every discount factor $\lambda \in [0, 1)$, compute the λ-discounted value and stationary optimal strategies of the two players.

8. The following game is a two-player zero-sum stochastic game with nondeterministic transitions and one absorbing state.

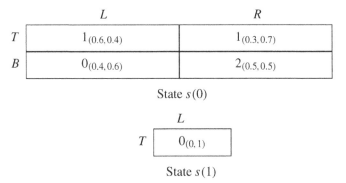

State $s(0)$

State $s(1)$

For every discount factor $\lambda \in [0, 1)$, compute the λ-discounted value of the game and stationary optimal strategies of the players.

9. Let σ^1 be a strategy of Player 1 that satisfies the following property: for every finite history $h_t = (s_1, a_1, \ldots, s_t) \in H$, the mixed action $\sigma^1(h_t)$ is an optimal strategy of Player 1 in the auxiliary game $G_{s_t, \lambda}(v_\lambda)$. Prove that σ^1 is a λ-discounted optimal strategy of Player 1.

10. Let $\Gamma = \langle\{1,2\}, S, (A^1(s), A^2(s))_{s \in S}, q, r\rangle$ be a two-player zero-sum stochastic game, and let σ^2 be a stationary strategy of Player 2. Prove that there exists a pure stationary strategy σ^1 such that

$$\gamma_\lambda^1(s; \sigma^1, \sigma^2) = \max_{\sigma'^1 \in \Sigma^2} \gamma_\lambda^1(s; \sigma'^1, \sigma^2),$$

that is, for every stationary strategy σ^2 of Player 2, Player 1 has a pure stationary best response.

11. Let $\Gamma = \langle\{1,2\}, S, (A^1(s), A^2(s))_{s \in S}, q, r\rangle$ be a two-player zero-sum stochastic game. Suppose that there exists a real number c such that the value of the one-stage game at all initial states is c, that is,

$$c = v_1(s), \quad \forall s \in S.$$

Prove that c is the λ-discounted value at all initial states $s \in S$ and every discount factor $\lambda \in (0, 1]$.

12. A two-player zero-sum stochastic game is *symmetric* if the following conditions are satisfied for every state $s \in S$:

  (a) The two players have the same set of actions in state s, that is, $A^1(s) = A^2(s)$.

  (b) The payoff matrix in state s is skew-symmetric:

  $$r(s, a^1, a^2) = -r(s, a^2, a^1), \quad \forall a^1 \in A^1(s), a^2 \in A^2(s).$$

  (c) The transition probabilities are symmetric:

  $$q(s, a^1, a^2) = q(s, a^2, a^1), \quad \forall a^1 \in A^1(s), a^2 \in A^2(s).$$

  Prove that for every discount factor $\lambda \in (0, 1]$, the λ-discounted value of a two-player zero-sum symmetric stochastic game is 0 for every initial state.

13. Let Γ be a two-player zero-sum stochastic game with *countably* many states and finitely many actions available to each player in each state. Suppose that the payoffs are between 0 and 1, that is, $r(s, a) \in [0, 1]$ for all pairs $(s, a) \in SA$. Prove that for every discount factor $\lambda \in (0, 1]$ the λ-discounted value exists at all initial states and both players have stationary λ-discounted optimal strategies.

14. Let $\Gamma = \langle\{1,2\}, S, (A^1(s), A^2(s))_{s \in S}, q, r\rangle$ be a two-player zero-sum stochastic game. For each player $i = 1, 2$ and every state $s \in S$, let $X^i(s) \subseteq \Delta(A^i(s))$ be a compact subset of mixed actions available to

player i in state s. Denote by Σ_X^i the set of strategies of player i such that the mixed action that is played when the game is in state s must belong to $X^i(s)$:

$$\Sigma_X^i := \{\sigma^i \in \Sigma^i : \sigma^i(h_t) \in X^i(s_t), \forall h_t \in H\}.$$

Denote the two players by i and j. Define

$$\underline{v}_\lambda^i(s) := \sup_{\sigma^i \in \Sigma_X^i} \inf_{\sigma^j \in \Sigma_X^j} \gamma_\lambda^i(s; \sigma^i, \sigma^j), \qquad (5.6)$$

and

$$\overline{v}_\lambda^i(s) := \inf_{\sigma^j \in \Sigma_X^j} \sup_{\sigma^i \in \Sigma_X^i} \gamma_\lambda^i(s; \sigma^i, \sigma^j). \qquad (5.7)$$

The quantity $\underline{v}_\lambda^i(s)$ is the max–min value of player i. It represents the maximum amount that player i can guarantee in the game. The quantity $\overline{v}_\lambda^i(s)$ is the min–max value of player i. It represents the maximum amount that player i can defend in the game.

(a) Show that the supremum in Eq. (5.6) and the infimum in Eq. (5.7) are attained by stationary strategies.

(b) Let $i \in \{1, 2\}$, and suppose that $X^i(s) = \Delta(A^i(s))$ for every state $s \in S$. Show that for every strategy $\sigma^j \in \Sigma_X^j$, there is a pure stationary strategy $\sigma^i \in \Sigma_X^i$ that attains the supremum in Eq. (5.6).

(c) Show that if all the sets $(X^i(s))_{s \in S}$, $i = 1, 2$, are convex, then $\underline{v}_\lambda^i(s) = \overline{v}_\lambda^i(s)$ for $i \in \{1, 2\}$.

15. Exercises 15 and 16 deal with stochastic games with perfect information. An infinite stochastic game has *perfect information* if in every state $s \in S$ at least one player has a single action, that is, $|A^1(s)| = 1$ or $|A^2(s)| = 1$.

Prove that in every two-player zero-sum stochastic game with perfect information, each player has a pure λ-discounted optimal strategy.

16. Consider a two-player zero-sum stochastic game with perfect information.

(a) Prove that for every pair of strategies $(\sigma^1, \sigma^2) \in \Sigma$ and for every initial state $s \in S$, the function $\lambda \mapsto \gamma_\lambda(s; \sigma^1, \sigma^2)$ is a *rational function*; that is, it is the ratio of two polynomials in λ.

(b) Prove that for every stationary strategy $\sigma^2 \in \Sigma^2$ of Player 2 and for every discount factor $\lambda \in (0, 1]$, Player 1 has a best reply that is a stationary pure strategy.

(c) Use the fact that Player 2 has a finite number of pure stationary strategies to conclude that there exist a pure stationary strategy σ^1 of

Player 1 and a discount factor $\lambda_0 \in [0,1)$ such that σ^1 is a λ-discounted optimal strategy for every $\lambda \in (\lambda_0, 1)$.

17. A (finite) *directed graph* is a pair (V, E), where V is a finite set of *vertices* (or *nodes*) and E is a finite set of *directed edges*. In other words, $E \subseteq V \times V$ and an element $e = (v_1, v_2) \in E$ is a directed edge from the vertex v_1 to the vertex v_2. A *game on a graph* is a two-player zero-sum game given by

    - A directed graph (V, E) with the property that $(v, v) \in E$ for each vertex $v \in V$. In other words, from every vertex v there is a directed edge to v itself.
    - A mapping $i \colon V \to \{1, 2\}$ assigning each vertex to one of the players.
    - A function $u \colon V \to \mathbb{R}$ assigning a payoff to each vertex.
    - An initial vertex $v_1 \in V$.

    The game is played in stages as follows:

    - At every stage the play is at one of the vertices, with v_1 being the initial vertex.
    - At stage $t \in \mathbb{N}$, the following takes place:
        - Player 2 pays Player 1 the amount of $u(v_t)$, where v_t is the vertex at stage t.
        - Player $i(v_t)$ chooses an edge starting at v_t, say the edge $(v_t, \hat{v})$.
        - Stage t is over, and the state for the next stage, stage $(t+1)$, is $v_{t+1} := \hat{v}$.

    Do the following:

    (a) Describe the game on a graph as a stochastic game.
    (b) Prove that for every $\lambda \in [0, 1)$, each player has a λ-discounted optimal strategy that is pure and stationary.
    (c) Assume that the two players follow optimal pure stationary strategies. Is it necessarily true that there exist a vertex v^* and a stage $t_0 \in \mathbb{N}$ such that $v_t = v^*$ for every $t \geq t_0$? Justify your answer.

6

Semi-Algebraic Sets and the Limit of the Discounted Value

In this chapter, we define semi-algebraic sets and study their basic properties. We then apply our findings to prove that for every initial state s the limit $\lim_{\lambda \to 0} v_\lambda(s)$ exists.

6.1 Semi-Algebraic Sets

Definition 6.1   A subset $A \subseteq \mathbb{R}^n$ is a *basic semi-algebraic set* if there is a polynomial $P \colon \mathbb{R}^n \to \mathbb{R}$ such that

$$A = \{x \in \mathbb{R}^n \colon P(x) = 0\},$$

or

$$A = \{x \in \mathbb{R}^n \colon P(x) > 0\}.$$

Note that $\emptyset$ and $\mathbb{R}^n$ are basic semi-algebraic sets for every $n \in \mathbb{N}$.

Recall that an *algebra* $\mathcal{A}$ is a family of sets that contains the empty set and is closed under finite unions and under complements: if $(A_i)_{i=1}^n$ is a finite collection of sets in $\mathcal{A}$, then $\bigcup_{i=1}^n A_i \in \mathcal{A}$, and if $A \in \mathcal{A}$, then $A^c \in \mathcal{A}$. From De Morgan's laws,[1] it follows that if $(A_i)_{i=1}^n$ is a finite collection of sets in $\mathcal{A}$, then $\bigcap_{i=1}^n A_i \in \mathcal{A}$.

Definition 6.2   A subset $A \subseteq \mathbb{R}^n$ of a Euclidean space is *semi-algebraic* if it belongs to the algebra generated by the family of basic semi-algebraic sets.

Example 6.3   Let $P_1, P_2, \ldots, P_K, P_{K+1}, \ldots, P_{K+L}, P_{K+L+1}, \ldots, P_{K+L+M}$ be $K + L + M$ polynomials on $\mathbb{R}^n$. Then the set

---

[1] Augustus De Morgan (Madurai, Madras Presidency, British Empire (present-day India), June 27, 1806 – London, UK, March 18, 1871) was a British mathematician and logician. He formulated De Morgan's laws and introduced the term mathematical induction, making its idea rigorous.

$$A := \left\{ x \in \mathbb{R}^n : P_1(x) > 0, \dots, P_K(x) > 0, P_{K+1}(x) \geq 0, \dots, P_{K+L}(x) \geq 0 \right.$$
$$\left. P_{K+L+1} = 0, \dots, P_{K+L+M} = 0 \right\}$$

is semi-algebraic. Indeed,

$$A = \left(\bigcap_{k=1}^{K} \{ x \in \mathbb{R}^n : P_k(x) > 0 \} \right)$$
$$\cap \left(\bigcap_{k=K+1}^{K+L} \left(\{ x \in \mathbb{R}^n : P_k(x) > 0 \} \cup \{ x \in \mathbb{R}^n : P_k(x) = 0 \} \right) \right)$$
$$\cap \left(\bigcap_{k=K+L+1}^{K+L+M} \{ x \in \mathbb{R}^n : P_k(x) = 0 \} \right). \qquad \blacklozenge$$

Let us give several examples of semi-algebraic sets.

1. The boundary of a disc in $\mathbb{R}^2$:
$$\left\{ (x, y) \in \mathbb{R}^2 : x^2 + y^2 = 1 \right\}.$$

2. The double solid cone:
$$\left\{ (x, y, z) \in \mathbb{R}^3 : x^2 + y^2 \leq z^2 \right\}.$$

3. The set of common zeros of two polynomials (the set of all points where two polynomials vanish): for every two polynomials $P, Q : \mathbb{R}^n \to \mathbb{R}$,
$$\left\{ x \in \mathbb{R}^n : P(x) = 0 \text{ and } Q(x) = 0 \right\} = \left\{ x \in \mathbb{R}^n : P^2(x) + Q^2(x) = 0 \right\}.$$

4. The graph of a polynomial: for every polynomial $P : \mathbb{R}^n \to \mathbb{R}$,
$$\left\{ (y_1, y_2, \dots, y_{n+1}) \in \mathbb{R}^{n+1} : P(y_1, y_2, \dots, y_n) = y_{n+1} \right\}.$$

5. For every two polynomials $P, Q : \mathbb{R}^n \to \mathbb{R}$ such that Q never vanishes, the graph of the rational function $\frac{P}{Q}$:
$$\left\{ (y_1, y_2, \dots, y_{n+1}) \in \mathbb{R}^{n+1} : \frac{P(y_1, y_2, \dots, y_n)}{Q(y_1, y_2, \dots, y_n)} = y_{n+1} \right\}.$$

6. An annulus in $\mathbb{R}^2$:
$$\left\{ (x, y) \in \mathbb{R}^2 : 1 \leq x^2 + y^2 \leq 2 \right\}.$$

Definition 6.4   Let $A \subseteq \mathbb{R}^n$ be a semi-algebraic set. A mapping $f : A \to \mathbb{R}^m$ is *semi-algebraic* if its graph is a semi-algebraic subset of $\mathbb{R}^{n+m}$.

As we have seen, every rational function whose denominator never vanishes is semi-algebraic.

Example 6.5   The function $\sqrt{x}$ is semi-algebraic.

Indeed,

$$\left\{(x,y) \in \mathbb{R}^2 \colon \sqrt{x} = y\right\}$$
$$= \left\{(x,y) \in \mathbb{R}^2 \colon x \geq 0, \ x = y^2\right\}$$
$$= \left\{(x,y) \in \mathbb{R}^2 \colon x \geq 0\right\} \cap \left\{(x,y) \in \mathbb{R}^2 \colon x - y^2 = 0\right\}.$$

As the two sets $\{(x,y) \in \mathbb{R}^2 \colon x \geq 0\}$ and $\{(x,y) \in \mathbb{R}^2 \colon x - y^2 = 0\}$ are semi-algebraic, so is their intersection. Consequently, the function $\sqrt{x}$ is semi-algebraic. ◆

Example 6.6   The value and optimal strategies of strategic-form games.

The space of all two-player zero-sum strategic-form games in which Player 1 has n actions and Player 2 has m actions is isomorphic to $\mathbb{R}^{nm}$. A mixed action of Player 1 is a probability distribution x on $\{1, 2, \ldots, n\}$, and a mixed action of Player 2 is a probability distribution y on $\{1, 2, \ldots, m\}$. Let B denote the set of all vectors $(u, v, x, y) \in \mathbb{R}^{nm+1+n+m}$ such that v is the value of the strategic-form game defined by the payoff function u; x is an optimal strategy of Player 1 in this game; and y is an optimal strategy of Player 2 in this game. The set B is a subset of $\mathbb{R}^{nm+1+n+m}$, and it is the set of all vectors (u, v, x, y) that satisfy the following polynomial equalities and inequalities:

$$x_i \geq 0, \quad \forall i \in \{1, 2, \ldots, n\},$$

$$\sum_{i=1}^{n} x_i = 1,$$

$$y_j \geq 0, \quad \forall j \in \{1, 2, \ldots, m\},$$

$$\sum_{j=1}^{m} y_j = 1,$$

$$\sum_{i=1}^{n} x_i u(i, j) \geq v, \quad \forall j \in \{1, 2, \ldots, m\},$$

$$\sum_{j=1}^{m} y_j u(i, j) \leq v, \quad \forall i \in \{1, 2, \ldots, n\}.$$

By Example 6.3, the set B is semi-algebraic. ◆

We now list three properties of semi-algebraic sets and semi-algebraic functions, which will be useful in the study of discounted games. We will not provide proofs for these results, because the proofs we are aware of are too lengthy for this book. The interested reader is referred to Benedetti and Risler (1990) or Bochnak et al. (2013).

- The projection of a semi-algebraic subset of $\mathbb{R}^{n+1}$ to the first n coordinates is a semi-algebraic subset of $\mathbb{R}^n$ (Theorem 6.7).
- Every semi-algebraic function can be expressed in a neighborhood of 0 as a Laurent series in fractional powers of λ (Theorem 6.9).
- Let A be a semi-algebraic subset of $\mathbb{R}^{n+1}$, and let B be its projection to the first coordinate. Then there is a semi-algebraic mapping $f : B \to \mathbb{R}^n$ such that the graph of f is a subset of A (Theorem 6.11).

The following theorem states that a projection of a semi-algebraic set is semi-algebraic.

Theorem 6.7 *Let $A \subseteq \mathbb{R}^{n+1}$ be a semi-algebraic set. Then the set*

$$B = \left\{ (x_1, x_2, \dots, x_n) \in \mathbb{R}^n : \exists x_{n+1} \in \mathbb{R} \text{ such that } (x_1, x_2, \dots, x_n, x_{n+1}) \in A \right\}$$

is semi-algebraic.

Every semi-algebraic function from $\mathbb{R}$ to $\mathbb{R}$ is locally a solution of a polynomial equation. This is the content of the next result, whose proof is left to the reader (Exercise 6.8).

Theorem 6.8 *Every semi-algebraic function $f : \mathbb{R} \to \mathbb{R}$ is a piecewise solution of a polynomial equation: there is a partition of $\mathbb{R}$ into a finite number of intervals $I_1, I_2, \dots, I_K$, and for each $k \in \{1, 2, \dots, K\}$ there is a polynomial $P_k : \mathbb{R}^2 \to \mathbb{R}$, such that $P_k(x, f(x)) = 0$ for every $x \in I_k$.*

As a conclusion, we obtain that every semi-algebraic function is locally a Laurent series in fractional powers.[2] Such a representation is called a *Puiseux series*.[3]

Theorem 6.9 *Let $f : (0, 1] \to \mathbb{R}$ be a semi-algebraic function. There exist a point $x_0 \in (0, 1]$, a positive integer L, an integer K, and real numbers $(a_k)_{k=K}^{\infty}$ such that:*

$$f(x) = \sum_{k=K}^{\infty} a_k x^{k/L}, \quad \forall x \in (0, x_0]. \tag{6.1}$$

---

[2] Pierre Alphonse Laurent (Paris, France, July 18, 1813 – Paris, France, September 2, 1854) was a French mathematician best known as the discoverer of the Laurent series. His work was not published until after his death.

[3] Victor Alexandre Puiseux (Argenteuil, France, April 16, 1820 – Frontenay, France, September 9, 1883) was a French mathematician and astronomer. He contributed to algebraic functions and uniformization.

That is, for every $x \in (0, x_0]$ the series in the right-hand side of Eq. (6.1) is summable,[4] and its sum is equal to $f(x)$.

The summability of the right-hand side of Eq. (6.1) implies that for every $k \in \mathbb{N}$, the term $a_k x^{k/M}$ dominates the tail $\sum_{l=k+1}^{\infty} a_l x^{l/M}$, that is,

$$\lim_{x \to 0} \frac{\sum_{l=k+1}^{\infty} a_l x^{l/M}}{a_k x^{k/M}} = 0$$

(see Exercise 6.15). In particular, we obtain the following (see Exercise 6.16).

Corollary 6.10   *Let $f: (0, 1] \to \mathbb{R}$ be a semi-algebraic function and let $f(x) = \sum_{k=K}^{\infty} a_k x^{k/M}$ be its Puiseux series representation.*

1. The limit $\lim_{x \to 0} f(x)$ exists and is given by

$$\lim_{x \to 0} f(x) = \begin{cases} 0, & \text{if } K > 0, \\ a_0, & \text{if } K = 0, \\ +\infty, & \text{if } K < 0, \ a_0 > 0, \\ -\infty, & \text{if } K < 0, \ a_0 < 0. \end{cases}$$

2. There is an $x_0 \in (0, 1]$ such that f is monotone in the interval $(0, x_0)$.

Example 4.4, continued   We have already calculated the λ-discounted value of the two-player zero-sum absorbing game that is depicted in Figure 6.1, and found out that it is given by

$$v_\lambda(s(0)) = \frac{1 - \sqrt{\lambda}}{1 - \lambda}.$$

Therefore,

$$v_\lambda(s(0)) = \left(1 - \sqrt{\lambda}\right)(1 + \lambda + \lambda^2 + \cdots) = 1 - \lambda^{\frac{1}{2}} + \lambda - \lambda^{\frac{3}{2}} + \lambda^2 - \cdots.$$

Thus, $v_\lambda(s(0))$ is a Puiseux series with $K = 0$, $M = 2$, and $a_k = (-1)^k$. Observe that the limit of the discounted value at $s(0)$ is $\lim_{\lambda \to 0} v_\lambda(s(0)) = 1$. ♦

	L	R
T	0	1 *
B	1 *	0 *

State $s(0)$

Figure 6.1 The game in Example 4.4.

---

[4] A sequence $(z_n)_{n \in \mathbb{N}}$ is *summable* if $\sum_{n=1}^{\infty} |z_n| < +\infty$.

The third property of semi-algebraic sets that we need is the following.

Theorem 6.11   *Let $A \subseteq \mathbb{R}^{n+1}$ be a semi-algebraic set and let $B := \{x \in \mathbb{R} : \exists y \in \mathbb{R}^n$ such that $(x,y) \in A\}$ be its projection on the first coordinate. Then there exists a semi-algebraic mapping $f : B \to \mathbb{R}^n$ such that the graph of f is a subset of A.*

6.2 Semi-Algebraic Sets and Zero-Sum Stochastic Games

In this section, we present some consequences of the theory of semi-algebraic sets for two-player zero-sum stochastic games. In Section 8.4, we will derive analogous results for multiplayer stochastic games.

Let $\Gamma = \langle \{1,2\}, S, (A^1(s), A^2(s))_{s \in S}, q, r \rangle$ be a two-player zero-sum stochastic game. Let $B(\Gamma)$ be the set of all vectors (λ, v, x^1, x^2), where $\lambda \in (0,1]$ is a discounted factor; $v = (v(s))_{s \in S}$ is the vector of λ-discounted values at all initial states; $x^1 = (x_s^1)_{s \in S}$ is a stationary λ-discounted optimal strategy of Player 1; and $x^2 = (x_s^2)_{s \in S}$ is a stationary λ-discounted optimal strategy of Player 2.

Theorem 6.12   *For every two-player zero-sum stochastic game Γ the set $B(\Gamma)$ is semi-algebraic.*

Proof   The set $B(\Gamma)$ is a subset of $\mathbb{R} \times \mathbb{R}^S \times \mathbb{R}^{\sum_{s \in S} |A^1(s)|} \times \mathbb{R}^{\sum_{s \in S} |A^2(s)|}$ and, by Theorem 5.10, it contains all vectors $(\lambda, v_\lambda, x, y)$ that satisfy the following finite list of polynomial equalities and inequalities:

$$\lambda > 0,$$

$$\lambda \leq 1,$$

$$x_s^1(a^1) \geq 0, \quad \forall s \in S, a^1 \in A^1(s),$$

$$\sum_{a^1 \in A^1(s)} x_s^1(a^1) = 1, \quad \forall s \in S,$$

$$x_s^2(a^2) \geq 0, \quad \forall s \in S, a^2 \in A^2(s),$$

$$\sum_{a^2 \in A^2(s)} x_s^2(a^2) = 1, \quad \forall s \in S,$$

$$v(s) \leq \sum_{a \in A^1(s)} x_s^1(a) \left(\lambda r(s, a^1, a^2) + (1 - \lambda) \sum_{s' \in S} q(s' \mid s, a^1, a^2) v(s') \right),$$

$$\forall s \in S, \forall a^2 \in A^2(s),$$

$$v(s) \geq \sum_{a \in A^2(s)} x_s^2(a) \left(\lambda r(s,a^1,a^2) + (1 - \lambda) \sum_{s' \in S} q(s' \mid s,a^1,a^2) v(s') \right),$$

$$\forall s \in S, \forall a^1 \in A^1(s).$$

By Example 6.3, the set $B(\Gamma)$ is semi-algebraic.   □

Since the set $B(\Gamma)$ is semi-algebraic, repeated use of Theorem 6.7 implies the following.

Corollary 6.13   *For every two-player zero-sum stochastic game* Γ, *the function* $\lambda \mapsto v_\lambda(s)$ *is semi-algebraic for every fixed initial state* $s \in S$.

From Corollary 6.13 and Theorem 6.9, we deduce that in a neighborhood of 0 the function $\lambda \mapsto v_\lambda(s)$ can be expressed as a Puiseux series.

Theorem 6.14   *Let* Γ *be a two-player zero-sum stochastic game. For every state* $s \in S$ *there exist a* $\lambda_0 \in (0,1]$, *a positive integer* M, *a nonnegative integer* K, *and real numbers* $(a_k)_{k=K}^\infty$ *with* $a_K \neq 0$, *such that*

$$v_\lambda(s) = \sum_{k=K}^\infty a_k \lambda^{k/M}, \quad \forall \lambda \in (0,\lambda_0].$$

Moreover, there exists a $\lambda_1 \in (0,\lambda_0]$ *such that for every* $s \in S$ *the function* $v_\lambda(s)$ *is monotone in the interval* $(0,\lambda_1)$.

Proof   The only point that requires explanation is why K can be chosen to be nonnegative. This follows from Corollary 6.10 and the fact that the function $\lambda \mapsto v_\lambda(s_1)$ is bounded by $\|r\|_\infty$.   □

In particular, we obtain that the limit of the discounted value as the discount factor goes to 0 exists.

Corollary 6.15   *In every two-player zero-sum stochastic game, the limit* $\lim_{\lambda \to 0} v_\lambda(s)$ *exists for every fixed initial state* $s \in S$.

Another corollary of Theorem 6.12 asserts that there is a semi-algebraic mapping that assigns a stationary λ-discounted optimal strategy for every discount factor $\lambda \in (0,1]$.

Corollary 6.16   *For every two-player zero-sum stochastic game and each player* $i \in \{1,2\}$ *there is a semi-algebraic mapping* $\lambda \mapsto x_\lambda^i$ *that assigns to every discount factor* $\lambda \in (0,1]$ *a stationary* λ-*discounted optimal strategy* x_λ^i *for player* i.

6.3 Comments and Extensions

In this chapter, we studied semi-algebraic properties of two-player zero-sum discounted stochastic games. As mentioned earlier, semi-algebraic properties of multiplayer discounted stochastic games will be discussed in Section 8.4.

The properties of semi-algebraic sets that we needed for the study of stochastic games are

- Every set that is defined by finitely many polynomial inequalities is semi-algebraic.
- The projection of a semi-algebraic set in $\mathbb{R}^{n+1}$ to $\mathbb{R}^n$ is a semi-algebraic set.
- If the projection of a semi-algebraic subset A of $\mathbb{R}^{n+1}$ to its first coordinate contains an interval (a,b), then there is a semi-algebraic mapping $f : (a,b) \to \mathbb{R}^n$ such that the graph of f is a subset of A.
- Every semi-algebraic subset of $\mathbb{R}$ is a finite union of intervals.

There are other families of sets that satisfy these properties. Suppose that for every $n \in \mathbb{N}$ we are given an algebra $\mathcal{A}_n$ of subsets of $\mathbb{R}^n$. The collection $(\mathcal{A}_n)_{n \in \mathbb{N}}$ is an *o-minimal structure* if the following conditions are satisfied:

- If $A \in \mathcal{A}_n$, then $A \times \mathbb{R}$ and $\mathbb{R} \times A$ are in $\mathcal{A}_{n+1}$.
- If $A \in \mathcal{A}_{n+1}$, then the natural projection of A to its first n coordinates is in $\mathcal{A}_n$.
- For every polynomial P in n real variables, the set of solutions (zero set) of P is in $\mathcal{A}_n$.
- A set is in $\mathcal{A}_1$ if and only if it is a finite unions of intervals.

The family of semi-algebraic sets is one example of an *o*-minimal structure. Stochastic games in which the sets of actions are members of an *o*-minimal structure and the graphs of the payoff functions are members of the same *o*-minimal structure were studied in Bolte et al. (2015).

We used the theory of semi-algebraic sets to prove that $\lim_{\lambda \to 0} v_\lambda(s)$ exists for every initial state $s \in S$. Alternative proofs that use different tools were given by Szczechla et al. (1997), and Oliu-Barton (2014).

When the set of states or the sets of actions of the players are not finite, the limit $\lim_{\lambda \to 0} v_\lambda(s)$ may fail to exist. This was shown by Vigeral (2013) for a game with four states and compact action sets, and by Ziliotto (2016c) for a game with a countable compact set of states and finitely many actions, see also Sorin and Vigeral (2015).

The concept of θ-evaluations, which generalizes the discounted evaluation and T-stage evaluations, was described in Section 2.3. Let $\theta = (\theta_t)_{t=1}^{\infty}$ be a sequence of nonnegative reals that sum to 1. Then the θ-payoff of a pair of strategies (σ^1, σ^2) at the initial state s in a two-player zero-sum stochastic game is the quantity

$$\gamma_\theta(s; \sigma^1, \sigma^2) := \mathbf{E}_{s, \sigma^1, \sigma^2} \left[\sum_{t=1}^\infty \theta_t r(s_t, a_t) \right],$$

and the θ-value at the initial state s is the quantity

$$v_\theta(s) := \min_{\sigma^2 \in \Sigma^2} \max_{\sigma^1 \in \Sigma^1} \gamma_\theta(s; \sigma^1, \sigma^2).$$

The relation between the limit $\lim_{\lambda \to 0} v_\lambda(s_1)$ and the limit of $v_\theta(s_1)$ as $\max_{t \in \mathbb{N}} \theta_t$ goes to 0 was studied by Ziliotto (2016b, 2018).

Exercise 6.17 is taken from Kocel-Cynk et al. (2014).

6.4 Exercises

Exercise 6.2 is used in Chapter 10. Exercises 6.3 and 6.4 are used in the solution of Exercise 6.11. Exercise 6.5 is used in the solution of Exercise 6.14. Exercise 6.6 is used in the solution of Exercise 6.7. Exercise 6.8 is used in the solution of Exercise 6.9. Exercise 6.10 is used in the proof of Theorems 10.4, 12.8, and 13.7, and in the solution of Exercise 6.16. Exercise 6.11 is used in the proof of Theorems 9.13 and 9.26, and in the solution of Exercise 6.16. Exercise 6.13 is used in the solution of Exercise 6.14. Exercise 6.15 is used in the solution of Exercise 6.16. Exercise 6.17 is used in the solution of Exercise 8.9.

1. Among the six examples of semi-algebraic sets provided after Example 6.3, which are basic semi-algebraic sets?

2. Prove that any composition of semi-algebraic mappings is a semi-algebraic mapping.

3. Let $A \subseteq \mathbb{R}^{n+1}$ be a semi-algebraic set. Prove that the set

   $$B := \left\{ (x_1, \ldots, x_n) \in \mathbb{R}^n : \forall x_{n+1} \in \mathbb{R} \text{ one has } (x_1, \ldots, x_n, x_{n+1}) \in A \right\}$$

   is semi-algebraic.

4. Show that if $f, g \colon X \to \mathbb{R}$ are semi-algebraic functions, then the function $h \colon X \to \mathbb{R}$ that is defined by

   $$h(x) := \max\{f(x), g(x)\}, \quad \forall x \in X$$

   is semi-algebraic as well.

5. Let X and Y be two semi-algebraic sets and let $f, g \colon X \to Y$ be two semi-algebraic mappings. Prove that the set $\{x \in X : f(x) = g(x)\}$ is semi-algebraic.

6. Let X and Y be two semi-algebraic sets, with Y compact. Let
   $f\colon X \times Y \to \mathbb{R}$ be continuous and semi-algebraic. Define two functions
   $g, h\colon X \to \mathbb{R}$ by

$$g(x) := \max_{y \in Y} f(x, y), \quad \forall x \in X,$$

$$h(x) := \min_{y \in Y} f(x, y), \quad \forall x \in X.$$

   Prove that the functions g and h are semi-algebraic.
7. In this exercise, we provide an alternative proof to Corollary 6.13. Use
   Exercise 6.6 to show that for every two-player zero-sum stochastic game
   and every fixed initial state $s \in S$, the function $\lambda \mapsto v_\lambda(s)$ is
   semi-algebraic.
8. Prove Theorem 6.8: for every semi-algebraic function $f\colon \mathbb{R} \to \mathbb{R}$ there is
   a partition of $\mathbb{R}$ into a finite number of intervals $I_1, I_2, \dots, I_K$, and for
   each $k \in \{1, 2, \dots, K\}$ there is a polynomial $P_k\colon \mathbb{R}^2 \to \mathbb{R}$ such that
   $P_k(x, f(x)) = 0$ for every $x \in I_k$.
   Hint: Use Bézout's Theorem.[5]
9. Show that the exponential function e^x is not semi-algebraic.
   Hint: Use Exercise 6.8.
10. Let $f\colon (0, 1) \to \mathbb{R}$ be a semi-algebraic function. Prove that there is an
    $x_0 > 0$ such that either $f(x) = 0$ for all $x \in (0, x_0)$, or $f(x) > 0$ for all
    $x \in (0, x_0)$, or $f(x) < 0$ for all $x \in (0, x_0)$.
11. Let $f\colon (0, 1) \to \mathbb{R}$ be a semi-algebraic function. Prove that
    f is differentiable everywhere, except possible at a finite number of
    points. Prove that the derivative of f is a semi-algebraic function.
12. Let G be a strategic-form game with finitely many players and finitely
    many actions for each player. Show that the set of Nash equilibria of G is
    semi-algebraic.
13. Let X and Y be two semi-algebraic sets, and let $f\colon X \times Y \to \mathbb{R}$ be a
    semi-algebraic function. Is the function $x \mapsto \sup_{y \in Y} f(x, y)$
    semi-algebraic? Prove or provide a counterexample.
14. Let X and Y be two semi-algebraic sets, and let $f\colon X \times Y \to \mathbb{R}$ be a
    semi-algebraic function. Define a correspondence (set-valued function)
    $G\colon X \to Y$ by

[5] Étienne Bézout (Nemours, France, March 31, 1730 – Avon, France, September 27, 1783) was a
French mathematician who contributed to the study of algebraic equations and to mathematics
education.

$$G(x) := \operatorname{argmax}_{y \in Y} f(x, y)$$
$$= \{y \in Y : f(x, y) \geq f(x, z) \ \forall z \in Y\}, \ \forall x \in X.$$

Prove that G is semi-algebraic.

15. Let $f : (0, 1] \to \mathbb{R}$ be a semi-algebraic function. Suppose that $f(x) = \sum_{k=K}^{\infty} a_k x^{k/M}$ for every $x \in (0, x_0]$. Prove that

$$\lim_{x \to 0} \frac{\sum_{l=k+1}^{\infty} a_l x^{l/M}}{a_k x^{k/M}} = 0.$$

16. Prove Corollary 6.10: Let $f : (0, 1] \to \mathbb{R}$ be a semi-algebraic function. Prove that

(a) The limit $\lim_{x \to 0} f(x)$ exists and is given by

$$\lim_{x \to 0} f(x) = \begin{cases} 0, & \text{if } K > 0, \\ a_0, & \text{if } K = 0, \\ +\infty, & \text{if } K < 0, \ a_0 > 0, \\ -\infty, & \text{if } K < 0, \ a_0 < 0. \end{cases}$$

(b) There is an $x_0 \in (0, 1]$ such that f is monotone in the interval $(0, x_0)$.

17. Let $A \subseteq (0, 1] \times \mathbb{R}^n$ be a bounded semi-algebraic set that is relatively closed[6] in $(0, 1] \times \mathbb{R}^n$. For every $\lambda \in (0, 1]$ denote $A_\lambda := \{x \in \mathbb{R}^n : (\lambda, x) \in A\}$. Define the set of all accumulation points of sequences in A_λ, as λ goes to 0, by

$$A_0 := \limsup_{\lambda \to 0} A_\lambda = \bigcap_{\mu > 0} \bigcup_{0 < \lambda \leq \mu} A_\lambda.$$

In other words, A_0 is the set of all points $x \in \mathbb{R}^n$ for which there exists a sequence $(\lambda_k)_{k \in \mathbb{N}}$ of real numbers in $(0, 1]$, and for each $k \in \mathbb{N}$ there exists $x_k \in A_{\lambda_k}$ such that $\lim_{k \to \infty} \lambda_k = 0$ and $\lim_{k \to \infty} x_k = x$.

(a) Prove that the set A_0 is compact.
(b) Prove that the set A_0 is semi-algebraic.
(c) Prove that the sets A_λ converge to the set A_0 in the Hausdorff distance:[7] for every $\epsilon > 0$, there is a $\lambda_0 > 0$ such that $d(A_0, A_\lambda) < \epsilon$ for all $\lambda \in (0, \lambda_0]$, where

[6] A set A is *relatively closed* in the set B if for every sequence $(x_k)_{k \in \mathbb{N}}$ of points in A that converges to a limit x in B we have $x \in A$. The set $(0, 1]$ is relatively closed in $(0, \infty)$, but not relatively closed in $\mathbb{R}$.

[7] Felix Hausdorff (Breslau, Kingdom of Prussia, November 8, 1868 – Bonn, Germany, January 26, 1942) was a German mathematician who is considered to be one of the founders of modern topology and who contributed significantly to set theory, descriptive set theory, measure theory, function theory, and functional analysis.

$$d(A_0, A_\lambda) := \max \left\{ \sup_{x \in A_0} \inf_{y \in A_\lambda} \|x - y\|_\infty, \sup_{y \in A_\lambda} \inf_{x \in A_0} \|x - y\|_\infty \right\}.$$

18. Denote by $\mathbb{R}[x]$ the set of all real-valued polynomials. A *polynomial game* is a vector $G = \langle I, (A^i)_{i \in I}, (u^i)_{i \in I} \rangle$, where I is a finite set of players, and for every player $i \in I$, A^i is a finite set of actions, and u^i is a function that assigns to each action profile $a \in A := \prod_{i \in I} A^i$ a polynomial $u^{i,a} \in \mathbb{R}[x]$. For each $x \in \mathbb{R}$ the polynomial game G determines a strategic-form game $G_x = \langle I, (A^i)_{i \in I}, (u^i_x)_{i \in I} \rangle$, where $u^i_x(a) := u^{i,a}(x)$ for every $a \in A$. Prove that the set

$$E := \left\{ (x, z) \in \mathbb{R} \times \mathbb{R}^{\sum_{i \in I} |A^i|} : x \in \mathbb{R}, z \text{ is an equilibrium of } G_x \right\}$$

is semi-algebraic in $\mathbb{R}^{1 + \sum_{i \in I} |A^i|}$.

19. Express the discounted value of the game in Exercise 5.5 as a Puiseux series.

20. Consider the following zero-sum absorbing game:

	A	B	C	D
a	0	0	0	1 *
b	0	0	1 *	0 *
c	0	1 *	0 *	0 *
d	1 *	0 *	0 *	0 *

State $s(0)$

Do the following.

(a) Prove that the λ-discounted value is positive for every $\lambda \in (0, 1]$.

(b) Prove that for every discount factor $\lambda \in (0, 1]$, the λ-discounted optimal strategy of Player 1 (the row player) gives positive probability to all actions.

(c) Calculate the discounted value.

(d) Express the discounted value as a Puiseux series. What is M?

(e) Express the stationary λ-discounted optimal strategy of Player 2 as a Puiseux series.

(f) Can you find for every positive integer M a game whose discounted value can be expressed as a Laurent series in $\lambda^{1/M}$?

7

B-Graphs and the Continuity of the Limit $\lim_{\lambda \to 0} v_\lambda(s; q, r)$

In Section 5.4, we proved that the discounted value is continuous in the parameters of the game, see Theorem 5.11. One weakness of this result is that it does not bound the Lipschitz constant of the value function $(\lambda, q, r) \mapsto v_\lambda(s; q, r)$. In this chapter, we will strengthen Theorem 5.11 and, using the concept of *B*-graphs, develop a bound on the Lipschitz constant of the value function. Our technique will allow us to study the continuity of the limit $\lim_{\lambda \to 0} v_\lambda(s; q, r)$ as a function of q and r.

7.1 *B*-Graphs

Definition 7.1   Let S be a finite set of states, let $B \subset S$ be a non-empty set, let $g: B \to S$ be a mapping, and let $s \in B$. The *g-path* that starts at s is the longest sequence $s = s^1, s^2, \ldots, s^k$ of states in S such that $s^{j+1} = g(s^j)$ for every $j \in \{1, 2, \ldots, k - 1\}$.

A g-path is finite if it ends outside B, that is, if $g(s^k) \notin B$, and it is infinite if it is a loop, that is, if there are $i, j \in \mathbb{N}$, $i \neq j$, such that $s^i = s^j$.

The fundamental concept that we will study in this section is that of *B*-graph.

Definition 7.2   Let S be a set of states and let $B \subset S$ be a non-empty set. A *B-graph* is a mapping $g: B \to S$ that contains no loops: every g-path ends outside B.

The set of all *B*-graphs is denoted by $\mathcal{G}(B)$. For a given pair of states $s \in B$ and $z \notin B$, the set of all *B*-graphs g with the property that the g-path that starts at s ends at z is denoted by $\mathcal{G}_{s \to z}(B)$.

Example 7.3 $|B| = 1$.

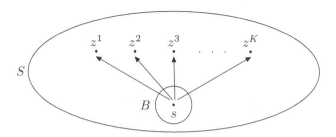

Figure 7.1 The case $|B| = 1$.

Suppose that B contains a single state, denoted by s. Denote by z^1, $z^2, \ldots, z^K$ the states outside of B (see Figure 7.1). The number of B-graphs is K; for every $k \in \{1, 2, \ldots, K\}$, the B-graph g_k is defined by $g_k(s) = z^k$. ◆

Example 7.4 $\quad |S| = 4, |B| = 2$.

Suppose that there are two states in B and two states outside B. Denote $B = \{s^1, s^2\}$ and $S \setminus B = \{z^1, z^2\}$ (see Figure 7.2). There are eight B-graphs:

$$g_1(s^1) = z^1, \qquad\qquad g_1(s^2) = z^2,$$
$$g_2(s^1) = z^1, \qquad\qquad g_2(s^2) = z^1,$$
$$g_3(s^1) = z^1, \qquad\qquad g_3(s^2) = s^1,$$
$$g_4(s^1) = s^2, \qquad\qquad g_4(s^2) = z^1,$$
$$g_5(s^1) = z^2, \qquad\qquad g_5(s^2) = z^1,$$
$$g_6(s^1) = s^2, \qquad\qquad g_6(s^2) = z^2,$$
$$g_7(s^1) = z^2, \qquad\qquad g_7(s^2) = s^1,$$
$$g_8(s^1) = z^2, \qquad\qquad g_8(s^2) = z^2,$$

The first four B-graphs are in $\mathcal{G}_{s^1 \to z^1}(B)$, and the last four B-graphs are in $\mathcal{G}_{s^1 \to z^2}(B)$. ◆

A *Markov chain* is a pair $\langle S, p \rangle$, where S is a finite set of states and $p \colon S \to \Delta(S)$ is a transition rule. In every stage, the chain is in one of its states. The chain starts at the initial state $s_1 \in S$, and, given the state s_t in stage t, the state s_{t+1} in stage $t + 1$ is chosen according to the probability distribution $p(\cdot \mid s_t)$. Thus, a Markov chain is equivalent to a stochastic game in which all players have a single action in all states.

Definition 7.5 $\quad$ The *weight* of a B-graph g in a Markov chain $\langle S, p \rangle$ is given by

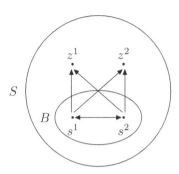

Figure 7.2 The case $|S| = 4$ and $|B| = 2$.

$$w(g) := \prod_{s \in B} p(g(s) \mid s).$$

Denote the sum of the weights of all B-graphs in a Markov chain $\langle S, p \rangle$ by

$$W(B) := \sum_{g \in \mathcal{G}(B)} w(g),$$

and the sum of the weights of all B-graphs in $\mathcal{G}_{s \to z}(B)$ by

$$W_{s \to z}(B) := \sum_{g \in \mathcal{G}_{s \to z}(B)} w(g).$$

Let $\langle S, p \rangle$ be a Markov chain, let $B \subset S$ be a non-empty set, let $s \in B$, and let $z \notin B$. Denote by $Q_s(z; B)$ the probability that, when the initial state of the Markov chain is s, the first state outside B that the process reaches is z.

The following result, which provides an expression for $Q_s(z; B)$ using the weights of B-graphs, follows from a more general result proved by Freidlin and Wentzell (1986).

Theorem 7.6   *Let $\langle S, p \rangle$ be a Markov chain and let B be a non-empty proper subset of S. If $\sum_{z \notin B} Q_s(z; B) > 0$ for every state $s \in B$ then*

$$Q_s(z; B) = \frac{W_{s \to z}(B)}{W(B)}. \tag{7.1}$$

The condition $\sum_{z \notin B} Q_s(z; B) > 0$ for every state $s \in S$ implies that in the Markov chain, given any initial state s, the process leaves B with positive probability. This in turn implies that the process leaves B with probability 1.

Proof of Theorem 7.6   We prove the result by induction on $|B|$, the number of states in B.

Step 1:   Without loss of generality, we can assume that $p(s \mid s) = 0$, for every state $s \in B$.

Let $B \subset S$ be a non-empty subset such that $\sum_{z \notin B} Q_s(z; B) > 0$ for each state $s \in B$. Let $\widehat{p}: B \to \Delta(S)$ be the normalized version of p: for every state $s \in S$,

$$\widehat{p}(z \mid s) := \begin{cases} \dfrac{p(z \mid s)}{1 - p(s \mid s)}, & \text{if } z \neq s, \\ 0, & \text{if } z = s. \end{cases} \tag{7.2}$$

Since $\sum_{t \notin B} Q_s(z; B) > 0$ for every state $s \in B$, it follows that $p(s \mid s) < 1$ for every state $s \in B$, and therefore the denominator in Eq. (7.2) is positive.

For every state $s \in B$ and every state $z \notin B$, denote by $\widehat{Q}_s(z; B)$ the probability that when the initial state is s, in the Markov chain $\langle S, \widehat{p} \rangle$ the first state outside B that is reached is z. The probability to stay in a state does not affect the probability distribution $Q_s(\cdot; B)$, and therefore

$$Q_s(z; B) = \widehat{Q}_s(z; B), \quad \forall s \in B, \forall z \in S. \tag{7.3}$$

The weight $\widehat{w}(g)$ of a B-graph g in the Markov chain $\langle S, \widehat{p} \rangle$ is given by

$$\widehat{w}(g) = \prod_{s \in B} \widehat{p}(g(s) \mid s) = \prod_{s \in B} \frac{p(g(s) \mid s)}{1 - p(s \mid s)} = \frac{w(g)}{\prod_{s \in B}(1 - p(s \mid s))}. \tag{7.4}$$

Denote the sum of the weights of all B-graphs in a Markov chain $\langle S, \widehat{p} \rangle$ by

$$\widehat{W}(B) := \sum_{g \in \mathcal{G}(B)} \widehat{w}(g),$$

and the sum of the weights of all B-graphs in $\mathcal{G}_{s \to z}(B)$ by

$$\widehat{W}_{s \to z}(B) := \sum_{g \in \mathcal{G}_{s \to z}(B)} \widehat{w}(g).$$

Since the denominator in the right-hand side of Eq. (7.4) is independent of g, it follows that

$$\frac{W_{s \to z}(B)}{W(B)} = \frac{\widehat{W}_{s \to z}(B)}{\widehat{W}(B)}. \tag{7.5}$$

Eqs. (7.3) and (7.5) show that if we prove that

$$\widehat{Q}_s(z; B) = \frac{\widehat{W}_{s \to z}(B)}{\widehat{W}(B)}, \quad \forall z, s \in B,$$

then it will follow that

$$Q_s(z; B) = \frac{W_{s \to z}(B)}{W(B)}, \quad \forall z, s \in B.$$

We thus assume from now on that $p(s \mid s) = 0$ for every state $s \in B$.

Step 2:   The case $|B| = 1$.

Suppose that $B = \{s\}$. The weight of a B-graph $g \in G(B)$ is $w(g) = p(g(s) \mid s)$. The number of B-graphs is $|S| - 1$, since $g(s)$ can have $|S| - 1$ values. The only B-graph in which the path from s leads to z is the one in which $g(s) = z$. We then have

$$Q_s(z; B) = p(z \mid s) = \frac{p(z \mid s)}{\sum_{s' \neq s} p(s' \mid s)} = \frac{W_{s \to z}(B)}{W(B)},$$

where the second equality holds by Step 1.

Step 3:   The general case.

Let $k \geq 2$. Assume by induction that Eq. (7.1) holds whenever B contains at most $k - 1$ states, and let B be a subset of S that contains k states.

Let $s \in B$ and $z \notin B$. By the induction hypothesis, for every state $s' \in B \setminus \{s\}$,

$$Q_{s'}(z; B \setminus \{s\}) = \frac{W_{s' \to z}(B \setminus \{s\})}{W(B \setminus \{s\})}.$$

To reach state z from state s in the Markov chain $\langle S, p \rangle$ (without reaching any other state not in B), we can either move from s directly to z; or move from s to some state $s' \in B \setminus \{s\}$, and then move from s' to z; or move from s to some state $s' \in B \setminus \{s\}$, then move back to s, and then move from s to z. This leads to the following expression for $Q_s(z; B)$:

$$Q_s(z; B) = p(z \mid s) + \sum_{s' \in B \setminus \{s\}} p(s' \mid s) Q_{s'}(z; B \setminus \{s\})$$

$$+ \sum_{s' \in B \setminus \{s\}} p(s' \mid s) Q_{s'}(s; B \setminus \{s\}) Q_s(z; B). \qquad (7.6)$$

It follows that

$$Q_s(z; B) = \frac{p(z \mid s) + \sum_{s' \in B \setminus \{s\}} p(s' \mid s) Q_{s'}(z; B \setminus \{s\})}{1 - \sum_{s' \in B \setminus \{s\}} p(s' \mid s) Q_{s'}(s; B \setminus \{s\})} \qquad (7.7)$$

$$= \frac{p(z \mid s) + \sum_{s' \in B \setminus \{s\}} p(s' \mid s) \frac{W_{s' \to z}(B \setminus \{s\})}{W(B \setminus \{s\})}}{1 - \sum_{s' \in B \setminus \{s\}} p(s' \mid s) Q_{s'}(s; B \setminus s)} \qquad (7.8)$$

$$= \frac{p(z \mid s) W(B \setminus \{s\}) + \sum_{s' \in B \setminus \{s\}} p(s' \mid s) W_{s' \to z}(B \setminus \{s\})}{\left(1 - \sum_{s' \in B \setminus \{s\}} p(s' \mid s) Q_{s'}(s; B \setminus s)\right) \cdot W(B \setminus \{s\})} \qquad (7.9)$$

$$= \frac{W_{s \to z}(B)}{\left(1 - \sum_{s' \in B \setminus \{s\}} p(s' \mid s) Q_{s'}(s; B \setminus s)\right) \cdot W(B \setminus \{s\})}, \qquad (7.10)$$

where Eq. (7.7) follows from Eq. (7.6), Eq. (7.8) follows from the induction hypothesis, and Eq. (7.10) holds because a g-path leads from s to z either if $g(s) = z$ (first summand in the numerator in Eq. (7.9)), or if $g(s) = s'$ and the g-path that starts at s' ends at z (second summand in the numerator in Eq. (7.9)).

Since the denominator in Eq. (7.10) is independent of z, it follows that

$$Q_s(z; B) = \frac{W_{s \to z}(B)}{\sum_{z' \notin B} W_{s \to z'}(B)} = \frac{W_{s \to z}(B)}{W(B)},$$

where the last equality holds since every path that leaves s ends in some state $z' \notin B$. We thus proved the induction step, which completes the proof of the theorem. □

7.2 The Mean Discounted Time

Let $\langle S, p \rangle$ be a Markov chain. For every discount factor $\lambda \in (0, 1]$ and every state $z \in S$, the *mean λ-discounted time* the process spends in state z, when the initial state is s, is given by

$$t_\lambda(s, p; z) := \mathbf{E}_{s, p} \left[\lambda \sum_{t \in \mathbb{N}} (1 - \lambda)^{t-1} \mathbf{1}_{\{s_t = z\}} \right], \tag{7.11}$$

where $\mathbf{1}_{\{s_t = z\}}$ is the indicator function.

Comparing Eq. (7.11) and Eq. (1.9), we see that $t_\lambda(s, p; z)$ is the λ-discounted payoff of a Markov decision problem with state space S, where the decision maker has a single action in each state, the transition is given by p, and the payoff is 1 in state z and 0 in all other states.

Our basic observation is that the function $p \mapsto t_\lambda(s, p; z)$ is a rational function of p, and that the two polynomials that define this rational function have nonnegative coefficients.

Proposition 7.7 *For every initial state s, every state $z \in S$, and every discount factor $\lambda \in (0, 1]$, there exist two polynomials $h_1(p)$ and $h_2(p)$ in the $|S|^2$ variables $(p(u \mid r))_{r, u \in S}$ such that* (i) *both $h_1(p)$ and $h_2(p)$ have degree at most $|S|$ and nonnegative coefficients, and* (ii) *$t_\lambda(s, p; z) = h_1(p)/h_2(p)$ for every transition rule p over S.*

Proof   Fix $\lambda \in (0, 1]$. Define an auxiliary Markov chain $\langle \widehat{S}, \widehat{p} \rangle$ as follows:

1. The state space is $\widehat{S} = S' \cup S''$, where S' and S'' are two disjoint copies of S. For every state $s \in S$, we denote by s' and s'' the corresponding states in S' and S'', respectively.
2. States in $s'' \in S''$ are absorbing: $\widehat{p}(s'' \mid s'') = 1$ for each state $s'' \in S''$.
3. The transition rule from each state $s' \in S'$ is given by

$$\widehat{p}(s'' \mid s') = \lambda,$$

$$\widehat{p}(z' \mid s') = (1 - \lambda)p(z \mid s), \quad \forall z \in S, \text{ and}$$

$$\widehat{p}(z'' \mid s') = 0 \quad \forall z \in S \setminus \{s\}.$$

In words, from state s' the process either moves with probability λ to state s'', where it is absorbed, or, with the complementary probability $1 - \lambda$, continues as in the original Markov chain. Another way to view this construction is that in every stage there is a fixed probability of λ that the process "terminates" and gets stuck in the current state; the transition from the current state $s' \in S'$ to its copy $s'' \in S''$, which is absorbing, corresponds to the termination of the process.

We claim that

$$t_\lambda(s, p; z) = Q_s(z''; S'), \quad \forall s, z \in S,$$

that is, the mean discounted time $t_\lambda(s, p; z)$ in the original Markov chain $\langle S, p \rangle$ is nothing but the probability that the Markov chain $\langle \widehat{S}, \widehat{p} \rangle$ is absorbed at state z''. Indeed, for a fixed state $z \in S$, both vectors $(t_\lambda(s, p; z))_{s \in S}$ and $(Q_s(z''; S'))_{s \in S}$ are solutions of the system of linear equations

$$x(s) = \lambda \mathbf{1}_{\{s = z\}} + (1 - \lambda) \sum_{r \in S} p(r \mid s) x(r), \quad \forall s \in S, \tag{7.12}$$

but this system has a unique solution. By Theorem 7.6,

$$t_\lambda(s, p; z) = \frac{W_{s \to z''}(S')}{W(S')}, \quad \forall s, z \in S,$$

and the result follows, since the weight $w(g)$ of every B-graph g is a product of the terms λ, $(1 - \lambda)$, and $(p(u \mid r))_{r, u \in S}$. $\square$

Comment 7.8   To prove that the function $p \mapsto t_\lambda(s, p; z)$ is a rational function of p, it is sufficient to show that it is the solution of the system of linear equations (7.12). We used B-graphs to show that the two polynomials that define this rational function have nonnegative coefficients.

Corollary 7.9 *For every initial state $s \in S$, every discount factor $\lambda \in (0,1]$, and every collection of nonnegative scalars $(\theta_z)_{z \in S}$, the function $p \mapsto \sum_{z \in S} \theta_z t_\lambda(s,p;z)$ is the ratio of two polynomials in the variables $(p(u \mid r))_{r,u \in S}$ of degree at most $|S|$ with nonnegative coefficients.*

Proof By Proposition 7.7, for every fixed pair of states $s, z \in S$, the function $p \mapsto t_\lambda(s,p;z)$ is the ratio of two polynomials in $(p(u \mid r))_{r,u \in S}$. By Theorem 7.6 and the proof of Proposition 7.7, all these ratios have the same denominator, which has nonnegative coefficients and degree at most $|S|$, as do each of the numerators. Since $(\theta_z)_{z \in S}$ are nonnegative scalars, the result follows. □

7.3 The Mean Discounted Time and Stochastic Games

Fix a two-player zero-sum stochastic game $\Gamma = \langle \{1,2\}, S, (A^1(s), A^2(s))_{s \in S}, q, r \rangle$. In this section, we will vary the transition rule and the payoff function, and denote by $\gamma_\lambda(s,q,r;x)$ the λ-discounted payoff at the initial state s under the stationary strategy profile x when the transition rule is q and the payoff function is r.

Every vector x of stationary strategies induces a Markov chain over the set of states S with the transition rule p defined by

$$p(z \mid s) := q(z \mid s, x(s)), \quad \forall s, z \in S;$$

that is, the probability to move from state s to state z in the Markov chain $\langle S, p \rangle$ is the probability that under the stationary strategy profile x the play moves from state s to state z. We denote by $t_\lambda(s,q,x;z)$ the λ-discounted time that the Markov chain induced by the stationary strategy profile x spends at state z when the initial state is s.

The concept of mean discounted time can be used to express the discounted payoff.

Lemma 7.10 *Let $\Gamma = \langle \{1,2\}, S, (A^1(s), A^2(s))_{s \in S}, q, r \rangle$ be a two-player zero-sum stochastic game. For every initial state $s \in S$, every stationary strategy profile x, and every discount factor $\lambda \in (0,1]$, we have*

$$\gamma_\lambda(s,q,r;x) = \sum_{z \in S} r(z, x(z)) t_\lambda(s,q,x;z).$$

Proof   The lemma follows from the following list of equalities.

$$\gamma_\lambda(s,q,r;x) = \mathbf{E}_{s,x}\left[\lambda \sum_{t \in \mathbb{N}}(1-\lambda)^{t-1}r(s_t,a_t)\right] \tag{7.13}$$

$$= \mathbf{E}_{s,x}\left[\lambda \sum_{t \in \mathbb{N}}(1-\lambda)^{t-1}r(s_t,x(s_t))\right] \tag{7.14}$$

$$= \sum_{z \in S} \mathbf{E}_{s,x}\left[\lambda \sum_{t \in \mathbb{N}}(1-\lambda)^{t-1}r(s_t,x(s_t))\mathbf{1}_{\{s_t=z\}}\right] \tag{7.15}$$

$$= \sum_{z \in S} r(z,x(z))\mathbf{E}_{s,x}\left[\lambda \sum_{t \in \mathbb{N}}(1-\lambda)^{t-1}\mathbf{1}_{\{s_t=z\}}\right] \tag{7.16}$$

$$= \sum_{z \in S} r(z,x(z))t_\lambda(s,q,x;z), \tag{7.17}$$

where Eq. (7.13) follows from the definition of the discounted payoff, Eq. (7.14) holds because $\mathbf{E}_{s,x}[r(s_t,a_t) \mid h_t] = r(s_t,x(s_t))$, Eqs. (7.15) and (7.16) hold because expectation is linear, and Eq. (7.17) follows from the definition of the mean discounted time.   □

7.4 A Distance between Transition Rules

In this section, we introduce and study a semi-metric $d(\cdot,\cdot)$ between transition rules in stochastic games. Recall that a *semi-metric* satisfies symmetry and separation ($d(q,q') = 0$ if and only if $q = q'$), but it does not necessarily satisfy the triangle inequality.

Let $\Gamma = \langle\{1,2\}, S, (A^1(s), A^2(s))_{s \in S}, q, r\rangle$ be a two-player zero-sum stochastic game. Denote by $\mathcal{Q}^* = (\Delta(S))^{S \times A_1 \times A_2}$ the set of all transition rules. Define a function $d: \mathcal{Q}^* \times \mathcal{Q}^* \to [0, +\infty]$ as follows: for every $q, q' \in \mathcal{Q}^*$,

$d(q,q')$

$$= \max\left\{\frac{q(z \mid s,a_1,a_2)}{q'(z \mid s,a_1,a_2)}, \frac{q'(z \mid s,a_1,a_2)}{q(z \mid s,a_1,a_2)} \;\middle|\; s,z \in S, a_1 \in A_1, a_2 \in A_2\right\} - 1,$$

where by convention $x/0 = +\infty$ for $x > 0$, and $0/0 = 1$. Let us list a few simple properties of the nonnegative valued function $d(\cdot,\cdot)$:

A.i) $d(q,q') = 0$ if and only if $q = q'$;
A.ii) Symmetry: $d(q,q') = d(q',q)$;

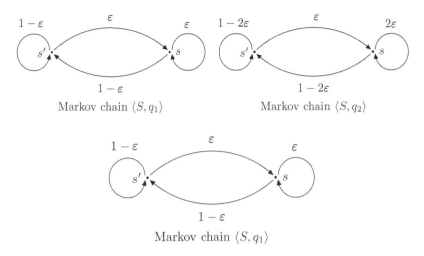

Figure 7.3 The Markov chains in Example 7.11.

A.iii) $d(q,q') < +\infty$ if and only if q and q' have the same supports, that is, $\text{supp}(q(\cdot \mid s,a) = \text{supp}(q'(\cdot \mid s,a))$ for every $(s,a) \in SA$; and

A.iv) $d(q_n,q) \to 0$ if and only if $q_n \to q$ in the Euclidean norm and q and q_n have the same support for all sufficiently large $n \in \mathbb{N}$.

As the following example shows, $d(\cdot,\cdot)$ does not necessarily satisfy the triangle inequality, hence it is *not* a metric.

Example 7.11 Let $S = \{s,s'\}$, and take $|A_1| = |A_2| = 1$; thus, the stochastic game reduces to a Markov chain. Fix $\epsilon \in \left(0, \frac{1}{4}\right)$, and consider the three Markov chains displayed in Figure 7.3. Formally, for $k = 1,2,4$, define a transition rule $q_k \colon S \to \Delta(S)$ by $q_k(s \mid z) = 1 - p_k(s' \mid z) = k\epsilon$ for each state $z \in S$. Since $\epsilon \in (0,\frac{1}{4})$ one has

$$d(q_1,q_2) = \max \left\{ \frac{\epsilon}{2\epsilon}, \frac{2\epsilon}{\epsilon}, \frac{1-\epsilon}{1-2\epsilon}, \frac{1-2\epsilon}{1-\epsilon} \right\} = 1,$$

$$d(q_2,q_4) = \max \left\{ \frac{2\epsilon}{4\epsilon}, \frac{4\epsilon}{2\epsilon}, \frac{1-2\epsilon}{1-4\epsilon}, \frac{1-4\epsilon}{1-2\epsilon} \right\} = 1,$$

and

$$d(q_1,q_4) = \max \left\{ \frac{\epsilon}{4\epsilon}, \frac{4\epsilon}{\epsilon}, \frac{1-\epsilon}{1-4\epsilon}, \frac{1-4\epsilon}{1-\epsilon} \right\} = 2.5.$$

Therefore, $d(q_1,q_2) + d(q_2,q_4) < d(q_1,q_4)$, and the triangle inequality does not hold. ♦

7.5 Continuity of the Value

The next result estimates the maximal change in the discounted value as one changes the transition rule and the payoff function.

Theorem 7.12   *Let* $\Gamma = \langle \{1, 2\}, S, (A^1(s), A^2(s))_{s \in S}, q, r \rangle$ *be a two-player zero-sum stochastic game, let* $q' \colon S \times A_1 \times A_2 \to \Delta(S)$ *be an arbitrary transition rule, and let* $r' \colon S \times A_1 \times A_2 \to \mathbb{R}$ *be an arbitrary payoff function. Then for every* $\lambda \in (0, 1]$ *and every initial state* $s \in S$,

$$|v_\lambda(s; q', r') - v_\lambda(s; q, r)| \le 4|S| d(q, q') \|r\|_\infty + \|r - r'\|_\infty.$$

Theorem 7.12 implies that the difference between the discounted values of two stochastic games that share the same set of states and actions can be bounded by a constant that is independent of the discount factor. To prove Theorem 7.12, we need the following technical result.

Lemma 7.13   *Let* $f(x_1, \ldots, x_k)$ *be a polynomial in* $x_1, \ldots, x_k$ *with non-negative coefficients and degree at most n, and let* $\epsilon \ge 0$. *Let* $y, y' \in \mathbb{R}^k$ *be two nonnegative vectors such that* $\dfrac{1}{1 + \epsilon} \le \dfrac{y_i}{y'_i} \le 1 + \epsilon$ *for every* $i = 1, \ldots, k$. *Then* $\dfrac{1}{1 + \epsilon^{-n}} \le \dfrac{f(y)}{f(y')} \le (1 + \epsilon)^n$.

Recall that $\frac{0}{0} = 1$, and therefore the condition $\dfrac{1}{1 + \epsilon} \le \dfrac{y_i}{y'_i} \le 1 + \epsilon$ implies that $y_i = 0$ if and only if $y'_i = 0$.

Proof   Denote $f(x) = \sum_{i=1}^{I} a_i \prod_{j=1}^{n_i} x_{k_{i,j}}$, where $I \in \mathbb{N}$, and for every $i = 1, \ldots, I$, $a_i \ge 0$, $0 \le n_i \le n$, and $1 \le k_{i,j} \le k$ for each $j = 1, \ldots, n_i$. Since y and y' are nonnegative vectors, the condition in the lemma implies that for every $i = 1, \ldots, I$ we have

$$\frac{1}{(1 + \epsilon)^n} \prod_{j=1}^{n_i} y'_{k_{i,j}} \le \prod_{j=1}^{n_i} y_{k_{i,j}} \le (1 + \epsilon)^n \prod_{j=1}^{n_i} y'_{k_{i,j}}. \tag{7.18}$$

Since $(a_i)_{i=1}^{I}$ are nonnegative, multiplying Eq. (7.18) by a_i and summing over $i = 1, \ldots, I$ we obtain

$$\frac{1}{(1 + \epsilon)^n} f(y') \le f(y) \le (1 + \epsilon)^n f(y'),$$

and the result follows since y, y', and the coefficients of f are nonnegative. $\square$

The next lemma states that if the payoff function is *nonnegative*, then one can uniformly bound the ratio $\dfrac{\gamma_\lambda(s_1,q',r;x_1,x_2)}{\gamma_\lambda(s_1,q,r;x_1,x_2)}$ over the set of stationary strategy pairs (x_1,x_2).

A payoff function $r\colon SA \to \mathbb{R}$ is *nonnegative* if $r(s,a) \geq 0$ for all $(s,a) \in SA$.

Lemma 7.14   *Let* $\Gamma = \langle \{1,2\}, S, (A^1(s), A^2(s))_{s\in S}, q, r \rangle$ *be a two-player zero-sum stochastic game with a nonnegative payoff function r, and let $q' \in \mathcal{Q}^*$ be a transition rule that satisfies $d(q,q') < \frac{1}{2|S|}$. Then for every stationary strategy profile x, one has*

$$(1 - d(q,q'))^{2|S|} \leq \frac{\gamma_\lambda(s_1,q',r;x)}{\gamma_\lambda(s_1,q,r;x)} \leq \frac{1}{(1 - d(q,q'))^{2|S|}}. \tag{7.19}$$

Proof   Fix a stationary strategy profile x. As mentioned before, x naturally defines a Markov chain over S with transition rule p defined by

$$p(z \mid s) = q(z \mid s, x(s)), \quad \forall z, s \in S.$$

By Lemma 7.10, for every $\lambda \in (0,1]$,

$$\gamma_\lambda(s,q,r;x) = \sum_{z\in S} r(z,x(z)) t_\lambda(s,q,x;z). \tag{7.20}$$

By Eq. (7.20) and Corollary 7.9, with $\theta_z = r(z,x(z))$ for every $z \in S$, and since r is nonnegative, for each state $s \in S$ there exist two polynomials h_1 and h_2 in the $|S|^2$ variables $(q(u \mid x(z)))_{u,z\in S}$ of degrees at most $|S|$ and nonnegative coefficients, such that

$$\frac{\gamma_\lambda(s,q',r;x)}{\gamma_\lambda(s,q,r;x)} = \frac{h_1(q)}{h_1(q')} \cdot \frac{h_2(q')}{h_2(q)}. \tag{7.21}$$

Set $\epsilon = d(q,q') < \frac{1}{2|S|} < 1$. By the definition of $d(q,q')$, for each $(z,s,a_1,a_2) \in S^2 \times A_1 \times A_2$ the two quantities

$$\frac{q(z \mid s,a_1,a_2)}{q'(z \mid s,a_1,a_2)} \quad \text{and} \quad \frac{q'(z \mid s,a_1,a_2)}{q(z \mid s,a_1,a_2)}$$

are between $\frac{1}{1+\epsilon}$ and $1+\epsilon$. Eq. (7.21) and Lemma 7.13, with $k = |S|^2 \cdot |A| \cdot |B|$, imply that

$$(1 + d(q,q'))^{-2|S|} \leq \frac{\gamma_\lambda(s_1,q',r;x)}{\gamma_\lambda(s_1,q,r;x)} \leq (1 + d(q,q'))^{2|S|}.$$

Since for every $y \in [0,1]$ one has $1 - y \leq 1/(1+y)$ and $1 + y \leq 1/(1-y)$, we obtain Eq. (7.19). $\qquad\square$

Lemma 7.15   *Let* $\Gamma = \langle \{1, 2\}, S, (A^1(s), A^2(s))_{s \in S}, q, r \rangle$ *be a two-player zero-sum stochastic game, and let* $q' \in \mathcal{Q}^*$ *satisfy* $d(q, q') < \frac{1}{2|S|}$. *Then for every initial state* $s \in S$,

$$|v_\lambda(s; q, r) - v_\lambda(s; q', r)| \leq 4|S| d(q, q') \|r\|_\infty. \tag{7.22}$$

Proof   Define a payoff function $r' : SA \to \mathbb{R}$ by

$$r'(s, a) := r(s, a) + \|r\|_\infty, \quad \forall (s, a) \in SA,$$

which is nonnegative. Fix an initial state $s \in S$. By Lemma 7.14,

$$(1 - d(q, q'))^{2|S|} \cdot \gamma_\lambda(s_1, q, r'; x)$$

$$\leq \gamma_\lambda(s, q', r'; x) \leq \frac{1}{(1 - d(q, q'))^{2|S|}} \cdot \gamma_\lambda(s, q, r'; x),$$

for every stationary strategy profile $x \in X$. For every $y \in \left(0, \frac{1}{2|S|}\right]$ we have

$$1 - 2|S|y \leq (1 - y)^{2|S|}$$

and

$$\frac{1}{(1 - y)^{2|S|}} \leq \frac{1}{1 - 2|S|y} \leq 1 + 4|S|y,$$

and since by assumption $d(q, q') < \frac{1}{2|S|}$, we deduce that

$$(1 - 2|S|d(q, q')) \cdot \gamma_\lambda(s_1, q, r'; x) \leq \gamma_\lambda(s, q', r'; x)$$
$$\leq (1 + 4|S|d(q, q')) \cdot \gamma_\lambda(s, q, r'; x), \quad \forall x \in X. \tag{7.23}$$

It follows that

$$(1 - 2|S|d(q, q')) \cdot \max_{x_1 \in X_1} \min_{x_2 \in X_2} \gamma_\lambda(s, q, r'; x)$$

$$\leq \max_{x_1 \in X_1} \min_{x_2 \in X_2} \gamma_\lambda(s, q', r'; x)$$

$$\leq (1 + 4|S|d(q, q')) \cdot \max_{x_1 \in X_1} \min_{x_2 \in X_2} \gamma_\lambda(s, q, r'; x).$$

This in turn implies that

$$(1 - 2|S|d(q, q')) \cdot v_\lambda(s; q, r')$$

$$\leq v_\lambda(s; q', r') \leq (1 + 4|S|d(q, q')) \cdot v_\lambda(s; q, r'). \tag{7.24}$$

By Exercise 3.3,

$$v_\lambda(s_1, q, r') = v_\lambda(s_1, q, r) + \|r\|_\infty,$$

$$v_\lambda(s_1, q', r') = v_\lambda(s_1, q', r) + \|r\|_\infty.$$

We deduce that

$$(1 - 2|S|d(q,q')) \cdot v_\lambda(s;q,r) - 2|S|d(q,q')\|r\|_\infty \leq v_\lambda(s;q',r)$$
$$\leq (1 + 4|S|d(q,q')) \cdot v_\lambda(s;q,r') + 4|S|d(q,q')\|r\|_\infty,$$

and Eq. (7.22) follows. □

We now turn to prove Theorem 7.12.

Proof of Theorem 7.12 By Theorem 3.15, for every initial state $s \in S$, every transition rule q, every pair of payoff functions (r,r'), and every $\lambda \in (0,1]$,

$$|v_\lambda(s,q,r) - v_\lambda(s,q,r')| \leq \|r - r'\|_\infty.$$

By the triangle inequality, it is therefore sufficient to prove that

$$|v_\lambda(s,q',r) - v_\lambda(s,q,r)| \leq 4|S|d(q,q')\|r\|_\infty.$$

This inequality holds trivially when $d(q,q') \geq 1/(2|S|)$. Indeed, in this case the right-hand side is at least $2\|r\|_\infty$, while the left-hand side is at most $2\|r\|_\infty$.

Assume then that $d(q,q') < 1/(2|S|)$. In this case, Lemma 7.15 yields the desired result. □

Corollary 7.16 *Let* $\Gamma = \langle\{1,2\}, S, (A^1(s), A^2(s))_{s \in S}, q, r\rangle$ *be a two-player zero-sum stochastic game, let* $(\lambda_n)_{n \in \mathbb{N}}$ *be a sequence of discount factors that converges to 0, let* $(q_n)_{n \in \mathbb{N}}$ *be a sequence of transition rules that satisfies* $\lim_{n \to \infty} d(q_n, q) = 0$, *and let* $(r_n)_{n \in \mathbb{N}}$ *be a sequence of payoff functions that satisfies* $\lim_{n \to \infty} \|r_n - r\|_\infty = 0$. *Then*

$$\lim_{n \to \infty} v_{\lambda_n}(s; q_n, r_n) = \lim_{\lambda \to 0} v_\lambda(s; q, r), \quad \forall s \in S.$$

Proof Fix an initial state $s \in S$. By the triangle inequality,

$$|v_{\lambda_n}(s; q_n, r_n) - \lim_{\lambda \to 0} v_\lambda(s; q, r)| \leq |v_{\lambda_n}(s; q_n, r_n) - v_{\lambda_n}(s; q, r)|$$
$$+ |v_{\lambda_n}(s; q, r) - \lim_{\lambda \to 0} v_\lambda(s; q, r)|.$$

The first term goes to 0 by Theorem 7.12, and the second term goes to 0 by Corollary 6.15. □

Let S be a set of states, and let $(A_1(s))_{s \in S}$ and $(A_2(s))_{s \in S}$ be the action sets of the two players. For every initial state $s \in S$, every transition rule $q: SA \to \Delta(S)$, and every payoff function $r: SA \to \mathbb{R}$, denote the limit of the λ-discounted value at the initial state s by

$$v_0(s; q, r) := \lim_{\lambda \to 0} v_\lambda(s; q, r).$$

As a conclusion of Corollary 7.16, we deduce that the function $(q,r) \mapsto v_0(s;q,r)$ is continuous (see Exercise 7.5).

Corollary 7.17   *Let* $\Gamma = \langle\{1,2\}, S, (A^1(s), A^2(s))_{s\in S}, q, r\rangle$ *be a two-player zero-sum stochastic game, let* $(q_n)_{n\in\mathbb{N}}$ *be a sequence of transition rules that satisfies* $\lim_{n\to\infty} d(q_n,q) = 0$, *and let* $(r_n)_{n\in\mathbb{N}}$ *be a sequence of payoff functions that satisfies* $\lim_{n\to\infty} \|r_n - r\|_\infty = 0$. *Then*

$$\lim_{n\to\infty} v_0(s;q_n,r_n) = v_0(s;q,r), \quad \forall s \in S.$$

7.6 Comments and Extensions

Theorem 7.6 provides an expression for the exit probability from a set using B-graphs. In Exercise 7.6, we provide an expression for the invariant distribution of a Markov chain using B-graphs. More general expressions of this type were obtained by Freidlin and Wentzell (1984, Section 6.3). They also provided an expression for the expected time until the process leaves a given set B using B-graphs. The first to introduce B-graphs to Game Theory was Vieille (2000b, 2000c), in his study of two-player stochastic games. Theorem 7.12 is due to Solan (2003). It was used in the literature to characterize the limit $v_0(s;q,r)$ or to compute it, see, for example, Chatterjee et al. (2008), Chatterjee et al. (2009), or Oliu-Barton (2014).

7.7 Exercises

1. Let $\langle S, p\rangle$ be a Markov chain with state space $S = \{s_1,s_2,s_3,s_4\}$, and let $B = \{s_1,s_2\}$. Using the tools developed in this chapter, calculate $Q_{s_1,p}(s_3 \mid B)$ in each of the following cases (all transitions from states in B that are not mentioned equal 0).

   (a)

   $$p(s_2 \mid s_1) = \tfrac{1}{4}, \quad p(s_3 \mid s_1) = \tfrac{3}{4},$$
   $$p(s_1 \mid s_2) = \tfrac{1}{7}, \quad p(s_3 \mid s_2) = \tfrac{2}{7}, \quad p(s_4 \mid s_2) = \tfrac{4}{7}.$$

   (b)

   $$p(s_2 \mid s_1) = \tfrac{1}{4}, \quad p(s_3 \mid s_1) = \tfrac{1}{2}, \quad p(s_4 \mid s_1) = \tfrac{1}{4},$$
   $$p(s_1 \mid s_2) = \tfrac{1}{5}, \quad p(s_3 \mid s_2) = \tfrac{2}{5}, \quad p(s_4 \mid s_2) = \tfrac{2}{5}.$$

2. Let $\langle S, p \rangle$ be a Markov chain with state space $S = \{s_1, s_2, s_3, s_4, s_5\}$, and let $B = \{s_1, s_2, s_3\}$. Using the tools developed in this chapter, calculate $Q_{s_1, p}(s_4 \mid B)$ in each of the following cases (all transitions from states in B that are not mentioned equal 0).

(a)

$$p(s_2 \mid s_1) = \tfrac{1}{9}, \quad p(s_3 \mid s_1) = \tfrac{3}{9}, \quad p(s_4 \mid s_1) = \tfrac{3}{9}, \quad p(s_5 \mid s_1) = \tfrac{2}{9},$$
$$p(s_1 \mid s_2) = \tfrac{5}{11}, \quad p(s_3 \mid s_2) = \tfrac{3}{11}, \quad p(s_4 \mid s_2) = \tfrac{2}{11}, \quad p(s_5 \mid s_2) = \tfrac{1}{11},$$
$$p(s_1 \mid s_3) = \tfrac{1}{3}, \quad p(s_2 \mid s_3) = \tfrac{2}{3}.$$

(b)

$$p(s_2 \mid s_1) = \tfrac{1}{9}, \quad p(s_3 \mid s_1) = \tfrac{3}{9}, \quad p(s_4 \mid s_1) = \tfrac{3}{9}, \quad p(s_5 \mid s_1) = \tfrac{2}{9},$$
$$p(s_1 \mid s_2) = \tfrac{5}{11}, \quad p(s_3 \mid s_2) = \tfrac{3}{11}, \quad p(s_4 \mid s_2) = \tfrac{2}{11}, \quad p(s_5 \mid s_2) = \tfrac{1}{11},$$
$$p(s_1 \mid s_3) = \tfrac{1}{7}, \quad p(s_2 \mid s_3) = \tfrac{2}{7}, \quad p(s_4 \mid s_3) = \tfrac{3}{7}, \quad p(s_5 \mid s_3) = \tfrac{1}{7}.$$

3. Let $\langle S, p \rangle$ be a Markov chain, and let $B \subset S$ be a non-empty set.

   (a) Can $W(B)$ be strictly larger than 1?
   (b) Can $W(B)$ be strictly smaller than 1?

   Justify your answers.

4. Explain where in the proof of Lemma 7.15 we used the assumption that $d(q, q') < 1/(2|S|)$.

5. Prove Corollary 7.17: Let $\Gamma = \langle \{1,2\}, S, (A^1(s), A^2(s))_{s \in S}, q, r \rangle$ be a two-player zero-sum stochastic game, let $(q_n)_{n \in \mathbb{N}}$ be a sequence of transition rule that satisfies $\lim_{n \to \infty} d(q_n, q) = 0$, and let $(r_n)_{n \in \mathbb{N}}$ be a sequence of payoff functions that satisfies $\lim_{n \to \infty} \|r_n - r\|_\infty = 0$. Show that

$$\lim_{n \to \infty} v_0(s; q_n, r_n) = v_0(s; q, r), \quad \forall s \in S.$$

6. In this exercise, we provide an expression for the invariant distribution of a Markov chain using B-graphs. Let $\langle S, p \rangle$ be a Markov chain. Assume that for every initial state $s \in S$ and every state $z \in S$ there is a positive probability that the process reaches z, that is, $\mathbf{P}_{s, p}(\exists t \in \mathbb{N} \text{ such that } s_t = z) > 0$. Let μ be the unique invariant measure of the Markov chain, that is, μ is the unique solution of the system of linear equations

$$\mu(s) = \sum_{s' \in S} \mu(s') p(s \mid s'), \quad \forall s \in S. \tag{7.25}$$

Prove that for every state $s \in S$, we have

$$\mu(s) = \frac{W(S \setminus \{s\})}{\sum_{s \in S} W(S \setminus \{s\})}. \tag{7.26}$$

Hint: Substitute Eq. (7.26) in Eq. (7.25).

8

Kakutani's Fixed-Point Theorem and Multiplayer Discounted Stochastic Games

In this chapter, we prove Kakutani's fixed point theorem, which is an extension of Brouwer's fixed point theorem to correspondences (set-valued functions). We then define the concept of λ-discounted equilibrium, and using Kakutani's fixed point theorem we prove that every multiplayer stochastic game admits a stationary λ-discounted equilibrium, for every discount factor $\lambda \in (0, 1]$.

8.1 Kakutani's Fixed-Point Theorem

We here state Kakutani's Fixed-Point Theorem (Kakutani, 1941),[1] and prove it using Brouwer's Fixed-Point theorem.[2]

Definition 8.1   A *correspondence* (or *set-valued mapping*) F between a set X and a set Y, denoted $F : X \rightrightarrows Y$, is a mapping that assigns to each point $x \in X$ a subset of Y. In other words, it is a mapping from X to the collection of all subsets of Y.

If F is a correspondence, a point $x \in X$ such that $x \in F(x)$ is called a *fixed point* of F.

Theorem 8.2 (Kakutani)   *Let $X \subseteq \mathbb{R}^n$ be a non-empty, compact, and convex set. Let $F : X \rightrightarrows X$ be a correspondence that satisfies the following conditions.*

---

[1] Shizuo Kakutani (Osaka, Japan, August 28, 1911 – New Haven, Connecticut, August 17, 2004) was a Japanese–American mathematician, best known for the fixed-point theorem that carries his name. Kakutani's other well-known mathematical contributions include the Kakutani skyscraper, a concept in ergodic theory, and his solution of the Poisson equation using methods of stochastic analysis.
[2] Luitzen Egbertus Jan Brouwer (Rotterdam, Netherlands, February 27, 1881 – Blaricum, Netherlands, December 2, 1966), was a Dutch mathematician who worked in topology, set theory, measure theory, and complex analysis.

1. *For every $x \in X$, the set $F(x)$ is non-empty and convex.*
2. *The graph of F, $\mathrm{Graph}(F) := \{(x, y) \in \mathbb{R}^{2n} \mid x \in X, y \in F(x)\}$, is compact.*

Then F has a fixed point.

Kakutani's Fixed-Point Theorem is an extension of Brouwer's Fixed-Point Theorem, which we state now. There are many proofs for Brouwer's Fixed-Point Theorem, see, for example, Border (1985).

Theorem 8.3 (Brouwer)   *Let $X \subset \mathbb{R}^n$ be a non-empty, compact, and convex set, and let $f : X \to X$ be a continuous mapping. Then there is point $x \in X$ such that $x = f(x)$.*

Comment 8.4   To see that Brouwer's Fixed-Point Theorem is a special case of Kakutani's Fixed-Point Theorem, let $f : X \to X$ be a continuous mapping and define a correspondence $F : X \rightrightarrows X$ by

$$F(x) := \{f(x)\}, \quad \forall x \in X.$$

Clearly, F has non-empty and convex values, and since f is continuous, F has a compact graph. By Kakutani's Fixed-Point Theorem, the correspondence F has a fixed point x, which is also a fixed point of f.

Carathéodory's Theorem states that if a point in $\mathbb{R}^n$ is a convex combination of $n + 2$ points, then it is a convex combination of $n + 1$ of those points.

Theorem 8.5 (Carathéodory)   *Let $x, x_1, \ldots, x_{n+2} \in \mathbb{R}^n$, and assume that $x = \sum_{i=1}^{n+2} \alpha_i x_i$, for some nonnegative real numbers $\alpha_1, \ldots, \alpha_{n+2}$ that sum up to 1. Then there are nonnegative real numbers $\beta_1, \ldots, \beta_{n+2}$ that sum up to 1 such that at least one of them is 0 and $x = \sum_{i=1}^{n+2} \beta_i x_i$.*

Proof   If there is an index $i \in \{1, \ldots, n+2\}$ such that $\alpha_i = 0$, then the conclusion is immediate. Assume then that $\alpha_i > 0$ for all $i \in \{1, 2, \ldots, n+2\}$.

Consider the following system of $n + 1$ linear equations with the $n + 2$ real unknowns $z_1, \ldots, z_{n+2}$:

$$\sum_{i=1}^{n+2} z_i x_i = x,$$

$$\sum_{i=1}^{n+2} z_i = 1. \tag{8.1}$$

Since $x_1, \ldots, x_{n+2} \in \mathbb{R}^n$, the first equation contributes n equations, one for each coordinate. Hence the system (8.1) is a system of $n + 1$ linear equations with $n + 2$ unknowns, and by assumption it has at least one solution in the nonnegative orthant $- \alpha_1, \ldots, \alpha_{n+2}$. Since the system is linear, its solution set is an affine subspace of $\mathbb{R}^{n+2}$. Since the number of variables is larger than the number of equations, if the system has at least one solution, then the dimension of the solution set is at least 1. It follows that there is a line ℓ of solutions that passes through the point $\alpha = (\alpha_1, \ldots, \alpha_{n+2})$.

The point $\alpha = (\alpha_1, \ldots, \alpha_{n+2})$ lies in the simplex

$$\Delta(n+2) := \left\{ z \in \mathbb{R}^{n+2} : z_1, \ldots, z_{n+2} = 0, \ \sum_{i=1}^{n+2} z_i = 1 \right\},$$

which is a compact set. Since the line ℓ is infinite, it must intersect the boundary of $\Delta(n+2)$ in some point (in fact, in two points). Denote one of the intersection points of ℓ and $\Delta(n+2)$ by β. Since β lies on ℓ, it solves the system (8.1). Since it lies on the boundary of $\Delta(n+2)$, it satisfies $\beta_i \geq 0$ for every $i \in \{1, \ldots, n+2\}$, and $\beta_i = 0$ for at least one i. Therefore, β satisfies the requirements of the theorem. □

Induction on k yields the following corollary of Theorem 8.5.

Corollary 8.6 *Let $k \geq 2$, and let $x, x_1, \ldots, x_{n+k} \in \mathbb{R}^n$. Assume that $x = \sum_{i=1}^{n+k} \alpha_i x_i$, for some nonnegative real numbers $\alpha_1, \ldots, \alpha_{n+k}$ that sum up to 1. Then there are nonnegative real numbers $\beta_1, \ldots, \beta_{n+k}$ that sum up to 1 such that at most $n+1$ of the β_i's are not 0 and $x = \sum_{i=1}^{n+k} \beta_i x_i$.*

We are now ready to prove Kakutani's Fixed-Point Theorem.

Proof of Theorem 8.2   Fix for the moment $m \in \mathbb{N}$. Since X is compact, there exist a positive integer $K = K_m \in \mathbb{N}$ and points $x_1^m, x_2^m, \ldots, x_K^m \in X$ such that the open balls with centers x_i^m and radius $1/m$ cover X:

$$X \subseteq \bigcup_{k=1}^{K} B(x_k^m, 1/m). \tag{8.2}$$

Denote by $B_k^c = \mathbb{R}^n \setminus B(x_k^m, 1/m)$ the complement of $B(x_k^m, 1/m)$. For each $k = 1, \ldots, K$, choose $y_k^m \in F(x_k^m)$. Since $F(x) \neq \emptyset$ for every $x \in X$, such a choice is possible.

Define a mapping $f^m : X \to \mathbb{R}^n$ by

$$f^m(x) := \sum_{k=1}^{K} \frac{d(x, B_k^c)}{\sum_{k=1}^{K} d(x, B_k^c)} \cdot y_k^m, \quad \forall x \in X. \tag{8.3}$$

Thus, $f^m(x)$ is a convex combination of $y_1^m, \ldots, y_K^m$. By Eq. (8.2), every $x \in X$ lies inside at least one of the open balls $(B(x_k^m, 1/m))_{k=1}^K$. It follows that the denominator in Eq. (8.3) is positive for every $x \in X$, and therefore the mapping f^m is well defined.

Let us verify that f^m satisfies the conditions of Brouwer's Fixed-Point Theorem. Since the function $x \mapsto d(x, B_k^c)$ is continuous on X, so is the mapping f^m. Further, since $y_k^m \in F(x_k^m) \subseteq X$, and since X is convex, $f^m(x) \in X$ for every $x \in X$. By Brouwer's Fixed-Point Theorem, the mapping f^m has a fixed point: there is a point $x^m \in X$ such that $x^m = f^m(x^m)$. That is,

$$x^m = f^m(x^m) = \sum_{k=1}^{K} \frac{d(x^m, B_k^c)}{\sum_{l=1}^{K} d(x^m, B_l^c)} \cdot y_k^m.$$

Thus, x^m is a convex combination of $y_1^m, \ldots, y_K^m$, and the coefficient of y_k^m is positive if and only if $d(x^m, x_k^m) < \frac{1}{m}$.

By Carathéodory's Theorem, there are numbers $k_1^m, \ldots, k_{n+1}^m \in \{1, \ldots, K\}$ and $\beta_1^m, \ldots, \beta_{n+1}^m$ such that

(A.1) $\beta_i^m \geq 0$ for $i = 1, \ldots, n+1$.
(A.2) $\sum_{i=1}^{n+1} \beta_i^m = 1$.
(A.3) $\beta_i^m > 0$ implies that $d\left(x^m, x_{k_i^m}^m\right) < 1/m$.
(A.4) $x^m = \sum_{i=1}^{n+1} \beta_i^m y_{k_i^m}^m$.

This construction is valid for every $m \in \mathbb{N}$. Since X is compact, there is a subsequence $(m_l)_{l \in \mathbb{N}}$ such that

- The sequence $(x^{m_l})_{l \in \mathbb{N}}$ converges to a limit x^*.
- For each $i = 1, \ldots, n+1$, the sequence $\left(y_{k_i^{m_l}}^{m_l}\right)_{l \in \mathbb{N}}$ converges to a limit y_i^*.
- For each $i = 1, \ldots, n+1$, the sequence $(\beta_i^{m_l})_{l \in \mathbb{N}}$ converges to a limit β_i^*.

Taking the limit $m \to \infty$, Conditions (A.1)–(A.4) have several consequences. By conditions (A.1) and (A.2), $\sum_{i=1}^{n+1} \beta_i^* = 1$ and $\beta_i^* \geq 0$ for every $i \in \{1, \ldots, n\}$. Since $d\left(x^m, x_{k_i^{m_l}}^{m_l}\right) < 1/m_l$ for every $l \in \mathbb{N}$, and since $(x^{m_l})_{l \in \mathbb{N}}$ converges to x^*, the sequence $\left(x_{k_i^{m_l}}^{m_l}\right)_{l \in \mathbb{N}}$ converges to x^* as well. Since $y_{k_i^{m_l}}^{m_l} \in F\left(x_{k_i^{m_l}}^{m_l}\right)$, and since the set Graph(F) is closed, $y_i^* \in F(x_i^*) = F(x^*)$ for each $i = 1, \ldots, n+1$. Condition (A.4) implies that $x^* = \sum_{i=1}^{n+1} \beta_i^* y_i^*$. Since $F(x^*)$ is convex, $x^* \in F(x^*)$, and therefore x^* is a fixed point of F. $\square$

8.2 Discounted Equilibrium

Let $\Gamma = \left\langle I, S, (A^i(s))_{s \in S}^{i \in I}, q, (r^i)_{i \in I} \right\rangle$ be a stochastic game. Recall that for every strategy profile $\sigma \in \Sigma$, every state $s \in S$, and every player $i \in I$, the λ-discounted payoff of player i under σ at the initial state s is

$$\gamma_\lambda^i(s;\sigma) := \mathbf{E}_{s,\sigma}\left[\lambda\sum_{t=1}^{\infty}(1-\lambda)^{t-1}r^i(s_t,a_t)\right]$$

$$= \lambda\sum_{t=1}^{\infty}(1-\lambda)^{t-1}\mathbf{E}_{s,\sigma}\left[r^i(s_t,a_t)\right].$$

The concept of λ-discounted equilibrium in stochastic games is the adaptation of the concept of λ-discounted optimal strategies in Markov decision problems that was introduced in Definition 1.21.

Definition 8.7   Let $\Gamma = \langle I, S, (A^i(s))_{s\in S}^{i\in I}, q, (r^i)_{i\in I}\rangle$ be a stochastic game, let $s \in S$, and let $\lambda \in (0,1]$. A strategy profile $\sigma_* = (\sigma_*^i)_{i\in I} \in \Sigma$ is a λ-discounted *equilibrium at the initial state s* if for every player $i \in I$ and every strategy $\sigma^i \in \Sigma^i$ we have

$$\gamma_\lambda^i(s;\sigma_*) \geq \gamma_\lambda^i(s;\sigma^i,\sigma_*^{-i}).$$

A strategy profile σ_* is a λ-*discounted equilibrium* if it is a λ-discounted equilibrium at all initial states.

Recall that the space of strategies of player i is the set Σ^i of all mappings

$$\sigma^i : (s_1,a_1,\ldots,a_t,s_{t+1}) \mapsto \Delta(A(s_{t+1})).$$

This set is compact in the product topology. In this topology, a sequence of stationary strategies $(\sigma_k^i)_{k\in\mathbb{N}}$ converges to σ^i if and only if

$$\lim_{k\to\infty}\sigma_k^i(h_t) = \sigma^i(h_t), \quad \forall h_t \in H.$$

Moreover, the λ-discounted payoff is continuous in this topology (Exercise 8.1): if $\lim_{k\to\infty}\sigma_k^i = \sigma^i$ for every $i \in I$, then

$$\lim_{k\to\infty}\gamma_\lambda(s;\sigma_k^1,\sigma_k^2,\ldots,\sigma_k^n) = \gamma_\lambda(s;\sigma^1,\sigma^2,\ldots,\sigma^n).$$

It follows that the set of λ-discounted equilibria is compact.

Theorem 8.8   *Let $(\sigma_k)_{k\in\mathbb{N}}$ be a sequence of λ-discounted equilibria in a stochastic game Γ that converges to a limit $\sigma \in \Sigma$. Then σ is a λ-discounted equilibrium.*

The proof of Theorem 8.8 is left for the reader as an exercise (Exercise 8.2). Since the image of a compact set under a continuous function is compact, we deduce the following result.

Theorem 8.9   *Let Γ be a stochastic game and let $E_\lambda(s) \subseteq \mathbb{R}^n$ denote the set of all λ-discounted equilibrium payoffs at the initial state s in Γ. Then $E_\lambda(s)$ is compact.*

In the next section, we prove that there always exists a λ-discounted equilibrium in stationary strategies.

8.3 Existence of Stationary Discounted Equilibria

Here we prove that any multiplayer stochastic game has at least one discounted equilibrium in stationary strategies, for every discount factor. Our approach is analogous to the one we used for zero-sum games. The main difference is that we use Kakutani's Fixed-Point Theorem (Theorem 8.2) instead of the fixed-point theorem for contracting mappings. This is done because while the value of a strategic-form game is a single number, the set of equilibrium payoffs of a strategic-form game usually contains more than one element.

We need the following technical Lemma.

Lemma 8.10   *Let $X, Y \subset \mathbb{R}^n$ be two compact sets, and let $f : X \times Y \to \mathbb{R}$ be a continuous function. Define a correspondence $F : X \rightrightarrows Y$ by*

$$F(x) := \mathrm{argmax}_{y \in Y} f(x, y) = \left\{ y \in Y \mid f(x, y) = \max_{z \in Y} f(x, z) \right\}.$$

Then

1. *$F(x)$ is non-empty for every $x \in X$.*
2. *Graph$(F) \subseteq X \times Y$ is a compact set.*
3. *If moreover $f(x, y)$ is linear in y for every fixed x, then the set $F(x)$ is convex for every $x \in X$.*

Proof   The first claim holds because f is continuous and Y is compact.

Let us prove the second claim. Since X and Y are compact and since Graph$(F) \subseteq X \times Y$, it is sufficient to prove that Graph(F) is closed. Let $(x_k, y_k)_{k \in \mathbb{N}}$ be a sequence that satisfies the following properties:

- For every $k \in \mathbb{N}$, $x_k \in X$, $y_k \in Y$, and $y_k \in F(x_k)$.
- The limits $x = \lim_{k \to \infty} x_k$, and $y = \lim_{k \to \infty} y_k$ exist.

We prove that $y \in F(x)$; that is, $f(x, y) \geq f(x, z)$ for every $z \in Y$. Let $z \in Y$ be arbitrary. For every k we have $y_k \in F(x_k)$, hence $f(x_k, y_k) \geq f(x_k, z)$. Since f is continuous, $f(x, y) \geq f(x, z)$.

Finally, we prove the third claim. Assume that $f(x, y)$ is linear in y for each fixed x. If $y_1, y_2 \in F(x)$, then $f(x, y_1) = f(x, y_2)$. By the linearity of f, we have $f(x, \alpha y_1 + (1 - \alpha) y_2) = f(x, y_1)$ for every $\alpha \in [0, 1]$. This implies that $\alpha y_1 + (1 - \alpha) y_2 \in F(x)$, for every $\alpha \in [0, 1]$, and therefore $F(x)$ is convex. □

As for the zero-sum case, we define for every state $s \in S$, every discount factor λ, and every mapping $w \colon S \to \mathbb{R}^n$ an auxiliary strategic-form game $G_{s,\lambda}(w)$, where the payoff is the sum of the stage payoff in state s (with weight λ) and a continuation payoff that is derived from w (with weight $1 - \lambda$). Let $\Gamma = \langle I, S, (A^i(s))_{s \in S}^{i \in I}, q, (r^i)_{i \in I} \rangle$ be a stochastic game, and let $w \colon S \to \mathbb{R}^n$ be an arbitrary mapping (recall that n is the number of players). For each state $s \in S$ consider the following strategic-form game $G_{s,\lambda}(w)$:

- The set of players is $I = \{1, 2, \ldots, n\}$.
- The set of actions of player i is $A^i(s)$.
- The payoff function of player i is

$$u^i(a) := \lambda r^i(s, a) + (1 - \lambda) \sum_{s' \in S} q(s' \mid s, a) w^i(s'), \quad \forall a \in A(s).$$

We can now state and prove the main result of this section.

Theorem 8.11 *Any stochastic game admits a stationary λ-discounted equilibrium, for every $\lambda \in (0, 1]$.*

Proof **Step 1:** Application of Kakutani's Fixed-Point Theorem.
Set

$$M := \max_{i \in I} \|r^i\|_\infty.$$

This is a bound on all the payoffs in the game. Consider the set

$$XW := \left(\prod_{i \in I} \prod_{s \in S} \Delta(A^i(s)) \right) \times [-M, M]^{n \times |S|} \subseteq \mathbb{R}^{\sum_{i \in I} \sum_{s \in S} |A^i(s)| + n|S|}.$$

The set XW is compact and convex. An element of X will be denoted by $\left((x^i(s))_{s \in S}^{i \in I}, (w^i(s))_{s \in S}^{i \in I} \right)$, where $x^i(s) \in \Delta(A^i(s))$ and $w^i(s) \in [-M, M]$ for every $i \in I$ and every $s \in S$. The coordinates $x^i = (x^i(s))_{s \in S}$ should be thought of as a stationary strategy of player i, and the coordinates $w^i = (w^i(s))_{s \in S}$ should be thought of as continuation payoffs for player i if the game reaches state s.

We now define a correspondence $F \colon XW \rightrightarrows XW$ as follows. The coordinates of F will be denoted by

$$F = (F_X^{i,s}, F_W^{i,s})_{s \in S}^{i \in I},$$

where $F_X^{i,s}(x, w) \subseteq \Delta(A^i(s))$ and $F_W^{i,s}(x, w)$ is a singleton in $[-M, M]$ for every $i \in I$, $s \in S$, and $(x, w) \in XW$. Since $F_W^{i,s}(x, w)$ is a singleton, we will abuse notation and denote this singleton by $F_W^{i,s}(x, w)$. The correspondence F is defined by:

$$F_X^{i,s}(x, w) := \operatorname{argmax}_{y^i \in \Delta(A^i(s))} \left(\lambda r^i(s, y^i, x^{-i}(s)) \right.$$

$$\left. + (1 - \lambda) \sum_{s' \in S} q(s' \mid s, y^i, x^{-i}(s)) w^i(s') \right),$$

and

$$F_W^{i,s}(x, w) := \lambda r^i(s, x(s)) + (1 - \lambda) \sum_{s' \in S} q(s' \mid s, x(s)) w^i(s').$$

Let us explain the motivation behind this definition. Consider the strategic-form game $G_{s,\lambda}(w)$. Then $(F_W^{i,s}(x, w))_{i \in I}$ is the expected payoff in this game when the players use the mixed action profile x. The quantity $F_X^{i,s}(x, w)$ is the set of all best replies of player i to the mixed action profile $x^{-i}(s)$ of the other players in this strategic-form game.

By Lemma 8.10, the correspondence F has non-empty and convex values and a compact graph. By the continuity of the functions $x(s) \mapsto r^i(x(s))$ and $x(s) \mapsto q(s' \mid s, x(s))$, the correspondence F has non-empty values. Hence, we can apply Kakutani's Fixed-Point Theorem (Theorem 8.2), which guarantees the existence of a fixed point $(x_*, w_*) \in XW$ for F. We prove that x_* is a stationary λ-discounted equilibrium, and that w_* is the corresponding λ-discounted equilibrium payoff.

Step 2: $\gamma_\lambda^i(s; x_*) = w_*^i(s)$ for every player $i \in I$ and every initial state $s \in S$.

Fix a player $i \in I$. Since (x_*, w_*) is a fixed point of F, we have

$$w_*^i(s) = \lambda r^i(s, x_*(s)) + (1 - \lambda) \sum_{s' \in S} q(s' \mid s, x_*(s)) w_*^i(s'), \quad \forall i \in I, \forall s \in S.$$

By Lemma 1.28, $\gamma_\lambda^i(s; x_*) = w_*^i(s)$ for every initial state $s \in S$.

Step 3: $\gamma_\lambda^i(s; \sigma^i, x_*^{-i}) \leq w_*^i(s)$ for every $i \in I$, every strategy σ^i of player i, and every initial state $s \in S$.

Fix a player $i \in I$ and a strategy $\sigma^i \in \Sigma^i$. By the definition of F, for every state $s \in S$ we have $w_*^i(s) = F_W^{i,s}(x_*, w_*)$. That is, $x_*^i(s)$ is a best reply of player i to x_*^{-i} in the strategic-form game $G_{s,\lambda}(w_*)$. It follows that for every mixed action $y^i \in \Delta(A^i(s))$ we have:

$$\lambda r^i(s, y^i, x_*^{-i}(s)) + (1 - \lambda) \sum_{s' \in S} q(s' \mid s, y^i, x_*^{-i}(s)) w_*^i(s')$$

$$\leq \lambda r^i(s, x_*(s)) + (1 - \lambda) \sum_{s' \in S} q(s' \mid s, x_*(s)) w_*^i(s') \qquad (8.4)$$

$$= F_W^{i,s}(x_*) = w_*^i(s).$$

For each history $h_t \in H$, substitute $s = s_t$ and $y^i = \sigma^i(h_t)$ in Eq. (8.4). We deduce that

$$\lambda r^i(s_t, \sigma^i(h_t), x_*^{-i}(s_t))$$

$$+ (1 - \lambda) \sum_{s' \in S} q(s' \mid s_t, \sigma^i(h_t), x_*^{-i}(s_t)) w_*^i(s') \leq w_*^i(s_t).$$

By Lemma 1.28, it follows that $\gamma_\lambda^i(s; \sigma^i, x_*^{-i}) \leq w_*^i(s)$ for every initial state $s \in S$. $\qquad\qquad\square$

8.4 Semi-Algebraic Sets and Discounted Stochastic Games

In this section, we will apply the results on semi-algebraic sets obtained in Chapter 6 to discounted stochastic games.

Let $\Gamma = \langle I, S, (A^i(s))_{s \in S}^{i \in I}, q, (r^i)_{i \in I} \rangle$ be a stochastic game. Let $B(\Gamma)$ be the set of all vectors $(\lambda, g, x) \in (0, 1] \times \mathbb{R}^{S \times I} \times \left(\prod_{s \in S, i \in I} \Delta(A^i(s)) \right)$ such that:

- λ is a discount factor and
- x is a stationary λ-discounted equilibrium in Γ with payoff $\gamma_\lambda(s; x) = g(s)$, for every state $s \in S$.

Theorem 8.12 *For every stochastic game Γ the set $B(\Gamma)$ is semi-algebraic.*

Proof The set B is a subset of the Euclidean space of dimension $1 + |S| \cdot |I| + \sum_{s \in S} \left(\prod_{i \in I} |A^i(s)| \right)$, and it contains all vectors (λ, g, x) that are solutions of the following finite list of polynomial equalities and inequalities:

$$\lambda > 0,$$

$$\lambda \leq 1,$$

$$x_s^i(a^i) \geq 0, \quad \forall s \in S, i \in I, a^i \in A^i(s),$$

$$\sum_{a^i \in A^i(s)} x^i_s(a^i) = 1, \quad \forall s \in S, i \in I,$$

$$g^i(s) = \sum_{a \in A(s)} \prod_{i \in I} x^i_s(a^i) \left(\lambda r^i(s,a) + (1-\lambda) \sum_{s' \in S} q(s' \mid s,a_1,a_2) g^i(s') \right),$$

$$\forall s \in S, i \in I,$$

$$g^j(s) \geq \sum_{a^{-j} \in A^{-j}(s)} \prod_{i \in I \setminus \{j\}} x^i_s(a^i)$$

$$\times \left(\lambda r(s,a^{-j},a^j) + (1-\lambda) \sum_{s' \in S} q(s' \mid s,a^{-j},a^j) g(s') \right), \quad \forall s \in S, i \in I.$$

Indeed, the first two inequalities imply that λ is a discount factor, the next inequality and equation imply that x is a stationary strategy profile, the fifth equality implies that $g(s) = \gamma_\lambda(s;x)$ for every initial state $s \in S$, and the last inequality implies that no player has a profitable deviation, so that x is a λ-discounted equilibrium. It follows that the set B is indeed semi-algebraic. $\qquad\square$

A repeated application of Theorem 6.7 yields the following corollary of Theorem 8.12.

Theorem 8.13   *For every stochastic game Γ and every discount factor λ, the set of all stationary λ-discounted equilibria of Γ is semi-algebraic.*

By Theorem 8.12, a repeated application of Theorem 6.7, and Theorem 6.11, there exists a semi-algebraic mapping $\lambda \mapsto x_\lambda$ that assigns to each discount factor λ a stationary λ-discounted equilibrium.

Corollary 8.14   *For every stochastic game $\Gamma = \langle I, S, (A^i(s))^{i \in I}_{s \in S}, q, (r^i)_{i \in I} \rangle$ there is a semi-algebraic mapping $\lambda \mapsto x_\lambda$ that assigns to every discount factor $\lambda \in (0,1]$ a stationary λ-discounted equilibrium $x_\lambda \in \prod_{s \in S} \prod_{i \in I} \Delta(A^i(s))$.*

8.5 Comments and Extensions

In this chapter, we proved that a discounted stationary equilibrium always exists. How does one find such an equilibrium? Methods to do this have been proposed by, for example, Nowak and Raghavan (1993), Herings and Peeters (2004), Govindan and Wilson (2010), Bourque and Raghavan (2014), and Eibelshäuser and Poensgen (2019).

Theorem 8.11, which was proved by Fink (1964) and Takahashi (1964), is valid whenever the sets of players, states, and actions are finite. When the number of players is countable, an equilibrium need not exist even in strategic-form games (see Peleg, 1969). When the number of states is countable and the sets of actions are finite, a stationary discounted equilibrium exists, see Exercise 8.16. When the number of states is finite and the sets of actions are compact, a stationary discounted equilibrium exists under suitable continuity conditions on the payoff function and transitions (Takahashi, 1964). When the set of states is general and the sets of actions are compact, the existence of a discounted equilibrium, not necessarily stationary, was established under suitable continuity conditions on the payoff function and transitions, see, for example, Nowak (1985a), Mertens and Parthasarathy (1987), Parthasarathy and Sinha (1989), Amir (1996), Solan (1998), Nowak (2003c), Horst (2005), or He and Sun (2017). Levy (2013a) and Levy and McLennan (2015) provided examples of multiplayer nonzero-sum stochastic games with general sets of states and actions that do not have measurable discounted equilibria. Levy's (2013a) example can be turned into a two-player nonzero-sum stochastic game with no measurable discounted equilibria.

When stationary discounted equilibrium is not known to exist, stationary discounted correlated equilibrium may still exist, see, for example, Nowak and Raghavan (1992), Duffie et al. (1994), and Harris, Reny, and Robson (1995).

In Chapter 6 we proved that the discounted value at a given initial state converges to a limit as the discount factor goes to 0. One can wonder whether the set of stationary discounted equilibria at a given initial state converges to a limit. We answer this question in the affirmative in Exercise 8.9. While the set of stationary discounted equilibria converges to a limit as the discount factor goes to 0, the set of (not necessarily stationary) discounted equilibrium payoffs does not necessarily converge to a limit as the discount factor goes to 0. For an example, see Renault and Ziliotto (2020a).

The game in Exercise 8.3 is taken from Flesch, Thuijsman, and Vrieze (1997). The game in Exercise 8.4 is taken from Sorin (1986). The game in Exercise 8.13 was proposed by Tristan Tomala.

8.6 Exercises

1. Prove that the λ-discounted payoff is continuous on the space of stationary strategies, endowed with the product topology.
2. Prove Theorem 8.8: Let $(\sigma_k)_{k\in\mathbb{N}}$ be a sequence of λ-discounted equilibria in a stochastic game Γ that converges to a limit $\sigma \in \Sigma$. Show that σ is a λ-discounted equilibrium.

3. Find the unique symmetric stationary λ-discounted equilibrium in the following three-player game,[3] where Player 1 chooses a row, Player 2 chooses a column, and Player 3 chooses a matrix. There is no need to calculate the strategy explicitly; it suffices to write down the polynomial that defines it.

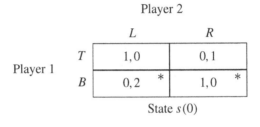

4. For every discount factor $\lambda \in (0, 1]$, find a λ-discounted equilibrium of the following two-player absorbing game.

Player 2

	L	R
T	1,0	0,1
B	0,2 *	1,0 *

Player 1

State $s(0)$

5. Let $\Gamma = \left\langle I, S, (A^i(s))_{s \in S}^{i \in I}, q, (r^i)_{i \in I} \right\rangle$ be a stochastic game. Suppose that there exists a $c \in \mathbb{R}^n$ such that for every state $s \in S$ the vector c is an equilibrium payoff in the strategic-form game with player set I, action set $A^i(s)$ for player i, and payoff $r(s, a)$. Prove that c is a λ-discounted equilibrium payoff at every initial state $s \in S$.

6. Consider the following repeated game:[4]

	D	C
D	1, 1	4, 0
C	0, 4	3, 3

(a) Find all stationary λ-discounted equilibria, for every discount factor $\lambda \in (0, 1]$.

(b) Show that if $\lambda \leq 2/3$, there exists a nonstationary λ-discounted equilibrium with payoff $(3, 3)$.

---

[3] A strategy profile is *symmetric* if all players use the *same* strategy.
[4] A *repeated game* is a stochastic game with a single state.

(c) Can you find a λ-discounted equilibrium with a different payoff than in the first two parts?

7. Let $\Gamma = \langle I, S, (A^i(s))_{s \in S}^{i \in I}, q, (r^i)_{i \in I} \rangle$ be a stochastic game. Denote by

$$E = \left\{ (\lambda, x, w) \in (0,1] \times \prod_{i \in I} \prod_{s \in S} \Delta(A^i(s)) \times \mathbb{R}^{n \times |S|} : \right.$$

$$\left. x \text{ is a } \lambda\text{-discounted stationary equilibrium with payoff } w \right\}.$$

Show that the set E is semi-algebraic.

8. Let $\Gamma = \langle I, S, (A^i(s))_{s \in S}^{i \in I}, q, (r^i)_{i \in I} \rangle$ be a stochastic game. Show that there exists a semi-algebraic mapping $w \colon (0,1] \to \mathbb{R}^{n \times |S|}$ such that $w(\lambda)$ is a λ-discounted equilibrium payoff for every $\lambda \in (0,1]$.

9. In this exercise, we prove that the limit of the sets of stationary discounted equilibria has a limit as the discount factor goes to 0.

   (a) Let X be a non-empty and compact subset of $\mathbb{R}^n$, and let $E \subseteq (0,1] \times X$ be a semi-algebraic set such that for every $\lambda \in (0,1]$ the set $E_\lambda := \{x \in X : (\lambda, x) \in E$ is non-empty. Prove that there exists a non-empty, semi-algebraic, and compact set $E_0 \subseteq X$ such that $\lim_{\lambda \to 0} d(E_\lambda, E_0) = 0$, where

   $$d(X,Y) := \max \left\{ \max_{x \in X} \min_{y \in Y} d(x,y), \max_{y \in Y} \min_{x \in X} d(x,y) \right\}$$

   is the Hausdorff distance between two sets.
   (b) Prove that the result in Part (a) does not hold when X is not bounded.
   (c) Let $\Gamma = \langle I, S, (A^i(s))_{s \in S}^{i \in I}, q, (r^i)_{i \in I} \rangle$ be a stochastic game, and let $s \in S$ be a state. Let $E_\lambda(s)$ be the set of λ-discounted stationary equilibria at the initial state s. Prove that there exists a non-empty, semi-algebraic, and compact set $E_0(s)$ such that $\lim_{\lambda \to 0} d(E_\lambda(s), E_0(s)) = 0$.

10. Let $\Gamma = \langle I, S, (A^i(s))_{s \in S}^{i \in I}, q, (r^i)_{i \in I} \rangle$ be a stochastic game. For every state $s \in S$ and every discount factor $\lambda \in (0,1]$ define

$$y(s; \lambda) := \max_{\sigma \in \Sigma} \sum_{i \in I} \gamma_\lambda^i(s; \sigma). \tag{8.5}$$

That is, the goal is to maximize the sum of discounted payoffs of the players.

    (a) Prove that the maximum in Eq. (8.5) is attained by a pure stationary strategy profile.

    (b) Show that the function $y(s; \lambda)$ is semi-algebraic.

    (c) Show that there is a semi-algebraic mapping $\lambda \mapsto x(\lambda)$ such that the stationary strategy $x(\lambda)$ attains the maximum in Eq. (8.5) for every λ.

    (d) Show that the mapping $\lambda \mapsto x(\lambda)$ can be chosen to be piecewise constant.

11. Consider the following two-player nonzero-sum stochastic game with two nonabsorbing states, where transition is such that at every nonabsorbing state, the play is either absorbed or moves to the other state.

	L	R
T	$0,0_{(0,1)}$	$0,0$ *
B	$0,0$ *	$3,1$ *

State $s(0)$

	L	R
T	$0,0_{(1,0)}$	$1,3$ *
B	$1,3$ *	$3,1$ *

State $s(1)$

    (a) For every discount factor $\lambda \in (0,1]$, find a stationary λ-discounted equilibrium in which in state $s(0)$ the players play (T, L).

    (b) For every discount factor $\lambda \in (0,1]$, find a stationary λ-discounted equilibrium in which in state $s(0)$ the players play (B, R).

    (c) For which discount factors $\lambda \in (0,1]$ does the game have a completely mixed stationary λ-discounted equilibrium? Find these equilibria.

12. Let $\Gamma = \langle \{1,2\}, S, (A^1(s), A^2(s))_{s \in S}, q, r \rangle$ be a two-player zero-sum stochastic game, and let σ^2 be a strategy of Player 2. Prove that there exists a pure strategy σ^1 such that

$$\gamma_\lambda^1(s; \sigma^1, \sigma^2) = \max_{\sigma'^1 \in \Sigma^2} \gamma_\lambda^1(s; \sigma'^1, \sigma^2),$$

that is, for every strategy σ^2 of Player 2, Player 1 has a pure best response.

13. A stochastic game $\Gamma = \langle I, S, (A^i(s))_{s \in S}^{i \in I}, q, (r^i)_{i \in I} \rangle$ has *perfect information* if in each state all players, except possibly for one, have one action; that is, for every state $s \in S$ there is a player $i_s \in I$ such that $|A^i(s)| = 1$ for every $i \in I \setminus \{i_s\}$.

    In this exercise, we show that in stochastic games with perfect information, a pure stationary discounted equilibrium need not exist. Consider the following two-player nonzero-sum stochastic game with two states:

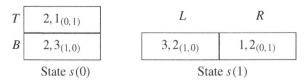

T	$2, 1_{(0,1)}$
B	$2, 3_{(1,0)}$

State $s(0)$

	L	R
	$3, 2_{(1,0)}$	$1, 2_{(0,1)}$

State $s(1)$

(a) Prove that for every $\lambda \in (0,1)$ there is no pure stationary λ-discounted equilibrium.

(b) Find the unique λ-discounted equilibrium in this game, for any $\lambda \in (0,1)$.

14. Let $\Gamma = \langle I, S, (A^i(s))_{s \in S}^{i \in I}, q, (r^i)_{i \in I} \rangle$ be a stochastic game. Let $(W_s)_{s \in S}$ be subsets of $\mathbb{R}^n$ with the following property: for every $s \in S$ and every $w \in W_s$ there are points $(w_{a,s'})_{a \in A(s), s' \in S}$ such that

- $w_{a,s'} \in W_{s'}$ for every $a \in A(s)$ and every $s' \in S$.
- w is an equilibrium of the strategic-form game with player set I, action set $A^i(s)$ for each player $i \in I$, and payoff function $\lambda r(s,a) + (1-\lambda) \sum_{s' \in S} q(s' \mid s,a) w_{a,s'}$, for $a \in A(s)$.

Prove that for every $s \in S$, every point $w \in W_s$ is a λ-discounted equilibrium payoff at the initial state s.

15. Let us generalize the model of stochastic games by allowing different discount factors for different players. That is, for every player $i \in I$, there is a corresponding discount factor $\lambda_i \in [0,1)$ and the payoff to player i, given the strategy profile σ and the initial state s_1 is $\gamma_i^{\lambda_i}(s^1, \sigma)$. Prove that in this generalized model, there exists a discounted equilibrium in stationary strategies for every vector of discount factors $(\lambda_i)_{i \in I}$.

16. In this exercise, we generalize Theorem 8.11 to the case where the state space is a countable set and each player has a finite number of actions in each state. Let $\Gamma = \langle I, S, (A^i(s))_{s \in S}^{i \in I}, q, (r^i)_{i \in I} \rangle$ be a stochastic game where the set S is countable, the sets I and $(A^i(s))_{s \in S}^{i \in I}$ are finite, and the payoff function r is bounded. Prove that the game admits a stationary λ-discounted equilibrium for every $\lambda \in (0,1]$.

17. Let $\Gamma = \langle I, S, (A^i(s))_{s \in S}^{i \in I}, q, (r^i)_{i \in I} \rangle$ be a stochastic game, let $i \in I$, let $s \in S$, and let $\lambda \in (0,1]$. The λ-*discounted min–max value of player i at the initial state s* is

$$\overline{v}_\lambda^i(s) := \inf_{\sigma^{-i} \in \Sigma^{-i}} \sup_{\sigma^i \in \Sigma^i} \gamma_\lambda^i(s; \sigma^i, \sigma^{-i}). \tag{8.6}$$

Prove that for every λ-discounted equilibrium σ_* we have

$$\gamma_\lambda^i(s; \sigma_*) \geq \overline{v}_\lambda^i(s), \quad \forall s \in S.$$

18. In this exercise, we characterize the set of λ-discounted equilibria in stochastic games.

    Let $\Gamma = \langle I, S, (A^i(s))_{s \in S}^{i \in I}, q, (r^i)_{i \in I} \rangle$ be a stochastic game, let $\lambda \in (0,1]$, and let σ be a strategy profile.

    For every history $h_t = (s_1, a_1, \ldots, a_{t-1})$ (without the state at stage t), the *continuation strategy at the history* h_t, denoted $\sigma^i_{|h_t}$, is the strategy given by the formula

    $$\sigma^i_{|h_t}(s'_1, a'_1, \ldots, a'_{j-1}, s'_j) := \sigma^i(s_1, a_1, \ldots, a_{t-1}, s'_1, a'_1, \ldots, a'_{j-1}, s'_j).$$

    This is the strategy that is played from stage t and on, conditioned that the play until stage t is h_t. We denote the *continuation strategy profile at* h_t by

    $$\sigma_{|h_t} = (\sigma^i_{|h_t})_{i \in I}.$$

    For every history $h_t = (s_1, a_1, \ldots, s_t)$, denote by $\mathbf{P}_\sigma(h_t)$ the probability that the history h_t occurs under σ when the initial state is s_1:

    $$\mathbf{P}_\sigma(h_t) := \left(\prod_{k=1}^{t} \prod_{i \in I} \sigma^i(a^i_t \mid h_k) \right) \cdot \left(\prod_{k=1}^{t-1} q(s_{k+1} \mid s_k, a_k) \right).$$

    Denote by $G_{h_t, \lambda}$ the following strategic-form game:

    - The set of players is I, and the set of actions of each player $i \in I$ is $A^i(s_t)$.
    - The payoff function of each player $i \in I$, denoted $R^i_{h_t, \lambda}$, is as follows. For every $a \in A(s_t)$, if $\prod_{i \in I} \sigma^i(a^i \mid h_t) > 0$,

    $$R^i_{h_t, \lambda}(a) := \lambda r^i(s_t, a) + (1 - \lambda) \sum_{s' \in S} q(s' \mid s_t, a) \gamma^i_\lambda(s'; \sigma_{|(h_t, s', a)}),$$

    while if $\prod_{i \in I} \sigma^i(a^i \mid h_t) = 0$,

    $$R^i_{h_t, \lambda}(a) := \lambda r^i(s_t, a) + (1 - \lambda) \sum_{s' \in S} q(s' \mid s_t, a) \overline{v}^i_\lambda(s').$$

    Thus, action profiles that can be played under $\sigma(h_t)$ lead to a continuation payoff given by the continuation strategy, while action profiles that cannot be played under $\sigma(h_t)$ lead to a continuation payoff that coincides with the min–max value, which is defined in Exercise 8.17.

    We say that two strategy profiles σ and σ' are *path-equivalent* if $\sigma(h) = \sigma'(h)$ for every history h with $\mathbf{P}_\sigma(h) > 0$.

Prove that the following two conditions are equivalent:

1. σ is path-equivalent to some λ-discounted equilibrium.
2. For every history h_t with $\mathbf{P}_\sigma(h_t) > 0$, the mixed action profile $\sigma(h_t)$ is an equilibrium of $G_{h_t,\lambda}$.

9

Uniform Equilibrium

In Chapter 2, we studied uniform ϵ-optimality in hidden Markov decision problems. In this chapter, we define the analogous concept for multiplayer stochastic games and prove its existence for two-player zero-sum stochastic games.

To motivate the concept of uniform ϵ-optimality, consider the two-player zero-sum absorbing game called the "Big Match," which was presented in Exercise 4.4 and is illustrated in Figure 9.1.

For every discount factor $\lambda \in (0, 1]$, the λ-discounted value at the initial state $s(0)$ is $v_\lambda(s(0)) = \frac{1}{2}$; the unique λ-discounted optimal strategy of Player 2 is $\left[\frac{1}{2}(L), \frac{1}{2}(R)\right]$; and the unique λ-discounted optimal strategy of Player 1 is $\left[\frac{1}{1+\lambda}(T), \frac{\lambda}{1+\lambda}(B)\right]$. The λ-discounted optimal strategy of Player 2 is independent of λ. This is not the case with Player 1: To play optimally, she must know the exact discount factor. Actually, if she does not properly evaluate the discount factor λ and plays a $\widehat{\lambda}$-optimal strategy with $\lambda \neq \widehat{\lambda}$, and if Player 2 exploits that error, then the λ-discounted payoff might be very low (Exercise 9.1).

Thus, in stochastic games there need not exist a uniformly optimal strategy. This example stands in sharp contrast to Markov decision problems in which a uniformly optimal strategy always exists. Later in this chapter, we will see that

	L	R
T	1	0
B	0 *	1 *

State $s(0)$

Figure 9.1 The "Big Match."

uniformly ϵ-optimal strategies exist in every two-player zero-sum stochastic game.

9.1 Definition of Uniform Equilibrium

We next introduce the concept of uniform ϵ-equilibrium, which we already studied in the setup of hidden Markov decision problems (see Definition 2.4).

Definition 9.1   Let $\Gamma = \langle I, S, (A^i(s))_{s \in S}^{i \in I}, q, (r^i)_{i \in I} \rangle$ be a stochastic game, let $\epsilon \geq 0$, let $s \in S$, and let $T \in \mathbb{N}$. A strategy profile σ_* is a *T-stage ϵ-equilibrium* at the initial state s if for each player $i \in I$ and every strategy $\sigma^i \in \Sigma^i$,

$$\gamma_T^i(s; \sigma_*) \geq \gamma_T^i(s; \sigma^i, \sigma_*^{-i}) - \epsilon.$$

A strategy profile is a T-stage ϵ-equilibrium if no player can gain more than ϵ in the T-stage game by deviating. We can also define ϵ-equilibria in the context of discounted games.

Definition 9.2   Let $\epsilon \geq 0$, let $s \in S$, and let $\lambda \in (0,1]$. A strategy profile σ_* is a *λ-discounted ϵ-equilibrium* at the initial state s if for each player $i \in I$ and every strategy $\sigma^i \in \Sigma^i$ of player i,

$$\gamma_\lambda^i(s; \sigma_*) \geq \gamma_\lambda^i(s; \sigma^i, \sigma_*^{-i}) - \epsilon.$$

Definition 9.3   Let $\Gamma = \langle I, S, (A^i(s))_{s \in S}^{i \in I}, q, (r^i)_{i \in I} \rangle$ be a stochastic game, let $s \in S$, and let $\epsilon > 0$. A strategy profile $\sigma_* = (\sigma_*^i)_{i \in I}$ is a *uniform ϵ-equilibrium at the initial state s* if there exist $\lambda_0 \in (0, 1]$ and $T_0 \in \mathbb{N}$ such that the following conditions hold:

(UE1) *For every $\lambda \in (0, \lambda_0)$, the strategy profile σ_* is a λ-discounted ϵ-equilibrium at the initial state s.*

(UE2) *For every $T \geq T_0$, the strategy profile σ_* is a T-stage ϵ-equilibrium at the initial state s.*

A strategy profile σ_* that is uniformly ϵ-optimal at all initial states is a *uniform ϵ-equilibrium*.

The interest in the concept of uniform equilibrium stems from its robustness. If a uniform equilibrium σ_* exists, then by playing this strategy profile the players ensure that no player can gain more than ϵ by deviating, regardless of the length of the game (provided it is sufficiently long), and regardless of the value of the discount factor (provided it is sufficiently low). It is also not necessary that all players have the same discount factor, or use the same criterion to evaluate the stream of payoffs (some of the players may try to maximize the discounted payoff, while others may try to maximize the T-stage payoff). As long as all players use a sufficiently low discount factor or a sufficiently large T, the strategy σ_* is an ϵ-equilibrium.

We first prove that Condition (UE2) implies Condition (UE1). Example 9.5 shows that the converse is false.

Theorem 9.4   *If the strategy profile σ satisfies Condition* (UE2) *with a given $\epsilon > 0$, then it also satisfies Condition* (UE1) *with 3ϵ.*

Proof   Fix $\epsilon > 0$ and an initial state $s \in S$. Denote the expected payoff at stage t under σ by

$$x_t := \mathbf{E}_{s,\sigma}[r(s_t, a_t)].$$

Then

$$\gamma_T(s;\sigma) = \frac{1}{T}\sum_{t=1}^{T} x_t,$$

and

$$\gamma_\lambda(s;\sigma) = \lambda \sum_{t=1}^{T}(1-\lambda)^{t-1}x_t.$$

Since the strategy profile σ satisfies Condition (UE2), there is a $T_0 \in \mathbb{N}$ such that for every player $i \in I$, every strategy $\sigma'^i \in \Sigma^i$, and every $T \geq T_0$ we have

$$\gamma_T^i(s;\sigma) \geq \gamma_T^i(s;\sigma'^i,\sigma^{-i}) - \epsilon.$$

Now fix a strategy $\sigma'^i \in \Sigma^i$ and set

$$y_t := \mathbf{E}_{s,\sigma'^i,\sigma^{-i}}[r(s_t,a_t)], \quad \forall t \in \mathbb{N}.$$

Condition (UE2) implies that

$$\frac{1}{T}\sum_{t=1}^{T}x_t = \gamma_T^i(s;\sigma) \geq \gamma_T^i(s;\sigma'^i,\sigma^{-i}) - \epsilon \geq \frac{1}{T}\sum_{t=1}^{T}y_t - \epsilon, \quad \forall T \geq T_0.$$

Let $\lambda_0 > 0$ be sufficiently small so that $\lambda \cdot \left(1 - (1-\lambda)^{T_0}\right) \cdot \|r\|_\infty < \epsilon$ for every $\lambda \in (0, \lambda_0]$. From Eqs. (2.2)–(2.8), we deduce that

$$\gamma_\lambda^i(s;\sigma) = \lambda \sum_{t=1}^{\infty}(1-\lambda)^{t-1}x_t \geq \lambda \sum_{t=1}^{\infty}(1-\lambda)^{t-1}y_t - 3\epsilon$$
$$= \gamma_\lambda^i(s;\sigma'^i,\sigma^{-i}) - 3\epsilon, \quad \forall \lambda \in (0,\lambda_0].$$

Since this inequality holds for every player $i \in I$ and every strategy $\sigma'^i \in \Sigma^i$, it follows that σ is a λ-discounted 3ϵ-equilibrium for every $\lambda \in (0,\lambda_0]$, as we wanted to show. $\square$

The following example shows that a strategy profile that satisfies Condition (UE1) with a given $\epsilon > 0$ need not satisfy Condition (UE2) with $C\epsilon$, for every fixed $C > 0$. It is based on Example 13.33 in Maschler et al. (2020).

	D	E	F
A	0, 2	1, 2	c, 2
B	0, 2	1, 2	c, 2

Figure 9.2 The payoff matrix of the game in Example 9.5.

Example 9.5   Let $(x_t)_{t=1}^{\infty}$ be a sequence of zeros and ones, satisfying

$$\limsup_{T \to \infty} \frac{\sum_{t=1}^{T} x_t}{T} > \limsup_{\lambda \to 0} \lambda \sum_{t=1}^{\infty} (1 - \lambda)^{t-1} x_t.$$

For details on how to construct such a sequence, see Exercise 9.19. Let c be a real number, satisfying

$$\limsup_{T \to \infty} \frac{\sum_{t=1}^{T} x_t}{T} > c > \limsup_{\lambda \to 0} \lambda \sum_{t=1}^{\infty} (1 - \lambda)^{t-1} x_t. \tag{9.1}$$

Consider the two-player repeated game displayed in Figure 9.2. In this game, the payoff to Player 2 is 2, under every action profile. Consequently, to prove that a pair of strategies is a λ-discounted ϵ-equilibrium or a T-stage ϵ-equilibrium, it is sufficient to show that Player 1 cannot profit more than ϵ by deviating.

Define the following strategy σ^2 of Player 2:

- In the first stage, play F.
- If in the first stage Player 1 played A, play F in all the remaining stages of the game.
- If in the first stage Player 1 played B, play D or E in all the remaining stages of the game, according to the sequence $(x_t)_{t=1}^{\infty}$: if $x_t = 0$, play D in stage t, and if $x_t = 1$, play E in stage t.

The strategy σ^2 does not depend on Player 1's actions after the first stage. Moreover, the stage payoff of Player 1 is determined only by the action of Player 2. In particular, the only part of the strategy of Player 1 that affects her payoffs in the game is her action in the first stage. Let σ_A^1 be a strategy of Player 1 where she plays the action A in the first stage, and let σ_B^1 be a strategy of Player 2 where she plays the action B in the first stage.

The reader can verify that

$$\gamma_T^1(\sigma_A^1,\sigma^2) = c, \quad \forall T \in \mathbb{N}, \tag{9.2}$$

$$\gamma_T^1(\sigma_B^1,\sigma^2) = \frac{1}{T}\left(c + \sum_{t=1}^{T-1} x_t\right), \quad \forall T \in \mathbb{N}, \tag{9.3}$$

$$\gamma_\lambda^1(\sigma_A^1,\sigma^2) = c, \quad \forall \lambda \in (0,1], \tag{9.4}$$

$$\gamma_\lambda^1(\sigma_B^1,\sigma^2) = \lambda\left(c + \sum_{t=1}^{\infty}(1-\lambda)^t x_t\right), \quad \forall \lambda \in (0,1]. \tag{9.5}$$

Eqs. (9.1), (9.4), and (9.5) imply that for every discount factor λ sufficiently close to 1, one has $\gamma_\lambda^1(\sigma_A^1,\sigma^2) > \gamma_\lambda^1(\sigma_B^1,\sigma^2)$. It follows that (σ_A^1,σ^2) is a λ-discounted equilibrium with payoff $(c,2)$, for every discount factor λ sufficiently close to 1. In particular, the strategy pair (σ_A^1,σ^2) satisfies Condition (UE1).

We next show that (σ_A^1,σ^2) does not satisfy Condition (UE2). Set $\epsilon_0 := \frac{1}{2}\left(\limsup_{T\to\infty}\frac{\sum_{t=1}^T x_t}{T} - c\right)$, By Eqs. (9.1)–((9.3)), for every $T \in \mathbb{N}$ such that $\frac{1}{T}\left(c + \sum_{t=1}^{T-1} x_t\right) > c + \epsilon_0$ we have $\gamma_T^1(\sigma_B^1,\sigma^2) > \gamma_T^1(\sigma_A^1,\sigma^2) + \epsilon_0$, and therefore (σ_A^1,σ^2) is not a T-stage ϵ-equilibrium for $\epsilon \in (0,\epsilon_0)$.   ♦

When σ is a uniform ϵ-equilibrium, the T-stage payoff for various T's and the λ-discounted payoff for various λ's may differ. When all those payoffs are within ϵ of some vector $w = (w^i(s))_{s\in S}^{i\in I} \in \mathbb{R}^{S\times I}$, we call w a *uniform ϵ-equilibrium payoff*. When a sequence of uniform ϵ-equilibrium payoffs converges to a limit, we call the limit a *uniform equilibrium payoff*.

Definition 9.6   The payoff vector $w = (w^i(s))_{s\in S}^{i\in I} \in \mathbb{R}^{S\times I}$ is a *uniform equilibrium payoff* if for every $\epsilon > 0$ there exist a uniform ϵ-equilibrium σ, $\lambda_0 \in (0,1]$, and $T_0 \in \mathbb{N}$ such that

$$\|\gamma_\lambda(s;\sigma_\epsilon) - w(s)\|_\infty \le \epsilon, \quad \forall s \in S, \forall \lambda \in (0,\lambda_0),$$

and

$$\|\gamma_T(s;\sigma_\epsilon) - w(s)\|_\infty \le \epsilon, \quad \forall s \in S, \forall T \ge T_0.$$

When a strategic-form game is a two-player zero-sum game, the notion of equilibrium reduces to optimal strategies, and all equilibrium payoffs are equal to the value of the game. Similarly, when a stochastic game is a two-player zero-sum game, we will talk about uniformly ϵ-optimal strategies and the uniform value.

Definition 9.7 Let $\Gamma = \left\langle \{1,2\}, S, (A^i(s))_{s\in S}^{i=1,2}, q, (r^i)_{i\in I}\right\rangle$ be a two-player zero-sum stochastic game. The vector $w = (w(s))_{s\in S} \in \mathbb{R}^S$ is the *uniform value* if for every $\epsilon > 0$, there exist strategies $\sigma_\epsilon^1 \in \Sigma^1$ and $\sigma_\epsilon^2 \in \Sigma^2$, $\lambda_0 \in (0,1]$ and $T_0 \in \mathbb{N}$ such that

- $\gamma_\lambda(s;\sigma_\epsilon^1,\sigma'^2) \geq w(s) - \epsilon$ for every initial state $s \in S$, every $\lambda \in (0,\lambda_0)$, and every strategy $\sigma'^2 \in \Sigma^2$.
- $\gamma_\lambda(s;\sigma'^1,\sigma_\epsilon^2) \leq w(s) + \epsilon$ for every initial state $s \in S$, every $\lambda \in (0,\lambda_0)$, and every strategy $\sigma'^1 \in \Sigma^1$.
- $\gamma_T(s;\sigma_\epsilon^1,\sigma'^2) \geq w(s) - \epsilon$ for every initial state $s \in S$, every $T \geq T_0$, and every strategy $\sigma'^2 \in \Sigma^2$.
- $\gamma_T(s;\sigma'^1,\sigma_\epsilon^2) \leq w(s) + \epsilon$ for every initial state $s \in S$, every $T \geq T_0$, and every strategy $\sigma'^1 \in \Sigma^1$.

The strategy σ_ϵ^i is said to be *uniformly ϵ-optimal* for player i. The quantity $w(s)$ is called *the uniform value at the initial state s*.

If the uniform value at the initial state s exists, it is equal to $\lim_{\lambda\to 0} v_\lambda(s)$ and to $\lim_{T\to\infty} v_T(s)$ (Exercise 9.3).

An alternative formulation of the uniform value is given by the amount a strategy can guarantee.

Definition 9.8 Let $\Gamma = \left\langle \{1,2\}, S, (A^i(s))_{s\in S}^{i=1,2}, q, (r^i)_{i\in I}\right\rangle$ be a two-player zero-sum stochastic game, let $c \in \mathbb{R}$, and let $s \in S$. A strategy $\sigma^1 \in \Sigma^1$ *uniformly guarantees c at the initial state s* if there exist $\lambda_0 \in (0,1]$ and $T_0 \in \mathbb{N}$ such that

$$\gamma_\lambda(s;\sigma^1,\sigma^2) \geq c, \quad \forall \sigma^2 \in \Sigma^2, \forall \lambda \in (0,\lambda_0),$$

and

$$\gamma_T(s;\sigma^1,\sigma^2) \geq c, \quad \forall \sigma^2 \in \Sigma^2, \forall T \geq T_0.$$

A strategy $\sigma^2 \in \Sigma^2$ *uniformly guarantees c at the initial state s* if there exist $\lambda_0 > 0$ and $T_0 \in \mathbb{N}$ such that

$$\gamma_\lambda(s;\sigma^1,\sigma^2) \leq c, \quad \forall \sigma^1 \in \Sigma^1, \forall \lambda \in (0,\lambda_0),$$

and

$$\gamma_T(s;\sigma^1,\sigma^2) \leq c, \quad \forall \sigma^1 \in \Sigma^1, \forall T \geq T_0.$$

Using the last definition, the uniform value can be defined as follows.

Definition 9.9 Let $\Gamma = \left\langle \{1,2\}, S, (A^1(s),A^2(s))_{s\in S}, q, (r^i)_{i\in I}\right\rangle$ be a two-player zero-sum stochastic game, and let $s \in S$. The real number $v(s)$ is the *uniform value* at the initial state s if for every $\epsilon > 0$, Player 1 has a strategy that uniformly guarantees $v(s) - \epsilon$ at s, and Player 2 has a strategy that uniformly guarantees $v(s) + \epsilon$ at s.

9.2 The "Big Match"

Before studying the uniform value in the general setup, we will analyze the "Big Match," see Figure 9.1.

This game was presented by Gillette (1957) and was studied by Blackwell and Ferguson (1968). As mentioned before (see Page 128), the strategy $\left[\frac{1}{2}(L), \frac{1}{2}(R)\right]$ of Player 2 guarantees that the expected payoff at every stage is $\frac{1}{2}$, and therefore it uniformly guarantees $\frac{1}{2}$.

As we show now, Player 1 cannot guarantee any positive payoff using stationary strategies.

Lemma 9.10   *Let $x^1 \in \Delta(A^1(s(0)))$ be a stationary strategy of Player 1 in the "Big Match." Then there is a strategy $\sigma^2 \in \Sigma^2$ of Player 2 such that*

$$\lim_{\lambda \to 0} \gamma_\lambda(x^1, \sigma^2) = 0.$$

Proof   Denote by $x^1(T)$ and $x^1(B)$ the probability by which Player 1 plays the actions T and B, respectively, under the stationary strategy x^1. Denote by σ_L^2 the stationary strategy of Player 2 where she always plays the action L, and by σ_R^2 the stationary strategy of Player 2 where she always plays the action R. By Theorem 5.2,

$$\gamma_\lambda(x^1, \sigma_L^2) = x^1(T) \cdot (\lambda + (1 - \lambda)\gamma_\lambda(x, \sigma_L^2)),$$

which solves to

$$\gamma_\lambda(x^1, \sigma_L^2) = \frac{\lambda x^1(T)}{1 - x^1(T) + \lambda x^1(T)}.$$

Applying again Theorem 5.2, we obtain

$$\gamma_\lambda(x^1, \sigma_R^2) = x^1(T) \cdot (1 - \lambda)\gamma_\lambda(x, \sigma_R^2) + 1 - x^1(T),$$

which solves to

$$\gamma_\lambda(x^1, \sigma_R^2) = \frac{1 - x^1(T)}{1 - x^1(T) + \lambda x^1(T)}.$$

If $x^1(T) = 1$, then $\gamma_\lambda^1(x^1, \sigma_R^2) = 0$ for every $\lambda \in (0, 1]$. If $x^1(T) < 1$, then for every $\lambda \in (0, 1]$,

$$\gamma_\lambda^1(x^1, \sigma_L^2) = \frac{\lambda x^1(T)}{1 - x^1(T) + \lambda x^1(T)} < \frac{\lambda}{1 - x^1(T)},$$

which goes to 0 as λ goes to 0. The claim follows.   □

We will prove the following.

Theorem 9.11   *For every $\epsilon > 0$ Player 1 has a strategy that uniformly guarantees $\frac{1}{2} - \epsilon$.*

As a corollary, we will deduce that the uniform value of the "Big Match" exists, and is equal to $\frac{1}{2}$.

Proof   Fix throughout $M \in \mathbb{N}$. In the proof, we construct a history-dependent strategy σ_M^1 that uniformly guarantees $\frac{M}{2(M+1)} - \delta$, for every $\delta > 0$. For every $t \in \mathbb{N}$ define three random variables, l_t, r_t, and k_t as follows:

- l_t is the number of times before stage t (not including stage t) in which Player 2 played L.
- r_t is the number of times before stage t (not including stage t) in which Player 2 played R.
- $k_t := l_t - r_t$.

Note that $l_t + r_t = t - 1$ for every $t \in \mathbb{N}$, so

$$k_t = l_t - (t - 1 - l_t) = 2l_t - t + 1. \tag{9.6}$$

Suppose that the game was not absorbed by stage t. If Player 2 played mainly L, then Player 1's average payoff so far is high, and k_t is large; Player 1 is quite happy from playing T, hence she wants to increase the probability of playing T. If Player 2 played mainly R, then Player 1's average payoff so far is low, and k_t is low (negative) as well; Player 1 wants to increase the probability of playing B, in the hope that when the game is absorbed, the absorbing payoff is 1.

Define a strategy σ_M^1 that reflects this idea as follows: at stage t, if the play has not been absorbed yet, play T with probability $1 - \frac{1}{(k_t + M + 1)^2}$, and B with probability $\frac{1}{(k_t + M + 1)^2}$. Observe that once $k_t = -M$, Player 1 plays B with probability 1, and the play is absorbed. In particular, all subsequent choices of the players do not affect the payoff.

It is not clear at all that this σ_M^1 uniformly guarantees an amount close to $\frac{1}{2}$, because Player 1 bases her decision on the actions played before stage t, while the absorbing payoff that will be realized if the game is absorbed at stage t depends on Player 2's action at stage 2. Surprisingly, though, the strategy σ_M^1 does uniformly guarantee $\frac{M}{2M+1} - \delta$ for every $\delta > 0$.

We will prove below that for every sequence of actions $\vec{a}^2 = (a_1^2, a_2^2, \ldots)$ of Player 2, one has

$$\gamma_t^i(s(0); \sigma_M^1, \vec{a}^2) \geq \frac{M}{2(M+1)} - \frac{M+1}{2t}, \quad \forall t \in \mathbb{N}.$$

Since every (not necessarily pure) strategy of Player 2 is a mixture of pure strategies, this implies that

$$\gamma_t^i(s(0); \sigma_M^1, \sigma^2) \geq \frac{M}{2(M+1)} - \frac{M+1}{2t}, \quad \forall \sigma^2 \in \Sigma^2, \forall t \in \mathbb{N},$$

and therefore σ_M^1 uniformly guarantees $\frac{M}{2(M+1)} - \delta$ for every $\delta > 0$, as claimed.

A useful property of the strategy σ_M^1 relates the play under σ_M^1 from the second stage on to either σ_{M-1}^1 or σ_{M+1}^1:

- If $a_1^2 = R$ (Player 2 plays R at the first stage), then the play under σ_M^1 from the second stage on coincides with σ_{M-1}^1.
- If $a_1^2 = L$ (Player 2 plays L at the first stage), then the play under σ_M^1 from the second stage on coincides with σ_{M+1}^1.

For example, suppose that $a_1^2 = L$. Then $l_2 = 1$, $r_2 = 0$, and $k_2 = 1$. Under σ_M^1, Player 1 plays T at the second stage with probability $1 - \frac{1}{(M+2)^2}$. But this is exactly what she plays at the first stage under σ_{M+1}^1.

Let t_* denote the first stage in which Player 1 plays B. From that stage on, the actions of the players do not affect the payoffs, so the game essentially ends at stage t_*.

Define for every $t \in \mathbb{N}$ the random variable X_t as follows.

$$X_t = \begin{cases} \frac{1}{2}, & \text{if } t_* > t, \\ 1, & \text{if } t_* \leq t,\ a_{t_*}^2 = R, \\ 0, & \text{if } t_* \leq t,\ a_{t_*}^2 = L. \end{cases}$$

If the game has ended before or at stage t, X_t represents the stage payoff to Player 1. Otherwise, it is $\frac{1}{2}$. As we will prove, the value of the game for the initial state $s(0)$ is $\frac{1}{2}$, and therefore X_t represents the value of the state at stage t.

Claim 9.12 *For every sequence of actions $\vec{a}^2 = (a_1^2, a_2^2, \ldots)$ of Player 2 and every $t \in \mathbb{N}$, one has*

$$\mathbf{E}_{s(0)\sigma_M^1, \vec{a}^2}[X_t] \geq \frac{M}{2(M+1)}.$$

Proof We proceed by induction on m.

For $t = 1$:

- If $a_1^2 = R$, then

$$\mathbf{E}_{s(0), \sigma_M^1, \vec{a}^2}[X_1] = \frac{1}{2}\left(1 - \frac{1}{(M+1)^2}\right) + \frac{1}{(M+1)^2} > \frac{1}{2} > \frac{M}{2(M+1)}.$$

- If $a_1^2 = L$, then

$$\mathbf{E}_{s(0), \sigma_M^1, \vec{a}^2}[X_1] = \frac{1}{2}\left(1 - \frac{1}{(M+1)^2}\right) = \frac{M^2 + 2M}{2(M+1)^2}$$
$$= \frac{M(M+2)}{2(M+1)^2} > \frac{M}{2(M+1)}.$$

Suppose now that the claim holds for $t = t_0$; let us prove that it holds for $t = t_0 + 1$ as well.

- If $a_1^2 = R$, then

$$\mathbf{E}_{s(0), \sigma_M^1, \vec{a}^2}[X_{t_0+1}] = \frac{1}{(M+1)^2} + \left(1 - \frac{1}{(M+1)^2}\right)\mathbf{E}_{\sigma_{M-1}^1, \vec{a}^2}[X_{t_0}]$$
$$\geq \frac{1}{(M+1)^2} + \left(1 - \frac{1}{(M+1)^2}\right)\frac{M-1}{2M}$$
$$= \frac{M}{2(M+1)}.$$

- If $a_1^2 = L$, then

$$\mathbf{E}_{s(0), \sigma_M^1, \vec{a}^2}[X_{t_0+1}] = \left(1 - \frac{1}{(M+1)^2}\right)\mathbf{E}_{\sigma_{M+1}^1, \vec{a}^2}[X_{t_0}]$$
$$\geq \left(1 - \frac{1}{(M+1)^2}\right)\frac{M+1}{2(M+2)}$$
$$= \frac{M}{2(M+1)}.$$

▲

For $t \in \mathbb{N}$ denote

$$t_\blacksquare := \min\{t_*, t\}.$$

The random variable $t_\blacksquare$ is the stage in which the game is absorbed if $t_* \leq t$, and t otherwise. Since once $k_t = -M$, Player 1 plays B with probability 1 and the game is absorbed; $k_{t_\blacksquare}$ is always at least $-M$. In view of Eq. (9.6),

$$l_{t_\blacksquare} \geq \frac{t_\blacksquare - 1 - M}{2}, \quad \forall t \in \mathbb{N}. \tag{9.7}$$

We note that

$$\gamma^i_{t-1}(s(0); \sigma^1_M, \vec{a}^2) = \mathbf{E}_{s(0), \sigma^1_M, \vec{a}^2} \left[\frac{l_{t_\blacksquare} + (t - t_\blacksquare)\mathbf{P}(a^2_{t_\blacksquare} = R)}{t - 1} \right]. \tag{9.8}$$

Indeed, if $t_\blacksquare = t$, then the first term in the numerator on the right-hand side of Eq. (9.8) counts the number of stages up to stage $t - 1$ in which the payoff was 1 and the second term vanishes. If, on the other hand, $t_\blacksquare = t_*$, then the first term in the numerator on the right-hand side of Eq. (9.8) counts the number of stages up to stage $t_* - 1$ in which the payoff was 1 and the second term is equal to the expected payoff between stages t_* and $t - 1$. Therefore,

$$
\begin{aligned}
\gamma^i_{t-1}(s(0); \sigma^1_M, \vec{a}^2) &= \mathbf{E}_{s(0), \sigma^1_M, \vec{a}^2} \left[\frac{l_{t_\blacksquare} + (t - t_\blacksquare)\mathbf{P}(a^2_{t_\blacksquare} = R)}{t - 1} \right] \\
&\geq \mathbf{E}_{s(0), \sigma^1_M, \vec{a}^2} \left[\frac{\frac{t_\blacksquare - M - 1}{2} + (t - t_\blacksquare)\mathbf{P}(a^2_{t_\blacksquare} = R)}{t - 1} \right] \\
&= \mathbf{E}_{s(0), \sigma^1_M, \vec{a}^2} \left[\frac{t_\blacksquare \cdot \frac{1}{2} + (t - t_\blacksquare)\mathbf{P}(a^2_{t_\blacksquare} = R)}{t - 1} \right] - \frac{M + 1}{2(t - 1)} \\
&= \frac{1}{t - 1} \sum_{j=1}^{t-1} \mathbf{E}_{s(0)}[X_j] - \frac{M + 1}{2(t - 1)} \\
&\geq \frac{M}{2(M + 1)} - \frac{M + 1}{2(t - 1)},
\end{aligned}
$$

where the first inequality follows from Eq. (9.7) and the last inequality follows from Claim 9.12. $\qquad\square$

9.3 Existence of the Uniform Value in Two-Player Zero-Sum Stochastic Games

In this section, we prove the following result due to Mertens and Neyman (1981).

Theorem 9.13 *Every two-player zero-sum stochastic game has a uniform value.*

Proof As already mentioned, if the uniform value at the initial state s exists, then it must be equal to $v_0(s) := \lim_{\lambda \to 0} v_\lambda(s)$. We will prove that for every initial state $s \in S$ and every $\epsilon > 0$ Player 1 can guarantee $v_0(s) - \epsilon$ in the T-stage game, for all sufficiently large T. Analogous arguments show that for every $\epsilon > 0$ Player 2 can uniformly guarantee $v_0(s) + \epsilon$ in the T-stage game that starts in state s, for all sufficiently large T. By Theorem 9.4, it will then follow that $v_0(s)$ is the uniform value at the initial state s.

By Shapley's Theorem (Theorem 5.9), for every discount factor $\lambda \in (0, 1]$ Player 1 has a stationary λ-discounted optimal strategy $x_\lambda^1 = (x_\lambda^1(s))_{s \in S}$. In fact, by Corollary 8.14, a stronger result holds: there is a semi-algebraic mapping $\lambda \mapsto x_\lambda^1$ such that x_λ^1 is a stationary λ-discounted optimal strategy, for every $\lambda \in (0, 1]$. In particular, the limit $x_0^1 = \lim_{\lambda \to 0} x_\lambda^1$ exists.

Since for every $\lambda \in (0, 1]$ the stationary strategy x_λ^1 is λ-discounted optimal, we have

$$\lambda r(s, x_\lambda^1(s), x^2(s)) + (1 - \lambda) \sum_{s' \in S} q(s' \mid s, x_\lambda^1(s), x^2(s)) v_\lambda(s') \geq v_\lambda(s),$$

$$\forall s \in S, \ \forall x^2(s) \in \Delta(A^2(s)), \ \forall \lambda \in (0, 1]. \tag{9.9}$$

Letting $\lambda \to 0$, we obtain

$$\sum_{s' \in S} q(s' \mid s, x_0^1(s), y) v_0(s') \geq v_0, \quad \forall s \in S, \ \forall y \in \Delta(A^2(s)).$$

In other words, if Player 1 plays the stationary strategy $x_0^1 = (x_0^1(s))_{s \in S}$, then for every strategy of Player 2, the process $(v_0(s_t))_{t \in \mathbb{N}}$ does not decrease in expectation: it is a submartingale. That is, as long as Player 1 plays this stationary strategy, his expected potential gain, measured by $v_0(s_t)$, does not decrease:

$$\mathbf{E}_{s, x_0^1, \sigma^2}[v_0(s_t)] \geq v_0(s), \quad \forall \sigma^2 \in \Sigma^2.$$

This is what happens in the "Big Match": as long as Player 1 plays T, the play remains in state $s(0)$, where $v_0(s(0)) = \frac{1}{2}$. Unfortunately for Player 1, while her potential gain remains high, there is no guarantee that her actual payoff will be high.

The uniform ϵ-optimal strategy for Player 1 that we will construct will not play the stationary strategy x_0^1. Rather, at every stage t, when the play is in state s_t, Player 1 will play the mixed action $x_{\lambda_t}^1(s_t)$, where $\lambda_t \in (0, 1]$ is history dependent and close to 0. One delicate aspect of the proof is the definition of the process $(\lambda_t)_{t \in \mathbb{N}}$, which on the one hand will always be close to 0, so that $\mathbf{E}_{s,\sigma^1,\sigma^2}[v_0(s_t)]$ will be close to $v_0(s)$ for every $\sigma^2 \in \Sigma^2$, and on the other hand will allow Player 1 to obtain a high payoff. To obtain a high payoff, at every stage t Player 1 will consider her past average payoff. If this past average payoff is high, Player 1 can think about future opportunities, which translates into lowering λ, because in the discounted payoff the weight of the payoff in the current stage is λ, and the payoff from tomorrow on is $1 - \lambda$. If the past average payoff is low, Player 1 should think about her short-term gains, which translates into increasing λ. This idea is a generalization of the construction of Blackwell and Ferguson (1968) that we have seen in Section 9.2.

We have already mentioned that the discount factor measures the importance of short-term gains versus long-term gains. When studying the uniform value, the discount factor may be arbitrarily low, hence future opportunities always outweigh present gains. However, if a player does not collect high stage payoffs, she will never utilize her future opportunities, and therefore her overall payoff will be low. The mechanism we described, adapting the discount factor dynamically as a function of past performance, allows the player to properly balance between stage payoffs and future opportunities.

We now turn to the formal proof. Set

$$M := \|r\|_\infty,$$

and fix throughout $\epsilon \in (0, M)$.

By Theorem 6.14, for every state $s \in S$ the function $\lambda \mapsto v_\lambda(s)$ is semi-algebraic, and can therefore be represented as a Puiseux series in a small neighborhood of 0. That is, there exists a $\lambda_* \in (0, 1]$ such that

$$v_\lambda(s) = \sum_{k=0}^\infty a_k(s)\lambda^{k/L}, \quad \forall \lambda \in (0, \lambda_*), \ \forall s \in S.$$

Since the set of states is finite, the constant L can be chosen to be independent of $s \in S$. By Corollary 6.10, the function $\lambda \mapsto v_\lambda(s)$ is monotone in a neighborhood of 0, and, by Exercise 6.11, its derivative is also semi-algebraic. In particular, there exists a $\lambda_{**} \in (0, \lambda_*]$ such that

$$v'_\lambda(s) = \sum_{k=1}^{\infty} \frac{ka_k(s)}{L} \cdot \lambda^{\frac{k-L}{L}}, \quad \forall \lambda \in (0, \lambda_{**}).$$

The series on the right-hand side has a positive radius of convergence, and therefore there is a constant $C > 0$ such that

$$v'_\lambda(s) < \frac{C}{L} \cdot \lambda^{-\frac{L-1}{L}}, \quad \forall \lambda \in (0, \lambda_{**}). \tag{9.10}$$

Define a function $\psi \colon (0, 1] \to \mathbb{R}$ by

$$\psi(\lambda) := \frac{C}{L} \cdot \lambda^{-\frac{L-1}{L}}. \tag{9.11}$$

Note that the function ψ is integrable:

$$\int_0^1 \psi(\lambda) d\lambda = \frac{C}{L} \int_0^1 \lambda^{-\frac{L-1}{L}} d\lambda = C\lambda^{1/L} \Big|_0^1 = C.$$

By the fundamental theorem of calculus, Eqs. (9.10) and (9.11) imply that for every $\lambda_1, \lambda_2 \in (0, \lambda_{**}]$ such that $\lambda_1 < \lambda_2$ and every state $s \in S$,

$$|v_{\lambda_1}(s) - v_{\lambda_2}(s)| \le \int_{\lambda_1}^{\lambda_2} \psi(\lambda) d\lambda.$$

Consequently, there exists a $\lambda_0 \in (0, \lambda_{**}]$ such that $|v_{\lambda_1}(s) - v_{\lambda_2}(s)| \le \epsilon$ for every $\lambda_1, \lambda_2 \in (0, \lambda_0]$ and every $s \in S$. Define a function $D \colon (0, \lambda_0] \to \mathbb{R}$ by the formula

$$D(y) := \frac{12M}{\epsilon} \int_y^{\lambda_0} \frac{\psi(\lambda)}{\lambda} d\lambda + \frac{1}{\sqrt{y}}. \tag{9.12}$$

Note that for every $y \in (0, \lambda_0]$ the integral in Eq. (9.12) is finite, hence $D(y)$ is well defined. We next exhibit several properties of D.

Claim 9.14   *The function D is integrable: $\int_0^{\lambda_0} D(y) dy < \infty$.*

Proof   We show that the two terms in the expression of D are integrable. We start with the second term:

$$\int_0^{\lambda_0} y^{-1/2} dy = 2\sqrt{y} \Big|_0^{\lambda_0} = 2\sqrt{\lambda_0}.$$

We now turn to the first term. By Tonelli's Theorem,[1] we can exchange the order of integration, and obtain

$$\int_0^{\lambda_0} \left(\int_y^{\lambda_0} \frac{\psi(\lambda)}{\lambda} d\lambda \right) dy = \int_0^{\lambda_0} \left(\int_0^{\lambda} \frac{\psi(\lambda)}{\lambda} dy \right) d\lambda$$

$$= \int_0^{\lambda_0} \psi(\lambda) d\lambda = C(\lambda_0)^{1/L} \leq C.$$

The result follows.   ▲

Claim 9.15   *The function D is decreasing in the range $[0, \lambda_0]$.*

Proof D is the sum of two terms, the first of which is nonincreasing in y, while the second is strictly decreasing in y. The claim follows.   ▲

Claim 9.16   *For every $a \in (0, 1)$,*

$$\lim_{y \to 0} (D(ay) - D(y)) = +\infty.$$

Proof   By the definition of D,

$$D(ay) - D(y)$$
$$= \int_{ay}^y \frac{\psi(\lambda)}{\lambda} d\lambda + \frac{1}{\sqrt{ay}} - \frac{1}{\sqrt{y}}$$
$$= \int_{ay}^y \frac{\psi(\lambda)}{\lambda} d\lambda + \frac{1}{\sqrt{y}} \left(\frac{1}{\sqrt{a}} - 1 \right).$$

The first term is positive for every $y > 0$, while the second term goes to $+\infty$ when $y \to 0$. The claim follows.   ▲

Recalling that D is decreasing, we define a function $\varphi(\lambda)$ as the shaded area in Figure 9.3.

Formally,

$$\varphi(\lambda) := \int_0^{\lambda} D(y) dy - \lambda D(\lambda).$$

By Claim 9.14, the function D is integrable, which implies

Claim 9.17 $\lim_{\lambda \to 0} \varphi(\lambda) = 0$.

---

[1] Leonida Tonelli (Gallipoli, Italy, April 19, 1885 – Pisa, Italy, March 12, 1946) was an Italian mathematician, noted for creating a variation of Fubini's theorem that bears his name, and for introducing semicontinuity methods as a common tool for the direct method in the calculus of variations.

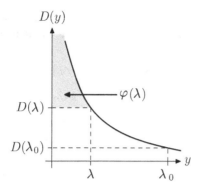

Figure 9.3 The functions D and φ.

Let $\lambda_1 \in (0, \lambda_0)$ be sufficiently small such that $\varphi(\lambda_1) < \epsilon$, and the following two inequalities hold:

$$D\left((1 - \frac{\epsilon}{6M})y\right) - D(y) > 6M, \quad \forall y \le \lambda_1, \qquad (9.13)$$

$$D(y) - D\left((1 + \frac{\epsilon}{6M})y\right) > 6M, \quad \forall y \le \lambda_1. \qquad (9.14)$$

By Claim 9.17, and by Claim 9.16 applied twice, with $a = 1 - \frac{\epsilon}{6M}$ to obtain Eq. (9.13) and with $a = \frac{1}{1+\frac{\epsilon}{6M}}$ to obtain Eq. (9.14), such λ_1 exists.

Define for every $t \in \mathbb{N}$ two random variables d_t and λ_t as follows:

$$d_1 := D(\lambda_1),$$
$$d_{t+1} := \max\{D(\lambda_1), d_t + r(s_t, a_t^1, a_t^2) - v_{\lambda_t}(s_{t+1}) + 4\epsilon\}, \qquad (9.15)$$
$$\lambda_{t+1} := D^{-1}(d_{t+1}).$$

Observe that $d_t \ge D(\lambda_0)$ for all $t \in \mathbb{N}$, and therefore $\lambda_t \le \lambda_0$.

Recall that x_λ^1 is a λ-discounted stationary optimal strategy of Player 1 for every $\lambda \in (0, \lambda_1]$. Define a strategy σ^1 for Player 1 as follows: at stage t, Player 1 plays the mixed action $x_{\lambda_t}^1(s_t)$. That is, she plays a mixed action that is optimal in the strategic-form game $G_{\lambda_t, s_t}(v_{\lambda_t})$. Set

$$T_0 = \max\left\{\frac{D(\lambda_1)}{\epsilon}, \frac{2M^2}{\epsilon^2 \lambda_1}\right\}.$$

We will prove that the strategy σ^1 guarantees $v_0(s) - 9\epsilon$ in the T-stage game, for every $T \ge T_0$:

$$\gamma_T(s; \sigma^1, \sigma^2) \ge v_0(s) - 9\epsilon, \quad \forall T \ge T_0, \forall \sigma^2 \in \Sigma^2. \qquad (9.16)$$

To prove that Eq. (9.16) holds, we will need to study some properties of the sequences $(\lambda_t)_{t \in \mathbb{N}}$ and $(d_t)_{t \in \mathbb{N}}$. To this end, we will define additional random variables.

Define for every $t \in \mathbb{N}$ a random variable Z_t by

$$Z_t := v_{\lambda_t}(s_t) - \varphi(\lambda_t).$$

When λ_t is small, Z_t is close to $v_0(s_t)$. As we will prove below, the strategy σ^1 ensures that, for every initial state $s \in S$ and every $\sigma^2 \in \Sigma^2$ we have $\lim_{t \to \infty} \lambda_t = 0$ with probability 1, and $\mathbf{E}_{s,\sigma^1,\sigma^2}[Z_{t+1} \mid \mathcal{H}_t] > Z_t$, where $\mathcal{H}_t$ is the algebra over the set H_∞ of plays that is spanned by the cylinder sets that correspond to histories of length t. This will imply in particular that $\lim_{t \to \infty} v_{\lambda_t}(s_t)$ exists with probability 1.

We now show that d_{t+1} cannot be too far from d_t. This follows from Eq. (9.15) since $|r(s_t, a_t^1, a_t^2)|$, $|v_{\lambda_t}(s_{t+1})|$, and ϵ are all at most M.

Claim 9.18 $|d_{t+1} - d_t| \le 6M$.

As a corollary, we bound the difference between λ_{t+1} and λ_t.

Claim 9.19 $|\lambda_{t+1} - \lambda_t| \le \frac{\epsilon \lambda_t}{6M}$.

Proof   Suppose first that $\lambda_{t+1} \le \lambda_t$. We need to prove that $\lambda_{t+1} \ge \lambda_t \left(1 - \frac{\epsilon}{6M}\right)$. Suppose that this is not the case, that is, $\lambda_{t+1} < \lambda_t \left(1 - \frac{\epsilon}{6M}\right)$. Then, by Eq. (9.13), $|d_{t+1} - d_t| > 6M$, which contradicts Claim 9.18.

Suppose now that $\lambda_{t+1} > \lambda_t$. We need to prove that $\lambda_{t+1} \le \lambda_t \left(1 + \frac{\epsilon}{6M}\right)$. Suppose this is not the case, that is, $\lambda_{t+1} > \lambda_t \left(1 + \frac{\epsilon}{6M}\right)$. Then, by Eq. (9.14), $|d_{t+1} - d_t| > 6M$, which again contradicts Claim 9.18.   ▲

Define:

$$
\begin{aligned}
C_1^t &:= \varphi(\lambda_t) - \varphi(\lambda_{t+1}), \\
C_2^t &:= v_{\lambda_{t+1}}(s_{t+1}) - v_{\lambda_t}(s_{t+1}), \\
C_3^t &:= \lambda_t(r(s_t, a_t^1, a_t^2) - v_{\lambda_t}(s_{t+1})).
\end{aligned}
$$

The following result relates the sequence $(Z_t)_{t \in \mathbb{N}}$ to the sequences $(C_1^t, C_2^t, C_3^t)_{t \in \mathbb{N}}$.

Claim 9.20   *For every state $s \in S$, every strategy $\sigma^2 \in \Sigma^2$, and every $t \in \mathbb{N}$,*

$$\mathbf{E}_{s,\sigma^1,\sigma^2}[Z_{t+1} - Z_t \mid \mathcal{H}_t] \ge \mathbf{E}_{s,\sigma^1,\sigma^2}[C_1^t + C_2^t - C_3^t \mid \mathcal{H}_t].$$

Proof   Since at every stage $t \in \mathbb{N}$ Player 1 plays a λ_t-discounted optimal strategy, by Eq. (9.9) we have

$$\mathbf{E}_{s,\sigma^1,\sigma^2}[\lambda_t r(s_t, a_t^1, a_t^2) + (1 - \lambda_t)v_{\lambda_t}(s_{t+1}) \mid \mathcal{H}_t] \ge v_{\lambda_t}(s_t).$$

Therefore,

$$\mathbf{E}_{s,\sigma^1,\sigma^2}[\lambda_t r(s_t, a_t^1, a_t^2) + (1 - \lambda_t)v_{\lambda_t}(s_{t+1}) - v_{\lambda_t}(s_t) \mid \mathcal{H}_t] \geq 0.$$

We add and subtract the terms $v_{\lambda_{t+1}}(s_{t+1})$, $\varphi(\lambda_t)$, and $\varphi(\lambda_{t+1})$, and obtain

$$
\begin{aligned}
\mathbf{E}_{s,\sigma^1,\sigma^2}[\lambda_t r(s_t, a_t^1, a_t^2) &+ (1 - \lambda_t)v_{\lambda_t}(s_{t+1}) - v_{\lambda_t}(s_t) \\
&+ v_{\lambda_{t+1}}(s_{t+1}) - v_{\lambda_{t+1}}(s_{t+1}) \\
&+ \varphi(\lambda_t) - \varphi(\lambda_t) \\
&+ \varphi(\lambda_{t+1}) - \varphi(\lambda_{t+1}) \mid \mathcal{H}_t] \geq 0.
\end{aligned}
\tag{9.17}
$$

Substituting C_1^t, C_2^t, C_3^t, Z_t, and Z_{t+1} in Eq. (9.17), we obtain

$$\mathbf{E}_{s,\sigma^1,\sigma^2}[C_3^t - C_2^t - C_1^t + Z_{t+1} - Z_t \mid \mathcal{H}_t] \geq 0,$$

or equivalently,

$$\mathbf{E}_{s,\sigma^1,\sigma^2}[Z_{t+1} - Z_t \mid \mathcal{H}_t] \geq \mathbf{E}_{s,\sigma^1,\sigma^2}[C_1^t + C_2^t - C_3^t \mid \mathcal{H}_t],$$

as stated. ▲

We will now bound C_1^t, C_2^t, and C_3^t.

Claim 9.21 $C_1^t \geq \lambda_t(d_{t+1} - d_t) - \epsilon\lambda_t$ (see Figure 9.4).

Proof   We consider the case $\lambda_{t+1} < \lambda_t$. As can be seen in Figure 2,

$$
\begin{aligned}
C_1^t &= \varphi(\lambda_t) - \varphi(\lambda_{t+1}) \\
&\geq \lambda_{t+1}(d_{t+1} - d_t) \\
&= \lambda_t(d_{t+1} - d_t) - (\lambda_t - \lambda_{t+1})(d_{t+1} - d_t) \\
&\geq \lambda_t(d_{t+1} - d_t) - \epsilon\lambda_t,
\end{aligned}
$$

where the last inequality follows from Claims 9.18 and 9.19. The case $\lambda_{t+1} > \lambda_t$ is analogous. ▲

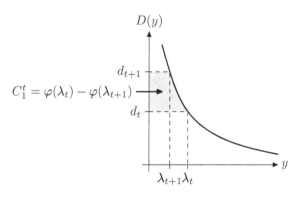

Figure 9.4 The quantity C_1^t.

Claim 9.22 $|C_2^t| \leq \epsilon \lambda_t$.

Proof By Claim 9.19, for every y between λ_t and λ_{t+1} we have $2\lambda_t \geq y$, and therefore

$$\frac{\lambda_t}{y} \geq \frac{1}{2}.$$

Hence,

$$|C_2^t| = |v_{\lambda_{t+1}}(s_{t+1}) - v_{\lambda_t}(s_{t+1})| \leq \left| \int_{\lambda_t}^{\lambda_{t+1}} \psi(\lambda) d\lambda \right| \leq 2\lambda_t \cdot \left| \int_{\lambda_t}^{\lambda_{t+1}} \frac{\psi(\lambda)}{\lambda} d\lambda \right|. \tag{9.18}$$

By the definition of the function D (see Eq. (9.12)),

$$d_{t+1} - d_t = D(\lambda_{t+1}) - D(\lambda_t) \tag{9.19}$$

$$= \frac{12M}{\epsilon} \int_{x=\lambda_{t+1}}^{\lambda_t} \frac{\psi(\lambda)}{\lambda} d\lambda + \frac{1}{\sqrt{\lambda_{t+1}}} - \frac{1}{\sqrt{\lambda_t}}.$$

Eq. (9.19) implies that

$$\int_{\lambda_{t+1}}^{\lambda_t} \frac{\psi(\lambda)}{\lambda} d\lambda = \frac{\epsilon}{12M} \left(D(\lambda_{t+1}) - D(\lambda_t) - \frac{1}{\sqrt{\lambda_{t+1}}} + \frac{1}{\sqrt{\lambda_t}} \right) \tag{9.20}$$

$$= \frac{\epsilon}{12M} \left(d_{t+1} - d_t - \frac{1}{\sqrt{\lambda_{t+1}}} + \frac{1}{\sqrt{\lambda_t}} \right).$$

Assume that $\lambda_t > \lambda_{t+1}$; in particular, $-\frac{1}{\sqrt{\lambda_{t+1}}} + \frac{1}{\sqrt{\lambda_t}} < 0$ and $d_t < d_{t+1}$. By Eqs. (9.18) and (9.20),

$$|C_2^t| \leq 2\lambda_t \int_{\lambda_{t+1}}^{\lambda_t} \frac{\psi(\lambda)}{\lambda} d\lambda$$

$$= \frac{\epsilon \lambda_t}{6M} \left(d_{t+1} - d_t - \frac{1}{\sqrt{\lambda_{t+1}}} + \frac{1}{\sqrt{\lambda_t}} \right)$$

$$\leq \frac{\epsilon \lambda_t}{6M} (d_{t+1} - d_t) \leq \epsilon \lambda_t,$$

where the last inequality holds by Claim 9.18. Assume now that $\lambda_t < \lambda_{t+1}$; in particular, $-\frac{1}{\sqrt{\lambda_t}} + \frac{1}{\sqrt{\lambda_{t+1}}} < 0$ and $d_t > d_{t+1}$. Hence, by Eqs. (9.18) and (9.20),

$$|C_2^t| \leq 2\lambda_t \int_{\lambda_t}^{\lambda_{t+1}} \frac{\psi(\lambda)}{\lambda} d\lambda$$

$$= \frac{\epsilon \lambda_t}{6M} \left(d_t - d_{t+1} - \frac{1}{\sqrt{\lambda_t}} + \frac{1}{\sqrt{\lambda_{t+1}}} \right)$$

$$\leq \frac{\epsilon \lambda_t}{6M} (d_t - d_{t+1}) \leq \epsilon \lambda_t. \qquad \blacktriangle$$

Claim 9.23 $C_3^t \leq \lambda_t(d_{t+1} - d_t) - 4\epsilon\lambda_t.$

Proof By the definition of d_{t+1},

$$d_{t+1} - d_t \geq r(s_t, a_t^1, a_t^2) - v_{\lambda_t}(s_{t+1}) + 4\epsilon.$$

Therefore,

$$C_3^t = \lambda_t(r(s_t, a_t^1, a_t^2) - v_{\lambda_t}(s_{t+1})) \leq \lambda_t(d_{t+1} - d_t) - 4\epsilon\lambda_t,$$

which is what we wanted to prove. ▲

The following result is a corollary of Claims 9.21–9.23.

Claim 9.24 *For every initial state $s \in S$, every strategy $\sigma^2 \in \Sigma^2$, and every $T \in \mathbb{N}$,*

$$\mathbf{E}_{s,\sigma^1,\sigma^2}[Z_T] \geq 2\epsilon\mathbf{E}_{s,\sigma^1,\sigma^2}\left[\sum_{t=1}^{T-1} \lambda_t\right] + Z_1. \tag{9.21}$$

Proof Consider the following chain of inequalities:

$$\mathbf{E}_{s,\sigma^1,\sigma^2}[Z_{t+1} - Z_t \mid \mathcal{H}_t]$$
$$\geq \mathbf{E}_{s,\sigma^1,\sigma^2}[C_1^t + C_2^t - C_3^t \mid \mathcal{H}_t]$$
$$\geq \mathbf{E}_{s,\sigma^1,\sigma^2}[\lambda_t(d_{t+1} - d_t) - \epsilon\lambda_t - \epsilon\lambda_t - \lambda_t(d_{t+1} - d_t) + 4\epsilon\lambda_t]$$
$$\geq 2\epsilon\lambda_t$$
$$> 0.$$

By summation over $t = 1, 2, \ldots, T - 1$ and the law of iterated expectation, we obtain Eq. (9.21). ▲

Comment 9.25 We note that the proof of Claim 9.24 implies in particular that the process $(Z_t)_{t\in\mathbb{N}}$ is a submartingale. We will not use this property in the proof.

Fix an initial state $s \in S$ and a strategy $\sigma^2 \in \Sigma^2$. From Claim 9.24, it follows that $\mathbf{E}_{s,\sigma^1,\sigma^2}[Z_t] \geq Z_1$ for every $t \in \mathbb{N}$. By substituting $Z_t = v_{\lambda_t}(s_t) - \varphi(\lambda_t)$ in Eq. (9.21), we deduce that the λ_t-discounted value at state s_t is high:

$$\mathbf{E}_{s,\sigma^1,\sigma^2}[v_{\lambda_t}(s_t)] \geq v_{\lambda_1}(s) + \mathbf{E}_{s,\sigma^1,\sigma^2}[\varphi(\lambda_t) - \varphi(\lambda_1)]$$
$$\geq v_{\lambda_1}(s) - \varphi(\lambda_1) \tag{9.22}$$
$$\geq v_0(s) - 2\epsilon.$$

Further, since $|Z_t| \leq 2M$ for every $t \in \mathbb{N}$, we deduce from Eq. (9.21) that

$$\mathbf{E}_{s,\sigma^1,\sigma^2}\left[\sum_{t=1}^{\infty} \lambda_t\right] \leq \frac{M}{\epsilon}. \tag{9.23}$$

Since $\lambda_t \leq \lambda_1$ for every $t \in \mathbb{N}$, Eq. (9.23) bounds the number of times that λ_t can be equal to λ_1:

$$\mathbf{E}_{s,\sigma^1,\sigma^2}\left[\sum_{t=1}^{\infty} \mathbf{1}_{\{\lambda_t=\lambda_1\}}\right] \leq \frac{M}{\epsilon\lambda_1}. \qquad (9.24)$$

We are now ready to bound the T-stage payoff under (σ^1,σ^2). By the definition of d_{t+1},

$$d_{t+1} - d_t \leq r(s_t,a_t^1,a_t^2) - v_{\lambda_t}(s_{t+1}) + 4\epsilon + 2M\mathbf{1}_{\{\lambda_{t+1}=\lambda_1\}}. \qquad (9.25)$$

Summing Eq. (9.25) over $t = 1, 2, \ldots, T$, and taking expectations, we obtain

$$\mathbf{E}_{s,\sigma^1,\sigma^2}\left[\sum_{t=1}^{T} r(s_t,a_t^1,a_t^2)\right]$$

$$\geq \mathbf{E}_{s,\sigma^1,\sigma^2}\left[\sum_{t=1}^{T} v_{\lambda_t}(s_{t+1})\right] + \mathbf{E}_{s,\sigma^1,\sigma^2}[d_{T+1} - d_1]$$

$$- 4T\epsilon - 2M\mathbf{E}_{s,\sigma^1,\sigma^2}\left[\sum_{t=1}^{\infty} \mathbf{1}_{\{\lambda_t=\lambda_1\}}\right]. \qquad (9.26)$$

Since $d_{T+1} \geq d_1$ for every $T \in \mathbb{N}$, the term $\mathbf{E}_{s,\sigma^1,\sigma^2}[d_{T+1}-d_1]$ is nonnegative, hence by removing it we decrease the right-hand side. Using Eqs. (9.24) and (9.22) we then obtain

$$\gamma_T(s;\sigma^1,\sigma^2) \geq v_{\lambda_1}(s) - 3\epsilon - 4\epsilon - \frac{2M^2}{T\epsilon\lambda_1} \geq v_0(s) - 9\epsilon,$$

provided T is larger than $\frac{D(\lambda_1)}{\epsilon}$ and $\frac{2M^2}{\epsilon^2\lambda_1}$. Since σ^2 is an arbitrary strategy of Player 2, we thus proved that σ^1 uniformly guarantees $v_0(s) - 9\epsilon$ at the initial state s, as desired. $\square$

9.4 The Average Cost Optimality Equation

Theorem 9.13 asserts that the uniform value exists. It does not indicate how to calculate the uniform value in specific games. In this section, we will develop a tool that will help us in calculating the uniform value, but is valid only when the uniform value is the same in all states.

To state the next result, we define for every state $s \in S$ and every function $w: S \rightarrow \mathbb{R}$ a strategic-form game $\widehat{G}_s(w)$, which is similar to the game

$G_{s,\lambda}(w)$, with weight 1 on both the stage payoff and the continuation payoff. Let $\widehat{G}_s(w)$ be the following two-player zero-sum strategic-form game:

- The set of players is I.
- The set of actions of player i is $A^i(s)$, for each $i \in I$.
- The payoff function is

$$r(s,a^1,a^2) + \sum_{s' \in S} q(s' \mid s,a^1,a^2)w(s'), \quad \forall(a^1,a^2) \in A^1 \times A^2.$$

The following result provides a way to calculate the uniform value in certain two-player zero-sum stochastic games.

Theorem 9.26 (The Average Cost Optimality Equation) *Let $\Gamma = \langle\{1,2\}, S, (A^1(s), A^2(s))_{s \in S}, q, r\rangle$ be a two-player zero-sum stochastic game. Assume that there exist a real number g and a function $w \colon S \to \mathbb{R}$ such that*

$$g + w(s) = val(\widehat{G}_s(w)), \quad \forall s \in S. \tag{9.27}$$

Then $v_0(s) = g$ for every state $s \in S$.

By Exercise 3.3, the value of a two-player zero-sum strategic-form game is invariant under the addition of a constant. It follows that the set of functions w that are part of a solution of Eq. (9.27) is invariant under the addition of a constant: If (g, w) is a solution of Eq. (9.27), then for every real number c the pair (g, w') is a solution of Eq. (9.27), where $w' \colon S \to \mathbb{R}$ is defined by $w'(s) := w(s) + c$, for each $s \in S$.

Example 9.27 Consider the two-player zero-sum stochastic game with two states that is displayed in Figure 9.5.

We will calculate the value of the game using Theorem 9.26. Since we can add a constant to w, we set $w(s(1)) = 0$. Denote $A := w(s(0))$. The games $\widehat{G}_{s(0)}(w)$ and $\widehat{G}_{s(1)}(w)$ are depicted in Figure 9.6. By Theorem 9.26,

$$g + A = val(\widehat{G}_{s(0)}(f)), \quad g = val(\widehat{G}_{s(1)}(w)). \tag{9.28}$$

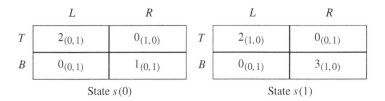

Figure 9.5 The stochastic game in Example 9.27.

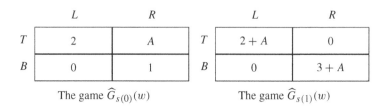

Figure 9.6 The games $G_{s(0)}(w)$ and $G_{s(1)}(w)$ in Example 9.27.

The values of the games $\widehat{G}_{s(0)}(w)$ and $\widehat{G}_{s(1)}(w)$ depend on A, and simple calculations show that:

- If $A \geq 2$, then $\mathrm{val}(G_{s(0)}(w)) = 2$.
- If $A \leq 2$, then $\mathrm{val}(G_{s(0)}(w)) = \frac{2}{3-A}$.
- If $A \geq -2$ or $A \leq -3$, then $\mathrm{val}(G_{s(1)}(w)) = \frac{(2+A)(3+A)}{5+2A}$.
- If $-3 \leq A \leq -2$, then $\mathrm{val}(G_{s(1)}(w)) = 0$.

To determine g we will solve Eq. (9.28) for each value of A, and we will see that there is a unique A for which the system has a solution.

- If $A \geq 2$, then (g, A) is a solution of the system

$$g + A = 2, \quad g = \frac{(2 + A)(3 + A)}{5 + 2A}.$$

However, this system has no solution with $A \geq 2$.
- If $-2 \leq A \leq 2$ or $A \leq -3$, then (g, A) is a solution of the system

$$g + A = \frac{2}{3 - A}, \quad g = \frac{(2 + A)(3 + A)}{5 + 2A},$$

which has three solutions:

$$A \approx -2.5427, \quad g = 2.9061,$$
$$A \approx -0.40165, \quad g = 0.9896,$$
$$A \approx 2.6111, \quad g = 2.5311.$$

among which only the second satisfies $-2 \leq A \leq 2$ or $A \leq -3$.
- If $-2 \leq A \leq -3$, then (g, A) is a solution of the system

$$g + A = \frac{2}{3 - A}, \quad g = 0,$$

which has no solution with $-2 \leq A \leq -3$.

It follows that the value of the game at both initial states is $v_0(s(0)) = v_0(s(1)) = g \approx 0.9896$. ◆

Proof of Theorem 9.26   By Eq. (9.27),

$$w(s) = -g + \text{val}(\widehat{G}_s(w)), \quad \forall s \in S. \tag{9.29}$$

Substituting this expression in itself, we see that for every initial state $s_1 \in S$,

$$w(s_1) = -g + \text{val}(\widehat{G}_{s_1}(-g + \text{val}(\widehat{G}_{s_2}(w)))),$$

where s_2 is the state in the second stage. Using the fact that adding a constant to all entries of the payoff matrix results in an increase of the value by the same constant (Exercise 3.3), we deduce that

$$w(s_1) = -2g + \text{val}(\widehat{G}_{s_1}(\text{val}(\widehat{G}_{s_2}(w)))). \tag{9.30}$$

The term $\text{val}(\widehat{G}_{s_1}(\text{val}(\widehat{G}_{s_2}(w))))$ is the value of the two-stage stochastic game that starts in state s_1, where the continuation payoff after the end of the second stage is $w(s_3)$, and the players are interested in the sum of payoffs (rather than the average payoff). Further substitution of Eq. (9.29) in Eq. (9.30) yields

$$w(s_1) = -3g + \text{val}(\widehat{G}_{s_1}(\text{val}(\widehat{G}_{s_2}(\text{val}(\widehat{G}_{s_3}(w)))))).$$

Thus, $w(s_1)$ is also the value of the three-stage game that starts in state s_1, where the continuation payoff after the end of the third stage is $w(s_4)$. Additional iterations of this substitution yield for every $T \in \mathbb{N}$,

$$w(s_1) = -Tg + \text{val}(\widehat{G}_{s_1}(\text{val}(\widehat{G}_{s_2}(\text{val}(\widehat{G}_{s_2}(\ldots (\text{val}(\widehat{G}_{s_T}(w)))))))). \tag{9.31}$$

The value of the T-stage game at the initial state s_1 when the players are interested in the sum of payoffs is $T v_T(s_1)$. This value is also equal to

$$\text{val}(\widehat{G}_{s_1}(\text{val}(\widehat{G}_{s_2}(\text{val}(\widehat{G}_{s_2}(\ldots (\text{val}(\widehat{G}_{s_T}(0)))))))). \tag{9.32}$$

Inductive use of Theorem 3.15 implies that

$$\begin{aligned}
&\left| \text{val}(\widehat{G}_{s_1}(\text{val}(\widehat{G}_{s_2}(\text{val}(\widehat{G}_{s_2}(\ldots (\text{val}(\widehat{G}_{s_T}(w))))))) \right. \\
&\left. - \text{val}(\widehat{G}_{s_1}(\text{val}(\widehat{G}_{s_2}(\text{val}(\widehat{G}_{s_2}(\ldots (\text{val}(\widehat{G}_{s_T}(0)))))))) \right| \leq \|w\|_\infty.
\end{aligned} \tag{9.33}$$

We thus deduce that

$$\begin{aligned}
\|w\|_\infty &\geq \left| \text{val}(\widehat{G}_{s_1}(\text{val}(\widehat{G}_{s_2}(\text{val}(\widehat{G}_{s_2}(\ldots (\text{val}(\widehat{G}_{s_T}(w))))))) \right. \\
&\quad \left. \text{val}(\widehat{G}_{s_1}(\text{val}(\widehat{G}_{s_2}(\text{val}(\widehat{G}_{s_2}(\ldots (\text{val}(\widehat{G}_{s_T}(0))))))) \right| \\
&= \left| T v_T(s_1) - w(s_1) - Tg \right|.
\end{aligned}$$

Dividing both sides of this equation by T, and letting T go to infinity, we obtain

$$v_0(s_1) = \lim_{T \to \infty} v_T(s_1) = g,$$

as claimed. $\qquad\qquad\qquad\qquad\qquad\qquad\qquad\qquad\qquad\qquad\qquad$ □

The following result has two consequences. First, it provides a condition which ensures that a solution to Eq. (9.27) exists. Second, it provides an interpretation to the function w in Eq. (9.27).

Theorem 9.28   *Let* $\Gamma = \langle \{1,2\}, S, (A^1(s), A^2(s))_{s \in S}, q, r \rangle$ *be a two-player zero-sum stochastic game, let* $s \in S$, *and suppose that*

$$\lim_{\lambda \to 0} \frac{|v_\lambda(s) - v_\lambda(s')|}{\lambda} < \infty, \quad \forall s' \in S. \tag{9.34}$$

Then the system of equations (9.27) has a solution (g, w) *that satisfies*

$$w(s) - w(s') = \lim_{\lambda \to 0} \frac{v_\lambda(s) - v_\lambda(s')}{\lambda}.$$

We note that the condition in Theorem 9.26 implies that $v_0(s) = v_0(s')$ for every state $s' \in S$. Indeed, $v_0(s') = \lim_{\lambda \to 0} v_\lambda(s')$ for every state $s' \in S$ and, by Eq. (9.34),

$$\begin{aligned}
|v_0(s) - v_0(s')| &= \lim_{\lambda \to 0} |v_\lambda(s) - v_\lambda(s')| \\
&= \lim_{\lambda \to 0} \lambda \cdot \frac{|v_\lambda(s) - v_\lambda(s')|}{\lambda} \\
&= \lim_{\lambda \to 0} \lambda \cdot \lim_{\lambda \to 0} \frac{|v_\lambda(s) - v_\lambda(s')|}{\lambda} = 0.
\end{aligned}$$

While under the conditions of Theorem 9.28, the uniform value at all states is the same, the discounted value need not be the same. According to Theorem 9.28, the quantity $w(s) - w(s')$ is equal to the limit of the difference between the λ-discounted value at the initial states s and s', divided by λ.

Proof of Theorem 9.28   Fix two distinct states $s, s' \in S$. By Theorem 5.10, for every discount factor $\lambda \in (0, 1]$ we have

$$v_\lambda(s) = \text{val}(G_{s,\lambda}(v_\lambda)), \tag{9.35}$$

where $G_{s,\lambda}(v_\lambda)$ is the two-player zero-sum game where the set of actions of each player i is $A^i(s)$ and the payoff function is

$$\lambda r(s, a) + (1 - \lambda) \sum_{\widehat{s} \in S} q(\widehat{s} \mid s, a) v_\lambda(\widehat{s}), \quad \forall a \in A^1 \times A^2.$$

Subtracting $v_\lambda(s')$ from both sides of Eq. (9.35), we find that

$$v_\lambda(s) - v_\lambda(s') = \text{val}(G_s')), \qquad (9.36)$$

where G_s' is the two-player zero-sum game in which the set of actions of each player i is $A^i(s)$ and the payoff function is

$$\lambda r(s,a) + \sum_{\widehat{s} \in S} q(\widehat{s} \mid s,a)(v_\lambda(\widehat{s}) - v_\lambda(s')) - \lambda \sum_{\widehat{s} \in S} q(\widehat{s} \mid s,a) v_\lambda(\widehat{s}),$$

$$\forall a \in A^1(s) \times A^2(s).$$

Dividing both sides of Eq. (9.36) by λ, we obtain

$$\frac{v_\lambda(s) - v_\lambda(s')}{\lambda} = \text{val}(G_s'')), \qquad (9.37)$$

where G_s'' is the two-player zero-sum game in which the set of actions of each player i is $A^i(s)$ and the payoff function is

$$r(s,a) + \sum_{\widehat{s} \in S} q(\widehat{s} \mid s,a) \frac{v_\lambda(\widehat{s}) - v_\lambda(s')}{\lambda} - \sum_{\widehat{s} \in S} q(\widehat{s} \mid s,a) v_\lambda(\widehat{s}),$$

$$\forall a \in A^1(s) \times A^2(s).$$

Set

$$w(\widehat{s}) := \lim_{\lambda \to 0} \frac{v_\lambda(\widehat{s}) - v_\lambda(s')}{\lambda}, \quad \forall \widehat{s} \in S,$$

which, by Eq. (9.34), is finite. Theorem 3.15 implies that the value operator is continuous. Therefore, letting $\lambda \to 0$ in Eq. (9.37) we get

$$w(s) = \text{val}(G_s'''), \qquad (9.38)$$

where G_s''' is the two-player zero-sum game in which the set of actions of each player i is $A^i(s)$ and the payoff function is

$$r(s,a) + \sum_{\widehat{s} \in S} q(\widehat{s} \mid s,a) w(\widehat{s}) - \sum_{\widehat{s} \in S} q(\widehat{s} \mid s,a) v_0(\widehat{s}), \quad \forall a \in A^1(s) \times A^2(s).$$

$$(9.39)$$

As mentioned before, Eq. (9.34) implies that $v_0(s) = v_0(\widehat{s})$ for every two states $s, \widehat{s} \in S$, hence the payoff function of G_s''' is

$$r(s,a) + \sum_{\widehat{s} \in S} q(\widehat{s} \mid s,a) w(\widehat{s}) - v_0(s), \quad \forall a \in A^1(s) \times A^2(s). \qquad (9.40)$$

Thus, the difference between the payoffs in $G_s(w)$ and G_s''' is the constant $v_0(s)$. Therefore,

$$w(s) = \text{val}(G_s''') = \text{val}(G_s(w)) - v_0(s),$$

and the result follows. $\qquad \qquad \square$

The following result identifies one class of stochastic games where the condition in Theorem 9.28 holds. For every state $s \in S$ denote by

$$\theta_s := \min\{t > 1 : s_t = s\}$$

the first stage after the initial stage in which the play visits the state s.

Theorem 9.29   *Let* $\Gamma = \langle\{1,2\}, S, (A^1(s), A^2(s))_{s \in S}, q, r\rangle$ *be a two-player zero-sum stochastic game. Suppose that for every two distinct states* $s, s' \in S$ *there is a strategy* $\sigma^1 \in \Sigma^1$ *and* $c > 0$ *such that for every strategy* $\sigma^2 \in \Sigma^2$ *we have*

$$\mathbf{E}_{s,\sigma^1,\sigma^2}[\theta_{s'}] \leq c. \tag{9.41}$$

Then

$$\lim_{\lambda \to 0} \frac{v_\lambda(s) - v_\lambda(s')}{\lambda} \leq 2c\|r\|_\infty, \quad \forall s, s' \in S.$$

Theorem 9.29 states that if for every two distinct states s, s', Player 1 can ensure that the expected time to reach state s' from state s is finite, then the condition in Theorem 9.28 holds. Plainly, Theorem 9.29 holds when the roles of the players are switched: Player 2 can ensure that the expected time to reach state s' from state s is finite.

Proof of Theorem 9.29   Fix two distinct states $s, s' \in S$, let $c > 0$, and let $\sigma^1 \in \Sigma^1$ be such that Eq. (9.41) holds for every strategy $\sigma^2 \in \Sigma^2$.

Let $\widehat{\sigma}^1$ be the strategy of Player 1 that follows the strategy σ^1 until stage $\theta_{s'}$, and a λ-discounted optimal strategy afterward. Let σ^2 be a stationary λ-discounted optimal strategy of Player 2. Then we have

$$v_\lambda(s) \geq \gamma_\lambda(s; \widehat{\sigma}^1, \sigma^2) \tag{9.42}$$

$$= \mathbf{E}_{s,\widehat{\sigma}^1,\sigma^2}\left[\lambda \sum_{t=1}^{\infty} (1-\lambda)^{t-1} r(s_t, a_t)\right]$$

$$= \mathbf{E}_{s,\widehat{\sigma}^1,\sigma^2}\left[\lambda \sum_{t=1}^{\theta_{s'}-1} (1-\lambda)^{t-1} r(s_t, a_t) + \lambda \sum_{t=\theta_{s'}}^{\infty} (1-\lambda)^{t-1} r(s_t, a_t)\right]$$

$$= \mathbf{E}_{s,\widehat{\sigma}^1,\sigma^2}\left[\lambda \sum_{t=1}^{\theta_{s'}-1} (1-\lambda)^{t-1} r(s_t, a_t) + (1-\lambda)^{\theta_{s'}} v_\lambda(s')\right], \tag{9.43}$$

$$\geq -\|r\|_\infty \cdot \mathbf{E}_{s,\widehat{\sigma}^1,\sigma^2}\left[\lambda \sum_{t=1}^{\theta_{s'}-1} (1-\lambda)^{t-1}\right] + v_\lambda(s')$$

$$\qquad - v_\lambda(s') \mathbf{E}_{s,\widehat{\sigma}^1,\sigma^2}\left[1 - (1-\lambda)^{\theta_{s'}}\right]$$

$$= -\|r\|_\infty \cdot \mathbf{E}_{s,\widehat{\sigma}^1,\sigma^2}\left[1-(1-\lambda)^{\theta_{s'}}\right] + v_\lambda(s')$$
$$- v_\lambda(s')\mathbf{E}_{s,\widehat{\sigma}^1,\sigma^2}\left[1-(1-\lambda)^{\theta_{s'}}\right]$$
$$\geq v_\lambda(s') - 2\|r\|_\infty \cdot \mathbf{E}_{s,\widehat{\sigma}^1,\sigma^2}\left[1-(1-\lambda)^{\theta_{s'}}\right]$$
$$\geq v_\lambda(s') - 2\|r\|_\infty \cdot \mathbf{E}_{s,\widehat{\sigma}^1,\sigma^2}\left[\lambda\theta_{s'}\right], \tag{9.44}$$

where Eq. (9.42) holds because the strategy σ^2 is λ-discounted optimal, Eq. (9.43) holds because after stage $\theta_{s'}$ both players follow λ-discounted optimal strategies, hence the λ-discounted payoff from stage $\theta_{s'}$ and on is $v_\lambda(s')$, and Eq. (9.44) holds because $1-\lambda t \leq (1-\lambda)^t$ for every $\lambda \in (0,1]$ and every $t \in \mathbb{N}$. We conclude that

$$\frac{v_\lambda(s) - v_\lambda(s')}{\lambda} \geq -2\|r\|_\infty \cdot \mathbf{E}_{s,\widehat{\sigma}^1,\sigma^2}\left[\theta_{s'}\right] \geq -2c\|r\|_\infty.$$

Switching the roles of s and s', we obtain that

$$\lim_{\lambda \to 0} \frac{|v_\lambda(s) - v_\lambda(s')|}{\lambda} \leq 2c\|r\|_\infty,$$

and the result follows.   □

9.5 Extensive-Form Correlated Equilibrium

In this section, we apply the technique developed in Section 9.3 to multiplayer stochastic games, and we derive the existence of a version of uniform correlated equilibrium.

We start by defining the concept of λ-discounted max–min value, encountered in the setup of two-player zero-sum games in Exercise 5.14.

Definition 9.30   Let $\Gamma = \langle I, S, (A^i(s))_{s\in S}^{i\in I}, q, (r^i)_{i\in I}\rangle$ be a stochastic game, let $i \in I$, let $s \in S$, and let $\lambda \in (0,1]$. The λ-discounted max–min value of player i at the initial state s is

$$\underline{v}_\lambda^i(s) := \sup_{\sigma^i \in \Sigma^i} \inf_{\sigma^{-i} \in \Sigma^{-i}} \gamma_\lambda^i(s;\sigma^i,\sigma^{-i}). \tag{9.45}$$

Let us show that the λ-discounted max–min value of player i is the λ-discounted value in an auxiliary two-player zero-sum stochastic game $\widehat{\Gamma}^i = \langle\{1,2\}, S, (\widehat{A}^1(s), \widehat{A}^2(s))_{s\in S}^{i\in I}, q, r^i\rangle$, defined as follows:

- The set of players is $\{1,2\}$.
- The set of states is S, the set of states in the original game Γ.
- The set of actions of Player 1 in state s is $\widehat{A}^1(s) := A^i(s)$, for every $s \in S$.

- The set of actions of Player 2 in state s is $\widehat{A}^2(s) := \prod_{j \neq i} A^j(s)$, for every $s \in S$.
- The payoff function is r^i.
- The transition rule is the transition rule in the original game Γ.

This stochastic game captures the situation in which the goal of the players in $I \setminus \{i\}$ is to minimize player i's payoff, and they can correlate their actions: if, for example, $I = \{i_1, i_2, i_3\}$, $i = i_1$, $A^{i_2}(s) = \{T, B\}$ and $A^{i_3}(s) = \{L, R\}$, then Player 1 in the game $\widehat{\Gamma}^i$ represents player i_1 in Γ, Player 2 in $\widehat{\Gamma}^i$ represents players i_2 and i_3 in Γ, and $\widehat{A}^2(s) = \{TL, TR, BL, BR\}$. In particular, at state s of the game $\widehat{\Gamma}^i$ Player 2 can choose the mixed action $\left[\frac{1}{2}(TL), \frac{1}{2}(BR)\right]$, a choice that is not available to players i_2 and i_3 in Γ.

The set $\widehat{\Sigma}^1$ of strategies of Player 1 in $\widehat{\Gamma}^i$ is identical to Σ^1, the set of strategies of player i in Γ. A strategy of Player 2 in $\widehat{\Gamma}^i$ is a function $\widehat{\sigma}^2$ that assigns to every history $h_t = (s_1, a_1, \ldots, s_t) \in H$ a probability distribution $\widehat{\sigma}^2(h_t) \in \Delta(\widehat{A}^2(s_t)) = \Delta\left(\prod_{j \neq i} A^i(s_t)\right)$. We note that the set $\widehat{\Sigma}^2$ of strategies of Player 2 in $\widehat{\Gamma}^i$ is convex and its extreme points are the pure strategies of Player 2 in $\widehat{\Gamma}^i$. Moreover, a pure strategy of Player 2 in $\widehat{\Gamma}^i$ is a pure strategy profile of the players in $I \setminus \{i\}$ in Γ.

Let $\widehat{\gamma}_\lambda(s; \widehat{\sigma}^1, \widehat{\sigma}^2)$ denote the λ-discounted payoff under strategy profile $(\widehat{\sigma}^1, \widehat{\sigma}^2)$ at the initial state s in the game $\widehat{\Gamma}^i$. Since $\widehat{\Gamma}^i$ is a two-player zero-sum stochastic game, it admits a discounted value at each initial state $s \in S$, which is given by

$$\widehat{v}_\lambda^i(s) = \max_{\widehat{\sigma}^1 \in \widehat{\Sigma}^1} \min_{\widehat{\sigma}^2 \in \widehat{\Sigma}^2} \widehat{\gamma}_\lambda(s; \widehat{\sigma}^1, \widehat{\sigma}^2) = \min_{\widehat{\sigma}^2 \in \widehat{\Sigma}^2} \max_{\widehat{\sigma}^1 \in \widehat{\Sigma}^1} \widehat{\gamma}_\lambda(s; \widehat{\sigma}^1, \widehat{\sigma}^2). \quad (9.46)$$

Since for every fixed strategy $\widehat{\sigma}^1 \in \Sigma^1$, state $s \in S$, and discounted factor $\lambda \in (0, 1]$, the function $\widehat{\sigma}^2 \mapsto \widehat{\gamma}_\lambda(s; \widehat{\sigma}^1, \widehat{\sigma}^2)$ is linear, and since the set $\widehat{\Sigma}^2$ is compact in the product topology, the minimum $\min_{\widehat{\sigma}^2 \in \widehat{\Sigma}^2} \widehat{\gamma}_\lambda(s; \widehat{\sigma}^1, \widehat{\sigma}^2)$ in Eq. (9.46) is attained at an extreme point of $\widehat{\Sigma}^2$, namely, a pure strategy of Player 2 in $\widehat{\Gamma}^i$. As mentioned before, the set of pure strategies of Player 2 in $\widehat{\Gamma}^i$ is Σ^{-i}, hence

$$\widehat{v}_\lambda^i(s) = \max_{\widehat{\sigma}^1 \in \widehat{\Sigma}^1} \min_{\widehat{\sigma}^2 \in \widehat{\Sigma}^2} \widehat{\gamma}_\lambda(s; \widehat{\sigma}^1, \widehat{\sigma}^2) = \max_{\sigma^1 \in \Sigma^1} \min_{\sigma^2 \in \Sigma^2} \gamma_\lambda(s; \sigma^1, \sigma^2) = \underline{v}_\lambda^i(s).$$

We thus obtained an interpretation for the λ-discounted max–min value of player i in the game Γ, as the value of an auxiliary two-player zero-sum game.

The function $\lambda \mapsto \underline{v}^i_\lambda(s)$ is semi-algebraic for every player $i \in I$ and every initial state $s \in S$ (Exercise 9.9). By Corollary 6.10, the limit $\underline{v}^i_0(s) := \lim_{\lambda \to 0} \underline{v}^i_\lambda(s)$ exists for every player $i \in I$ and every initial state $s \in S$. By Theorem 9.13, $\underline{v}^i_0(s)$ is the *uniform value* of the game $\widehat{\Gamma}^i$ at the initial state s. Moreover, in this game Player 2 has a uniformly ϵ-optimal strategy. Such a strategy is called a *uniform ϵ-punishment strategy against player i.* The quantity $\underline{v}^i_0(s)$ is called the *uniform max–min value* of player i at the initial state s in Γ.

Recall that on Page 117 we defined for every state $s \in S$, every discount factor $\lambda \in (0,1]$, and every function $w \colon S \to \mathbb{R}$, an auxiliary strategic-form game $G_{s,\lambda}(w)$, namely, the game played at state s when the continuation payoff is given by the function w. Here we will define an analogous game in which each player has a different discount factor. Formally, for every vector $\vec{\lambda} = (\lambda^i)_{i \in I} \in (0,1]^I$, every state $s \in S$, and every vector $w \in \mathbb{R}^{S \times I}$, the game $G_{s,\vec{\lambda}}(w)$ is the strategic-form game $\langle I, (A^i(s))_{i \in I}, (u^i_{s,\vec{\lambda}})_{i \in I}\rangle$, where the payoff function of each player i is given by

$$u^i_{s,\vec{\lambda}}(a) := \lambda^i r^i(s,a) + (1 - \lambda^i) \sum_{s' \in S} q(s' \mid s,a) w^i(s'), \quad \forall a \in A(s).$$

The max–min value of each player $i \in I$ in the game $G_{s,\vec{\lambda}}(w)$ depends only on player i's payoff function, hence it does not depend on $(\lambda^j)_{j \neq i}$. As in Theorem 5.10, the max–min value of player i in the auxiliary game $G_{s,\vec{\lambda}}(\underline{v}_{\vec{\lambda}})$ is $\underline{v}^i_{\lambda^i}(s)$, where $\underline{v}_{\vec{\lambda}}(s) = (\underline{v}^i_{\lambda^i}(s))_{i \in I}$ (Exercise 9.20). Denote the payoff in the game $G_{s,\vec{\lambda}}(\underline{v}_{\vec{\lambda}})$ under the mixed action profile α by $u_{s,\vec{\lambda}}(\alpha)$.

For every state $s \in S$ and every vector of discount factors $\vec{\lambda} \in (0,1]^I$ let $x_{\vec{\lambda}}(s)$ be a stationary equilibrium in the game $G_{s,\vec{\lambda}}(\underline{v}_{\vec{\lambda}}(s))$. The equilibrium payoff of each player i in a strategic-form game is always at least her min–max value in the game (Exercise 3.5), which is at least her max–min value in the game. Hence

$$\underline{v}^i_{\lambda^i}(s) \leq u^i_{s,\vec{\lambda}}(x_{\vec{\lambda}}(s))$$

$$\leq \mathbf{E}_{x_{\vec{\lambda}}(s)}\left[\lambda^i r^i(s, x_{\vec{\lambda}}(s)) + (1-\lambda^i)\sum_{s' \in S} q(s' \mid s, x_{\vec{\lambda}}(s)) \underline{v}^i_{\lambda^i}(s')\right].$$

We will now have each player individually apply the strategy from the proof of Theorem 9.13. Fix $\epsilon > 0$ sufficiently small. Let $L \in \mathbb{N}$ and $C > 0$ be such that

$$\frac{d\underline{v}^i_\lambda(s)}{d\lambda} < \frac{C}{L}\lambda^{-\frac{L-1}{L}},$$

for every $\lambda > 0$ sufficiently small, every state $s \in S$, and every player $i \in I'$, and define a function $\psi \colon (0,1] \to \mathbb{R}$ by

$$\psi(\lambda) := \frac{C}{L}\lambda^{-\frac{L-1}{L}}.$$

(Compare this definition with Eq. (9.11).)

Consider the following strategy σ_*^i of player i, where λ_1 is sufficiently small.

- Let $\vec{\lambda}_1 \in (0,1]^I$ be the vector with all coordinates equal to λ_1. Define for every $t \in \mathbb{N}$,

$$d_1^i := D(\lambda_1),$$
$$d_{t+1}^i := \max\{D(\lambda_1), d_t^i + r^i(s_t,a_t^1,a_t^2) - \underline{v}_{\lambda_t^i}^i(s_{t+1}^i) + 4\epsilon\},$$
$$\lambda_{t+1}^i := D^{-1}(d_{t+1}^i),$$

where $D \colon [0,\lambda_0] \to \mathbb{R}$ is the function defined in Eq. (9.12).
- In stage t, play the mixed action $x_{\vec{\lambda}_t}^i(s_t)$, where $\vec{\lambda}_t = (\lambda_t^i)_{i \in I}$.

Let $\sigma_* = (\sigma_*^i)_{i \in I}$ denote the strategy profile in which each player i plays the strategy σ_*^i.

Fix for the moment a player $i \in I$ and a state $s \in S$. By the proof in Section 9.3, there is a $T_0 \in \mathbb{N}$ such that

$$\gamma_T^i(s;\sigma_*) \geq \underline{v}_0^i(s) - 9\epsilon. \tag{9.47}$$

for all $T \geq T_0$. For every two positive integers $k < T$ and every history $h_k \in H$ denote by

$$\gamma_{k,T}^i(s;\sigma_* \mid h_k) := \frac{1}{T-k+1}\mathbf{E}_{s,\sigma_*}\left[\sum_{t=k}^T r(s_t,a_t^1,a_t^2) \mid h_k\right]$$

the expected payoff of player i under the strategy profile σ_* at the initial state s between stages k and T, conditioned that the history up to stage k is h_k. If in Eq. (9.26) we sum up over $t = k, k+1, \ldots, T$ (instead of $t = 1, 2, \ldots, T$) we deduce that for every sufficiently large T (in particular, T should be larger than $\frac{6k\|r\|_\infty}{\epsilon}$, $\frac{D(\lambda_1)}{\epsilon}$, and $\frac{2(\|r\|_\infty)^2}{\epsilon^2\lambda_1}$),

$$\gamma_{k,T}^i(s;\sigma_* \mid h_t) = \frac{1}{T-k+1}\mathbf{E}_{s,\sigma_*}\left[\sum_{t=k}^T r(s_t,a_t^1,a_t^2) \mid h_t\right]$$

$$\geq \underline{v}_0^i(s_t) - 8\epsilon - \frac{d_k}{T} \tag{9.48}$$

$$\geq \underline{v}_0^i(s_t) - 8\epsilon - \frac{6k\|r\|_\infty}{T}$$

$$\geq \underline{v}_0^i(s_t) - 9\epsilon.$$

By Eq. (9.48) the expected average payoff of player i between stages k and T is at least $\underline{v}_{\vec{\lambda}_1}^i(s) - 9\epsilon$, provided T is sufficiently large. Since in stage $k - 1$ the strategy σ^* plays an equilibrium in the auxiliary game $G_{s, \vec{\lambda}_{k-1}}(\underline{v}_{\vec{\lambda}_{k-1}}^-)$, and since $|\underline{v}_{\vec{\lambda}^i}^i(s_{k-1}) - \underline{v}_0^i(s_{k-1})| < \epsilon$, we have

$$\gamma_T^i(s; \sigma^* \mid h_{k-1}) \geq u^i(s, x_{\vec{\lambda}_1}^-(s)) - 11\epsilon.$$

for all sufficiently large T.

For each player $i \in I$, let σ_ϵ^{-i} be a uniform ϵ-punishment strategy against player i. In particular, for every strategy σ^i of player i and every sufficiently large T,

$$\gamma_T^i(s; \sigma^i, \sigma_\epsilon^{-i}) \leq \underline{v}_0^i(s) + \epsilon. \tag{9.49}$$

This strategy is a behavior strategy. By Kuhn's Theorem, the strategy σ_ϵ^{-i} is a mixed strategy, that is, a probability distribution over pure strategy profiles of players j with $j \in I \setminus \{i\}$.

Suppose we add to the model an impartial observer who observes the play and can privately and securely send messages to each player. Suppose also that the observer uses the following mechanism, which mimics the strategy profile σ.

- At the outset of the game, for each $i \in I$ the observer chooses one pure strategy profile $\sigma_{i,\epsilon} = (\sigma_{i,\epsilon}^j)_{j \neq i}$ according to the probability distribution σ_ϵ^{-i}, and sends to each player j her part in this pure strategy profile, namely, $\sigma_{i,\epsilon}^j$. The players will use this pure strategy if player i deviates.
- In stage 1, the observer chooses for each player i an action $\widehat{a}_1^i \in A^i(s_1)$ according to the probability distribution $\sigma_*^i(s_1)$, and informs each player privately which action $\widehat{a}_1^i$ was selected for him.
- At each stage $t > 1$, the observer chooses for each player i an action $\widehat{a}_t^i \in A^i(s_t)$ according to the probability distribution $\sigma_*^i(h_t)$, and informs each player privately which action $\widehat{a}_t^i$ was selected for him, as well as all the actions $(\widehat{a}_{t-1}^j)_{j \neq i}$ that she transmitted to the other players in the previous stage.

Suppose that each player $i \in I$ keeps a flag that indicates whether some player is punished, and if so, who. The flag is initialized to $\emptyset$ – at the outset of the game, no player is punished. Player i adopts the following[2] strategy τ_*^i:

- In stage 1, player i follows the action $\widehat{a}_1^i$ recommended to her by the observer.

- In stage $t > 1$, if the flag is $\emptyset$, player i compares the action a_{t-1}^j each player $j \neq i$ played in stage $t - 1$ to the action $\widehat{a}_{t-1}^j$ that the observer transmitted to player j at stage $t - 1$ (and which the observer sent to player i in stage t). If $a_{t-1}^j = \widehat{a}_{t-1}^j$ for every $j \neq i$, the value of the flag remains $\emptyset$. Otherwise, let j_* be the minimal index such that $a_{t-1}^{j_*} \neq \widehat{a}_{t-1}^{j_*}$. The flag is set to "Start punishing player j_* in stage t."

- In stage t, if the flag is $\emptyset$, player i plays the action $\widehat{a}_t^i$ transmitted to her by the observer.

- Otherwise, from stage t and on, player i forgets past play and starts following the pure strategy $\sigma_{j_*,\epsilon}^i$.

We argue that this construction is an ϵ-equilibrium in a suitable sense. If all players follow the strategy profile $\tau = (\tau^i)_{i \in I}$, then with the help of the observer the players implement the strategy profile σ_*, and by Eq. (9.47) we have

$$\gamma_T^i(s; \tau) \geq \underline{v}_0^i(s) - 11\epsilon, \quad \forall T \geq T_0. \tag{9.50}$$

Suppose that player i deviates, and in stage $t - 1$ after the history h_{t-1} she does not follow the action chosen for her by the observer. Since at each stage the observer reveals her recommendations in the previous stage, this deviation is detected in stage t, and from then on player i is punished by the other players at her uniform max–min value. Consequently, if we denote by σ^i the strategy taken by the deviator, we deduce from Eq. (9.49) that

$$\gamma_{t,t+T}^i(s; \sigma^i, \tau^{-i} \mid h_t) \leq \underline{v}_0^i(s_t) + \epsilon, \quad \forall T \geq T_0(\epsilon). \tag{9.51}$$

Eqs. (9.50) and (9.51) show that in the game with the observer, the strategy profile τ satisfies the following property: for every $\epsilon > 0$, every $t \in \mathbb{N}$, and every history $h_t \in H$ of length t, there is a $T_0 = T_0(\epsilon, t)$ such that for every player $i \in I$, every strategy τ'^i in this game, and every initial state $s \in S$, we have

$$\gamma_T^i(s; \tau \mid h_t) \geq \gamma_T^i(s; \tau'^i, \tau^{-i} \mid h_t) - 12\epsilon, \quad \forall T \geq T_0.$$

---

[2] We denote a strategy of player i by τ^i and not σ^i, because this is a strategy in an extended game that includes the observer, and not in the original stochastic game.

In this sense, the mechanism we provided together with the strategy τ is a correlated equilibrium in the game Γ.

The fact that T_0 depends on t means that the number of stages required to effectively punish a deviator depends on the stage in which the deviation occurs. By working harder, one can show that this bound can be made uniform in σ^i. This is beyond the scope of this book, and the reader is referred to Solan and Vieille (2002) for more details.

9.6 Comments and Extensions

The concept of uniform value was defined in Mertens and Neyman (1981). The "Big Match" was introduced by Gillette (1957) and solved by Blackwell and Ferguson (1968). The proof that every two-player zero-sum stochastic game admits a uniform value is due to Mertens and Neyman (1981), and the proof presented here follows Sorin (2002). The application of the technique of Mertens and Neyman (1981) to extensive-form correlated equilibrium in multiplayer stochastic games is taken from Solan and Vieille (2002).

An extensive literature is devoted to algorithms for calculating the uniform value of two-player zero-sum stochastic games. Algorithms that calculate the uniform value in certain classes of games were developed in, for example, Condon (1992), Zwick and Paterson (1996), Jurdziński et al. (2008), Andersson and Miltersen (2009), Hansen et al. (2011), and Etessami et al. (2019). Using algorithms devised for the theory of semi-algebraic sets, one can approximate the uniform value (see Chatterjee, Alfaro, and Henzinger, 2008) and uniformly ϵ-optimal strategies (see Solan and Vieille, 2010). The uniform value can be calculated efficiently for some classes of stochastic games, see, for example, Vrieze et al. (1983), Filar and Raghavan (1984), and Breton (1991), and the references therein. Attia and Oliu-Barton (2020) characterized the uniform value as a solution of a certain equation. Using this characterization, Oliu-Barton (2020) provided an algorithm to calculate the uniform value in complexity that is exponential in the size of the sets of states and actions.

When the players adopt uniformly ϵ-optimal strategies, they guarantee that the long-run average payoff is close to the value. This does not rule out the possibility that the payoff fluctuates along the play: during some long blocks of stages the payoff is high, in other long blocks of stages the payoff is low, and the blocks are arranged in such a way that the average payoff is close to the value. Catoni, Oliu-Barton, and Ziliotto (2021) proved that if both players adopt uniformly optimal strategies, then for every fixed sufficiently large positive integer m, the expected average payoff in stages $n, n+1, \ldots, n+m-1$ is close to the value. Thus, the short-run average payoff does not vary much along the play. This is an extension of the analogous result for Markov decision problems, proven by Sorin et al. (2010).

There are other concepts of value that were studied in the literature; see, for example, Maitra and Sudderth (1993, 1998, 2012). Two examples are the following: the value at the initial state s, denoted $v(s)$, is said to exist if

$$
v(s) = \sup_{\sigma^1 \in \Sigma^1} \inf_{\sigma^2 \in \Sigma^2} \limsup_{T \to \infty} \mathbf{E}_{s, \sigma^1, \sigma^2} \left[\frac{1}{T} \sum_{t=1}^{T} u(s_t, a_t) \right]
$$
$$
= \inf_{\sigma^2 \in \Sigma^2} \sup_{\sigma^1 \in \Sigma^1} \limsup_{T \to \infty} \mathbf{E}_{s, \sigma^1, \sigma^2} \left[\frac{1}{T} \sum_{t=1}^{T} u(s_t, a_t) \right],
$$

in one case, and

$$
v(s) = \sup_{\sigma^1 \in \Sigma^1} \inf_{\sigma^2 \in \Sigma^2} \mathbf{E}_{s, \sigma^1, \sigma^2} \left[\limsup_{T \to \infty} u(s_t, a_t) \right]
$$
$$
= \inf_{\sigma^2 \in \Sigma^2} \sup_{\sigma^1 \in \Sigma^1} \mathbf{E}_{s, \sigma^1, \sigma^2} \left[\limsup_{T \to \infty} u(s_t, a_t) \right],
$$

in the other.

According to its definition, a strategy is a mapping from past play to mixed actions. Thus, the amount of information the player needs to store so that she can follow a strategy is not bounded. One strand of literature studies players who have bounded memory. In Exercise 9.5, we explore this issue in the context of the "Big Match." Additional results regarding the "Big Match" can be found in Fortnow and Kimmel (1998). Uniformly ϵ-optimal strategies that have one bit of memory and in addition have access to a clock that counts the stage number were studied by Hansen et al. (2018, 2021).

We assumed that players observe the past states that the play visited and the past actions that the other player took. In a more general model, the players do not observe this history. Rather, at every stage, each player observes a random signal, whose distribution depends on the current state and on the actions chosen by the players. This model is difficult to analyze because of two reasons. First, since a player does not necessarily know the action of the other player, she cannot use Bayes' rule[3] to update her belief about the state of nature, as was done in the setup of Markov decision problems in Chapter 1. Second, if the signals that the players observe are different, then each player does not

---

[3] Thomas Bayes (1701 – Tunbridge Wells, England, April 7, 1761) was an English statistician, philosopher, and Presbyterian minister, who formulated a specific case of the theorem that nowadays bears his name.

necessarily know the information that the other player receives. Special classes of stochastic games with signals have been studied, and the convergence of the discounted value and the existence of the uniform min–max value, the uniform max–min value, or the limsup value, have sometimes been established, see, for example, Coulomb (2003), Rosenberg et al. (2003, 2004, 2009), Renault (2006, 2012), Hörner et al. (2010), Gensbittel and Renault (2015), and Gimbert et al. (2016). Ziliotto (2016c) and Renault and Ziliotto (2020b) provided examples that show that in general the discounted value need not converge as the discount factor goes to 0. Rosenberg et al. (2009) and Gimbert et al. (2016) provided examples where the limsup value need not exist.

Exercise 9.7 is adapted from Kohlberg (1974). Exercise 9.19 is taken from Maschler et al. (2020) and is based on Liggett and Lippman (1969). Exercise 9.18 is based on Thuijsman and Raghavan (1997).

9.7 Exercises

Exercise 9.7 is used in the proof of Theorem 10.4.

1. For every $\lambda \in (0, 1]$, denote by x_λ^1 the unique stationary λ-discounted optimal strategy of Player 1 in the "Big Match." Calculate
   $C(\lambda, \widehat{\lambda}) := \inf_{x^2 \in X^2} \gamma_{\widehat{\lambda}}(s(0), x_\lambda^1, x^2)$. The quantity $C(\lambda, \widehat{\lambda})$ is the lowest payoff that can be attained when the discount factor is $\widehat{\lambda}$, yet Player 1 mistakenly thinks that the discount factor is λ.

2. Prove that Definitions 9.7 and 9.9 for the uniform value are equivalent.

3. Prove that if the uniform value at the initial state s exists, then it is equal to $\lim_{\lambda \to 0} v_\lambda(s)$ and to $\lim_{T \to \infty} v_T(s)$.

4. Consider a stochastic game $\Gamma = \langle S, I, (A^i(s))_{s \in S}^{i \in I}, q, (r^i)_{i \in I} \rangle$ and let $\Gamma' = \langle S, I, (A^i(s))_{s \in S}^{i \in I}, q, (r'^i)_{i \in I} \rangle$ be the stochastic game that differs from Γ only in its payoff function, and let $\epsilon \geq 0$. Prove that if the strategy profile σ_* is a uniform ϵ-equilibrium in Γ, then σ_* is a uniform $(\epsilon + 2\|r - r'\|_\infty)$-equilibrium in Γ'.

5. Prove the following for the "Big Match" game.

   (a) Player 1 has no deterministic strategy that uniformly guarantees more than 0.

   (b) Player 1 has no bounded recall strategy[4] that uniformly guarantees more than 0.

---

[4] For $k \in \mathbb{N}$, a strategy is *k-bounded recall* if the mixed action that is played at stage t depends only on the actions that were played in the past k stages and on the states that were visited in those stages. A strategy is *bounded recall* if it is k-bounded recall for some $k \in \mathbb{N}$.

6. Prove that in the "Big Match," Player 1 does not have a strategy that guarantees $\frac{1}{2}$: for every strategy σ^1 of Player 1 there exist $T_0 \in \mathbb{N}$ and a strategy σ^2 of Player 2 such that

$$\gamma_T(\sigma^1, \sigma^2) < \frac{1}{2}, \quad \forall T \geq T_0.$$

7. In this exercise, we generalize the construction presented in Section 9.2 for a class of games that is more general than the "Big Match" in three respects. First, Player 2 may have more than two actions, second, there may be absorbing entries in the top row of the matrix, and third, the probability of absorption in the various entries may be strictly less than 1.

Since in two-player zero-sum absorbing games[5] the play effectively terminates once the play leaves the initial nonabsorbing state, it is convenient to shorten the notation as follows:

- The sets of actions of the two players at the initial state are A^1 and A^2.
- The payoff at the initial state when the players play the action profile $a \in A^1 \times A^2$ is $r(a)$.
- The probability of moving from the initial state to some absorbing state when the players play the action profile $a \in A^1 \times A^2$ is $p_*(a)$.
- The expected absorbing payoff if at the initial state the players play the action profile $a \in A^1 \times A^2$ is $r_*(a)$.

Denote the probability of absorption under strategy profile $x \in \Delta(A^1) \times \Delta(A^2)$ by

$$p_*(x) := \sum_{a \in A} \left(\prod_{i=1}^{2} x^i(a^i) \right) p_*(a),$$

and the expected absorbing payoff under strategy profile $x \in \Delta(A^1) \times \Delta(A^2)$ by

$$r_*(x) := \frac{\sum_{a \in A} \left(\prod_{i=1}^{2} x^i(a^i) \right) p_*(a) r_*(a)}{p_*(x)},$$

which is defined whenever $p_*(x) > 0$. By convention, we set $p_*(x)r_*(x) = 0$ whenever $p_*(x) = 0$. We denote by t_* the stage in which the play moves to an absorbing state.

Consider an absorbing game where $A^1 = \{T, B\}$ and $r_*(T, a^2) \leq 0$ whenever $p_*(T, a^2) > 0$. Suppose that $\|r_*\|_\infty \leq 1$, and that there is a

[5] An absorbing game is a stochastic game with one nonabsorbing state, which is the initial state, see Definition 4.6.

mixed action $\widehat{x}^2 \in \Delta(A^2)$ that satisfies (a) $p_*(T, \widehat{x}^2) = 0$, (b) $p_*(B, \widehat{x}^2) > 0$, and (c) $r_*(B, \widehat{x}^2) = 0$.

In this exercise, we will prove that for every $M \geq 1$ there is a strategy $\sigma_M^1 \in \Sigma^1$ that satisfies the following properties:

(a) For every strategy $\sigma^2 \in \Sigma^2$,

$$\mathbf{E}_{\sigma^1, \sigma^2}\left[r_*^2(a_{t_*}^1, a_{t_*}^2)\mathbf{1}_{\{t_* < \infty\}}\right] \leq \frac{2}{M+1}, \tag{9.52}$$

that is, the expected unconditional absorbing payoff is at most $\frac{2}{M+1}$ whatever Player 2 plays.

(b) If Player 2 plays the stationary strategy $\widehat{x}^2$, the game is absorbed with probability 1:

$$\mathbf{P}_{\sigma^1, \widehat{x}^2}(t_* < \infty) = 1.$$

Define two sequences of random variables $(X_t)_{t \in \mathbb{N}}$ and $(k_t)_{t \in \mathbb{N}}$ as follows:

$$k_1 := 0,$$

$$k_t := \sum_{l=1}^{t-1} p_*(B, a_l) r_*^2(B, a_l), \quad \forall t \geq 2,$$

$$X_t := \begin{cases} 0, & \text{if } t_* > t, \\ r_*^2(B, a_{t_*}^2), & \text{if } t_* \leq t, \end{cases} \quad \forall t \geq 1.$$

For every $M \geq 1$, define a strategy $\sigma_M^1 \in \Sigma^1$ as follows:

- If $M + 1 + k_t \geq 1$, then at stage t play the action B with probability $\frac{1}{(M+1+k_t)^2}$ (and play T with the complementary probability $1 - \frac{1}{(M+1+k_t)^2}$).
- If $M + 1 + k_t < 1$, then at stage T play the action B with probability 1.

Do the following.

(a) Let $\vec{a}^2 = (a_1^2, a_2^2, \ldots)$ be a sequence of actions of Player 2. Show that the play under the strategy pair $(\sigma_M^1, \vec{a}^2)$ from stage 2 and on is similar to the play under the strategy $(\sigma_{M+p_*(B,a_1^2)r_*^2(B,a_1^2)}^1, (a_2^2, a_3^2, \cdots))$ from stage 1 and on.
(b) Prove that $\mathbf{E}_{\sigma_M^1, \vec{a}^2}[X_t] \leq 1$ for every $\vec{a}^2 \in (A^2)^\infty$, every $M \geq 1$, and every $t \in \mathbb{N}$.

(c) Prove that for every $\vec{a}^2 \in (A^2)^\infty$, every $M \geq 1$, and every $t \in \mathbb{N}$ we have

$$\mathbf{E}_{\sigma_M^1, \vec{a}^2}[X_t] \leq \tfrac{2}{M+1}.$$

(d) Deduce that Eq. (9.52) holds.
(e) Prove that for every $M \geq 1$ we have $\mathbf{E}_{\sigma_M^1, x_*^2}(t_* < \infty) = 1$.
(f) Prove that for every $M \geq 1$,

$$\mathbf{E}_{\sigma_M^1, x_*^2}[r_*^2(a_{t_*}^1, a_{t_*}^2)\mathbf{1}_{\{t_* < \infty\}}] = 0.$$

(g) Can you relate the strategies $(\sigma_M^1)_{M \in \mathbb{N}}$ to the strategies developed in Section 9.2 for the "Big Match"?

8. What is the limit of the discounted values $v_0(s(0)) := \lim_{\lambda \to 0} v_\lambda(s(0))$ of the following two-player zero-sum absorbing game?

	L	R
T	1	0
B	0 *	2 *

State $s(0)$

9. Let $\Gamma = \langle I, S, (A^i(s))_{s \in S}^{i \in I}, q, (r^i)_{i \in I}\rangle$ be a stochastic game. For every player $i \in I$, every state $s \in S$, and every discount factor $\lambda \in (0, 1]$, define:

$$\bar{v}_\lambda^i(s) := \inf_{\sigma^{-i} \in \Sigma^{-i}} \sup_{\sigma^i \in \Sigma^i} \gamma_\lambda^i(s; \sigma^i, \sigma^{-i}).$$

This quantity is called the *λ-discounted min–max value of player i at the initial state s*. It represents the maximum amount that player i can defend in the game. By Exercise 3.6, the λ-discounted min–max value of player i at the initial state s may differ from the λ-discounted max–min value of player i at the same initial state s.

(a) Prove that the function $\lambda \mapsto \bar{v}_\lambda^i(s)$ is semi-algebraic for every player $i \in I$ and every initial state $s \in S$.
(b) Deduce that the limit $\bar{v}_0^i(s) := \lim_{\lambda \to 0} \bar{v}_\lambda^i(s)$ exists.
(c) Prove that for every $i \in I$, there exist a strategy profile $\sigma^{-i} \in \Sigma^{-i}$ and a $\lambda_0 > 0$ such that

$$\gamma_\lambda^i(s; \sigma^i, (\sigma^j)_{j \neq i}) \leq \bar{v}_\lambda^i(s) + \epsilon, \quad \forall \lambda \in (0, \lambda_0), \forall \sigma^i, \forall s \in S.$$

10. Calculate the uniform value of the game that is displayed in Example 4.4, and find stationary uniformly ϵ-optimal stationary strategies of the two players.

11. Find a *stationary* uniformly ϵ-optimal strategy of Player 1 in the game in Exercise 6.20.

12. Prove that the uniform value of the following game in both states is 1.

	L	C	R
T	$-1_{(1,0)}$	$0_{(0,1)}$	5 $*$
B	$-1_{(0,1)}$	1 $*$	-1 $*$

State $s(0)$

	L	R
T	$1_{(0,1)}$	$1_{(1,0)}$
B	$0_{(1,0)}$	1 $*$

State $s(1)$

13. Find the uniform value of the following game with two nonabsorbing states.

	L	R			L	R
T	$0_{(0,1)}$	$1_{(0,1)}$		T	$1_{(1,0)}$	$0_{(\frac{1}{2},\frac{1}{2})}$
B	1 $*$	0 $*$		B	0 $*$	2 $*$

State $s(0)$ State $s(1)$

14. A *quitting game* is an absorbing game $\Gamma = \langle I, (\{C^i, Q^i\})_{i \in I}, p_*, (r^i, r_*^i)_{i \in I} \rangle$, where each player has two actions, $A^i = \{C^i, Q^i\}$, interpreted as a continue action and a quit action, and the transition rule p_* satisfies $p_*(\vec{C}) = 0$ and $p_*(a) = 1$ for every action profile $a \in A \setminus \{\vec{C}\}$, where $\vec{C} = (C^i)_{i \in I}$.

Let Γ be a quitting game that satisfies the following property: the absorbing payoff of a player depends only on her action and on the number of players who quit; that is, there are two functions $a, b: \{0, 1, \ldots, n\} \to \mathbb{R}$ such that $r_*^i(Q^J, C^{J^c}) = a(|J|)$ if $i \notin J$ and $r_*^i(Q^J, C^{J^c}) = b(|J|)$ if $i \in J$.

Prove that there exists a pure stationary strategy profile that is uniformly ϵ-optimal for every $\epsilon > 0$.

15. In this exercise, we write down the strategy that is constructed in the proof of Theorem 9.13 for a specific game. Consider the following two-player zero-sum absorbing game.

	L	R
T	0	1 *
B	1 *	0 *

State $s(0)$

(a) Calculate the discounted optimal strategies of both players.
(b) Write down the functions ψ and D.
(c) Write down the optimal strategy of Player 1 that is constructed in the proof of Theorem 9.13, and explain why it is a stationary strategy.
(d) Explain why the optimal strategy of Player 2 that is constructed in the proof of Theorem 9.13 is not stationary.

16. A two-player zero-sum stochastic game is *recursive* if the payoff in all nonabsorbing states is 0. The game is *positive* if the payoff in all absorbing states is positive.

(a) Prove that in positive recursive games, if the uniform value in all nonabsorbing states is positive, then Player 1 has a stationary uniformly ϵ-optimal strategy for every $\epsilon > 0$.
(b) Prove that even when the uniform value in some states is 0, Player 1 has a stationary uniformly ϵ-optimal strategy for every $\epsilon > 0$.

Hint: For item (a), consider the strategy constructed in the proof of Theorem 9.13. For item (b), consider an auxiliary game that is similar to Γ, except that the nonabsorbing states in Γ whose uniform value is 0 are turned into absorbing states with absorbing payoff 0.

17. The goal of this exercise is to improve our understanding of the proof of Theorem 9.13, and uses the notations presented in Section 9.3. Let $\Gamma = \langle \{1,2\}, S, (A^1(s), A^2(s))_{s \in S}, q, r \rangle$ be a two-player zero-sum stochastic game. Let $\sigma = (\sigma^1, \sigma^2)$ be a strategy profile that satisfies the following condition:

$$\mathbf{E}_{s,\sigma}\left[\lambda_t r(s_t, a_t^1, a_t^2) + (1 - \lambda_t)v_{\lambda_t}(s_{t+1}) \mid \mathcal{H}_t\right] = v_{\lambda_t}(s_t),$$

$$\forall t \in \mathbb{N}, s \in S.$$

(a) Prove that Claim 9.20 holds with equality.

(b) Prove that $C_t^1 \leq \lambda_t(d_{t+1} - d_t)$. (Compare this with Claim 9.21.)

(c) Prove that $C_t^3 \geq \lambda_t(d_{t+1} - d_t) - 6\lambda_t M \cdot \mathbf{1}_{\{\lambda_{t+1}=\lambda_1\}} - 4\epsilon\lambda_t$.

(d) Prove that

$$\mathbf{E}_{s,\sigma^1,\sigma^2}\left[\sum_{t=1}^T r(s_t, a_t^1, a_t^2)\right]$$

$$\leq \mathbf{E}_{s,\sigma^1,\sigma^2}\left[\sum_{t=1}^T v_{\lambda_t}(s_{t+1})\right]$$

$$+ \mathbf{E}[d_{T+1} - d_1] + 4\,T\epsilon + 2M\mathbf{E}\left[\sum_{t=1}^\infty \mathbf{1}_{\{\lambda_t=\lambda_1\}}\right].$$

(Compare this equation with Eq. (9.26).)

(e) Explain why we cannot conclude that $\gamma_T(s;\sigma)$ is close to $v_0(s)$ for every T sufficiently large.

18. A stochastic game $\Gamma = \langle I, S, (A^i(s))_{s\in S}^{i\in I}, q, (r^i)_{i\in I}\rangle$ has *perfect information* if in all states the action set of all players but one contain one action; that is, for every state $s \in S$ there is a player i such that $|A^j(s)| = 1$ for every $j \in I \setminus \{i\}$. In this exercise, we prove that two-player nonzero-sum stochastic games admit a uniform equilibrium payoff.

(a) Prove that in stochastic games with perfect information for each player the uniform min–max value and the uniform max–min value coincide at all initial states: we have

$$\overline{v}_0^i(s) = \underline{v}_0^i(s), \quad \forall s \in S, \forall i \in I,$$

where the uniform max–min value is defined by

$$\underline{v}_0^i(s) := \lim_{\lambda \to 0} \underline{v}_\lambda^i(s), \quad \forall i \in I, \forall s \in S.$$

(b) Prove that for each player $i \in I$ there is a pure stationary strategy σ_0^i that guarantees $\underline{v}^i(s) - \epsilon$, for every $s \in S$ and every $\epsilon > 0$.

(c) Prove that for each player $i \in I$ there is a strategy $\widehat{\sigma}_{[i]}^{-i} \in \Sigma^{-i}$ that ensures that player i's payoff at every initial state s is at most $\overline{v}^i(s) + \epsilon$: for every $\epsilon > 0$ there is $T_0 \in \mathbb{N}$ such that

$$\gamma_T^i(s;\sigma^i, \sigma^{-i}) \leq \overline{v}_0^i(s) + \epsilon, \quad \forall s \in S, \forall T \geq T_0.$$

(d) Denote by τ^i the first stage in which player i deviates from σ_0^i; let $\tau := \min\{\tau^i, i \in I\}$ be the first stage in which some player deviates from $\sigma_0 := (\sigma_0^i)_{i \in I}$; and let $i_* := \min\{i \in I : \tau = \tau^i\}$ be a player who deviated first from σ_0. Define the following strategy $\sigma_*^i \in \Sigma^i$:

- Until stage τ follow the strategy σ_0^i.
- At stage $\tau + 1$ forget past play, and start following the strategy $\widehat{\sigma}_{[i]}^{-i_0}$.

Prove that the strategy profile $\sigma_* := (\sigma_*^i)_{i \in I}$ is a uniform 2ϵ-equilibrium.

19. In this exercise, we prove the existence of a sequence $(x_t)_{t=1}^{\infty}$ of zeros and ones satisfying

$$\liminf_{T \to \infty} \frac{\sum_{k=1}^{T} x_k}{T} < \liminf_{\lambda \to 0} \lambda \sum_{t=1}^{\infty} (1 - \lambda)^{t-1} x_t. \tag{9.53}$$

Let $(q_t)_{t \in \mathbb{N}}$ be a sequence of natural numbers. Define a sequence $(p_t)_{t \in \mathbb{N}}$ as follows:

$$p_1 := 0,$$
$$p_t := q_1 + q_2 + \cdots + q_{t-1}.$$

Define a sequence $(x_t)_{t \in \mathbb{N}}$ as follows:

$$x_t = \begin{cases} 1, & \text{if there exists } k \text{ such that } 2p_k < t \le 2p_k + q_k, \\ 0, & \text{otherwise.} \end{cases}$$

In words, in the sequence $(x_t)_{t \in \mathbb{N}}$ the first q_1 elements equal 1, the next q_1 elements equal 0, the next q_2 elements equal 1, the next q_2 elements equal 0, and so on.

(a) Prove that $\liminf_{T \to \infty} \frac{\sum_{k=1}^{T} x_k}{T} = \frac{1}{2}$.
(b) Denote $A(\lambda) = \lambda \sum_{t=1}^{\infty} (1 - \lambda)^{t-1} x_t$. Prove that

$$A(\lambda) = \sum_{k=1}^{\infty} (1 - \lambda)^{2p_k} (1 - (1 - \lambda)^{q_k}).$$

(c) Denote $\alpha_k = (1 - \lambda)^{p_k} - (1 - \lambda)^{p_{k+1}}$ for every $k \in \mathbb{N}$. Using item (b), prove that

$$A(\lambda) = \frac{1}{2} \left(\sum_{k=1}^{\infty} (\alpha_k)^2 + 1 \right).$$

(d) Let $\epsilon \in \left(0, \frac{1}{4}\right)$, and denote $c := \frac{\ln(\epsilon)}{\ln(1-\sqrt{\epsilon})}$. Prove that $c > 2$.

(e) Suppose that the sequence $(q_t)_{t \in \mathbb{N}}$ satisfies $q_k > \frac{2p_k}{c-2}$ for every $k \in \mathbb{N}$. Define

$$a_k := \frac{|\ln(1-\sqrt{\epsilon})|}{q_k}, \qquad b_k := \frac{|\ln(\epsilon)|}{2p_k}. \tag{9.54}$$

Prove that $\lim_{k \to \infty} b_k = 0$, and that for every $k \in \mathbb{N}$, (a) $b_{k+1} < b_k$, (b) $cq_k > 2p_k + 2q_k$, and (c) $a_k < b_{k+1}$.

(f) Prove, with the aid of item (e), that $\bigcup_{k \in \mathbb{N}}(a_k, b_k) = (0, \infty)$. Deduce that for every $\lambda \in (0, 1]$ there exists a $k(\lambda) \in \mathbb{N}$ such that $a_{k(\lambda)} \le |\ln(1-\lambda)| < b_{k(\lambda)}$.

(g) Using Eq. (9.54), prove that $\epsilon < (1-\lambda)^{2p_{k(\lambda)}}$, and $1 - \sqrt{\epsilon} \ge (1-\lambda)^{q_{k(\lambda)}}$. Deduce that

$$(\alpha_k)^2 = (1-\lambda)^{2p_{k(\lambda)}}(1 - (1-\lambda)^{q_{k(\lambda)}})^2 > \epsilon^2.$$

(h) Deduce, with the aid of item (c), that $\liminf_{\lambda \to 0} A(\lambda) \ge \frac{\epsilon^2+1}{2}$.

(i) Deduce that Eq. (9.53) holds for the sequence $(x_t)_{t \in \mathbb{N}}$ defined in item (d).

(j) Construct a sequence $(y_t)_{t=1}^{\infty}$ of zeros and ones satisfying

$$\limsup_{T \to \infty} \frac{\sum_{t=1}^{T} y_t}{T} > \limsup_{\lambda \to 0} \lambda \sum_{t=1}^{\infty} (1-\lambda)^{t-1} y_t.$$

Such a sequence was used in Example 9.5.

20. Prove that the min–max value of player i in the strategic-form game $G_{s,\lambda}(\overline{v}_\lambda)$ is $\overline{v}_\lambda^i(s)$.

21. Using the Average Cost Optimality Equation, calculate the uniform value at each state of the following two-player zero-sum stochastic game with two nonabsorbing states.

	L	R		L	R
T	$3_{(1,0)}$	$-1_{(0,1)}$	T	$2_{(1,0)}$	$-1_{(0,1)}$
B	$0_{(0,1)}$	$2_{(1,0)}$	B	$0_{(0,1)}$	$3_{(1,0)}$

State $s(0)$ State $s(1)$

22. Consider the following two-player zero-sum stochastic game with two states.

	L	R
T	$-1_{(1,0)}$	$0_{(1,0)}$
B	$0_{(1,0)}$	$-3_{(0,1)}$

State $s(0)$

	L	R
T	$1_{(0,1)}$	$0_{(0,1)}$
B	$0_{(0,1)}$	$3_{(1,0)}$

State $s(1)$

Do the following.

(a) Prove that $v_\lambda(s(0)) = -v_\lambda(s(1))$ for every $\lambda \in (0, 1]$.

(b) Prove that $v_0(s(0)) = v_0(s(1)) = 0$.

(c) For each $\lambda \in (0, 1]$, calculate the λ-discounted value at each initial state.

(d) Write down the Average Cost Optimality Equation for this game and solve it. Verify that the result that you obtained agrees with Eq. (9.34).

(e) Explain which of the results proved in Section 9.4 applies for this game.

10

The Vanishing Discount Factor
Approach and Uniform Equilibrium
in Absorbing Games

In this chapter, we present a technique to study uniform equilibria in stochastic games, called the *vanishing discount factor approach*. This approach was developed to prove the existence of a uniform ϵ-equilibrium in two-player nonzero-sum absorbing games using a mapping $\lambda \mapsto x_\lambda$, which assigns a stationary λ-discounted equilibrium x_λ to every $\lambda \in (0, 1]$, and analyzing the asymptotic properties of this mapping as λ goes to 0. We will use this approach to show that every absorbing game in which the probability of absorption is positive, whatever the players play, has a stationary uniform 0-equilibrium, and that every two-player absorbing game has a uniform ϵ-equilibrium, which need not be stationary, for every $\epsilon > 0$. To prove the second result, we will show how statistical tests are used in the construction of uniform ϵ-equilibria.

10.1 Preliminaries

Recall that a state $s \in S$ in a stochastic game is *absorbing* if $q(s \mid s, a) = 1$ for every action profile $a \in A(s)$: once the game reaches this state, it never leaves it, whatever the players play. Recall also that an *absorbing game* is a stochastic game with a single nonabsorbing state.

As mentioned on Page 58, to study the existence of a uniform equilibrium payoff in absorbing games we can assume that the payoff in each absorbing state is constant: once the game reaches an absorbing state, the players' payoff in every stage is independent of their actions. It is therefore convenient to present an absorbing game as a vector $\Gamma = \langle I, (A^i)_{i \in I}, p_*, (r^i, r_*^i)_{i \in I} \rangle$, where

- $I = \{1, 2, \ldots, n\}$ is the set of players.
- A^i is the finite set of actions of player i in the nonabsorbing state, for every $i \in I$. Denote by $A = \prod_{i \in I} A^i$ the set of action profiles at the nonabsorbing state.

- $p_*: A \to [0,1]$ represents the transition rule: $p_*(a)$ is the probability that the game is absorbed when the players adopt the action profile a.
- $r: A \to \mathbb{R}^n$ is the nonabsorbing payoff function: $r^i(a)$ is the payoff to player $i \in I$ in the nonabsorbing state when the players adopt the action profile a.
- $r_*: A \to \mathbb{R}^n$ is the absorbing payoff function: $r_*^i(a)$ is the payoff to player $i \in I$ in each future stage, given that the game is absorbed when the players adopt the action profile a.

Denote by M the maximal payoff (in absolute values) in the game:

$$M := \max_{i \in I} \max\{\|r^i\|_\infty, \|r_*^i\|_\infty\}.$$

For every mixed action profile $x \in \prod_{i \in I} \Delta(A^i)$ define the following three quantities: First,

$$r(x) := \sum_{a \in A} \left(\prod_{i \in I} x^i(a^i)\right) r(a);$$

this is the expected stage payoff to player $i \in I$ when the players adopt the mixed action profile x. Second,

$$p_*(x) := \sum_{a \in A} \left(\prod_{i \in I} x^i(a^i)\right) p_*(a);$$

this is the probability of absorption in a single stage when the players adopt the mixed action profile x. A mixed action profile $x \in \prod_{i \in I} \Delta(A^i)$ is called *absorbing* if $p_*(x) > 0$. And third,

$$r_*(x) := \frac{\sum_{a \in A} \left(\prod_{i \in I} x^i(a^i)\right) p_*(a) r_*(a)}{\sum_{a \in A} \left(\prod_{i \in I} x^i(a^i)\right) p_*(a)} = \frac{\sum_{a \in A} \left(\prod_{i \in I} x^i(a^i)\right) p_*(a) r_*(a)}{p_*(x)};$$

this is the expected absorbing payoff when the players adopt the mixed action profile x, conditional that absorption occurs. The quantity $r_*(x)$ is defined only for absorbing mixed action profiles $x \in \prod_{i \in I} \Delta(A^i)$.

Note that the mapping r and the function p_* are multilinear, and in particular continuous, and the mapping r_* is continuous on the set $\{x \in \prod_{i \in I} \Delta(A^i): p_*(x) > 0\}$ of absorbing mixed action profiles. Note also that the function $p_* \cdot r_*$ is multilinear whenever it is defined; that is, for every player $i \in I$, every two mixed actions $x^i, x'^i \in \Delta(A^i)$, every $\beta \in [0,1]$, and every mixed action profile $x^{-i} \in \prod_{j \neq i} \Delta(A^j)$ we have

$$\beta p_*(x^i, x^{-i}) r_*(x^i, x^{-i}) + (1 - \beta) p_*(x'^i, x^{-i}) r_*(x'^i, x^{-i})$$
$$= p_*(\beta x^i + (1 - \beta) x'^i, x^{-i}) r_*(\beta x^i + (1 - \beta) x'^i, x^{-i}),$$

whenever $p_*(x^i, x^{-i}) > 0$ and $p_*(x'^i, x^{-i}) > 0$.

We now express the discounted payoff under stationary strategy profiles using these mappings. When the players use the stationary strategy profile x, the λ-discounted payoff can be written as (see Theorem 5.2):

$$\gamma_\lambda(x) = \lambda r(x) + (1 - \lambda)\big(p_*(x) r_*(x) + (1 - p_*(x))\gamma_\lambda(x)\big).$$

Therefore,

$$\gamma_\lambda(x) = \frac{\lambda r(x) + (1 - \lambda) p_*(x) r_*(x)}{\lambda + (1 - \lambda) p_*(x)}. \tag{10.1}$$

Setting

$$\alpha_\lambda(x) := \frac{(1 - \lambda) p_*(x)}{\lambda + (1 - \lambda) p_*(x)}, \tag{10.2}$$

we obtain that

$$\gamma_\lambda(x) = (1 - \alpha_\lambda(x)) r(x) + \alpha_\lambda(x) r_*(x). \tag{10.3}$$

That is, the λ-discounted payoff is a convex combination of the expected stage payoff $r(x)$ and the expected absorbing payoff $r_*(x)$, with weights that depend on the discount factor λ and the per-stage probability of absorption $p_*(x)$.

Suppose now that we are given a semi-algebraic mapping $\lambda \mapsto x_\lambda$ that assigns a stationary strategy profile $x_\lambda \in \prod_{i \in I} \Delta(A^i)$ to every $\lambda \in (0, 1]$. Since this mapping is semi-algebraic, the limit $x_0 := \lim_{\lambda \to 0} x_\lambda$ exists. Thanks to the continuity of r and p_*,

$$r(x_0) = \lim_{\lambda \to 0} r(x_\lambda),$$
$$p_*(x_0) = \lim_{\lambda \to 0} p_*(x_\lambda).$$

If $p_*(x_0) > 0$, then $r_*(x_0)$ is well defined and

$$r_*(x_0) = \lim_{\lambda \to 0} r_*(x_\lambda).$$

The function $(\lambda, x) \mapsto \alpha_\lambda(x)$ is semi-algebraic, hence so is the function $\lambda \mapsto \alpha_\lambda(x_\lambda)$ (Exercise 6.2). Set

$$\alpha_0 := \lim_{\lambda \to 0} \alpha_\lambda(x_\lambda).$$

By Eq. (10.3),

$$\lim_{\lambda \to 0} \gamma_\lambda(x_\lambda) = (1 - \alpha_0) r(x_0) + \alpha_0 \lim_{\lambda \to 0} r_*(x_\lambda). \tag{10.4}$$

	L	R
T	1	0
B	0 *	1 *

Figure 10.1 The "Big Match."

Thus, the limit $\lim_{\lambda \to 0} \gamma_\lambda(x_\lambda)$ of the discounted payoff is equal to $r(x_0)$ if $\alpha_0 = 0$, to the limit of the absorbing payoff $\lim_{\lambda \to 0} r_*(x_\lambda)$ if $\alpha_0 = 1$, and to a convex combination of these two quantities if $\alpha_0 \in (0, 1)$.

Since the function $\lambda \mapsto p_*(x_\lambda)/\lambda$ is semi-algebraic, the limit $\lim_{\lambda \to 0} p_*(x_\lambda)/\lambda$ exists (it may be ∞). The following result relates this limit to α_0. Its proof follows from the definition of α_λ (Eq. (10.2)).

Theorem 10.1   *Let* $\Gamma = \langle I, (A^i)_{i \in I}, p_*, (r^i, r^i_*)_{i \in I} \rangle$ *be an absorbing game and let* $\lambda \mapsto x_\lambda$ *be a semi-algebraic mapping that assigns a stationary strategy profile to every* $\lambda \in (0, 1]$. *The following statements hold:*

1. $\alpha_0 = 0$ *if and only if* $\lim_{\lambda \to 0} \frac{p_*(x_\lambda)}{\lambda} = 0$.
2. $\alpha_0 = 1$ *if and only if* $\lim_{\lambda \to 0} \frac{p_*(x_\lambda)}{\lambda} = \infty$.
3. $\alpha_0 \in (0, 1)$ *if and only if* $\lim_{\lambda \to 0} \frac{p_*(x_\lambda)}{\lambda} \in (0, \infty)$.

Example 10.2   Consider the "Big Match" game that is displayed in Figure 10.1.

Here we have

$$r(T, L) = r(B, R) = 1, \quad r(T, R) = r(B, L) = 0,$$
$$p_*(T, L) = p_*(T, R) = 0, \quad p_*(B, L) = p_*(B, R) = 1,$$
$$r_*(B, L) = 0, \quad r_*(B, R) = 1.$$

Suppose that $x^1_\lambda = \left[\frac{2}{2+\lambda}(T), \frac{\lambda}{2+\lambda}(B)\right]$ and $x^2_\lambda = \left[\frac{1}{3}(L), \frac{2}{3}(R)\right]$. Then

$$r(x_\lambda) = \frac{2}{3(2+\lambda)} + \frac{2\lambda}{3(2+\lambda)} = \frac{2(1+\lambda)}{3(2+\lambda)},$$
$$p_*(x_\lambda) = \frac{\lambda}{2+\lambda},$$
$$r_*(x_\lambda) = \frac{2}{3}.$$
$$\alpha_\lambda(x_\lambda) = \frac{(1-\lambda)\frac{\lambda}{2+\lambda}}{\lambda + (1-\lambda)\frac{\lambda}{2+\lambda}} = \frac{1-\lambda}{3},$$
$$\gamma_\lambda(x_\lambda) = \frac{2+\lambda}{3} \cdot \frac{2(1+\lambda)}{3(2+\lambda)} + \frac{1-\lambda}{3} \cdot \frac{2}{3} = \frac{4}{9}.$$

Note that $x_0^1 = [1(T)]$ and $x_0^2 = \left[\frac{1}{3}(L), \frac{2}{3}(R)\right]$, and therefore $p_*(x_0) = 0$ and $r_*(x_0)$ is not defined. Also, $\gamma_\lambda(x_0) = \frac{1}{3}$ for every $\lambda \in (0, 1]$, and therefore $\lim_{\lambda \to 0} \gamma_\lambda(x_0) \neq \lim_{\lambda \to 0} \gamma_\lambda(x_\lambda)$.   ♦

In the rest of the chapter, we will use the following notation: t_* is the stage in which the play moves to an absorbing state.

10.2 Uniform Equilibrium in Absorbing Games with Positive Absorbing Probability

In this section, we prove that in every absorbing game in which the probability of absorption is positive regardless of what the players play, there exists a stationary uniform 0-equilibrium.

Theorem 10.3   *Let* $\Gamma = \langle I, (A^i)_{i \in I}, p_*, (r^i, r_*^i)_{i \in I} \rangle$ *be an absorbing game, satisfying* $p_*(a) > 0$ *for every action profile* $a \in A$. *Then the game admits a stationary uniform* ϵ-*equilibrium, for every* $\epsilon > 0$.

Proof   **Step 1:**   Definition of a stationary strategy profile x_0.

By Corollary 8.14, there is a semi-algebraic mapping $\lambda \mapsto x_\lambda$ that assigns a stationary λ-discounted equilibrium x_λ to every discount factor $\lambda \in (0, 1]$. This semi-algebraic mapping will remain fixed throughout the proof. Set

$$x_0 := \lim_{\lambda \to 0} x_\lambda.$$

We will prove that x_0 is a stationary uniform ϵ-equilibrium in Γ, for every $\epsilon > 0$.

Step 2:   Calculation of the payoff under x_0.

Because the action sets are finite, there is a $\delta > 0$ such that $p_*(a) \geq \delta$ for every action profile $a \in A$. This implies that $p_*(x) \geq \delta$ for every mixed action profile $x \in \prod_{i \in I} \Delta(A^i)$. That is, the per-stage probability of absorption is bounded from below. It follows that the game is absorbed at a geometric rate:

$$\mathbf{P}_\sigma(t_* > T) \leq (1 - \delta)^{T-1}, \quad \forall T \in \mathbb{N}, \forall \sigma \in \prod_{i \in I} \Sigma^i. \tag{10.5}$$

By Theorem 10.1, this implies that $\lim_{\lambda \to 0} \alpha_\lambda(x_\lambda') = 1$ for every semi-algebraic function $\lambda \mapsto x_\lambda'$ that assigns a stationary strategy profile to each discount factor. In other words, for every such semi-algebraic function,

$$\lim_{\lambda \to 0} \gamma_\lambda(x_\lambda') = r_*(x_0'). \tag{10.6}$$

Denote

$$w := \lim_{\lambda \to 0} \gamma_\lambda(x_\lambda) \in \mathbb{R}^I.$$

By Eq. (10.6) applied to the semi-algebraic mapping $\lambda \mapsto x_\lambda$,

$$w = \lim_{\lambda \to 0} \gamma_\lambda(x_\lambda) = r_*(x_0).$$

Thus, when the players play the stationary strategy profile x_0, the game is absorbed at every stage with probability at least δ, and the expected absorbing payoff is $r_*(x_0) = w$. Fix $\epsilon > 0$, and let $T_\epsilon \in \mathbb{N}$ be sufficiently large such that

$$(1 - \delta)^{T_\epsilon - 1} \leq \epsilon.$$

Eq. (10.5) implies that

$$\left| \gamma_T^i(x_0) - w^i \right| \leq 2M\epsilon, \quad \forall T \geq T_\epsilon / \epsilon,$$

see Exercise 3.

Step 3: The stationary strategy profile x_0 is a uniform ϵ-equilibrium for every $\epsilon > 0$.

We will show that no player can profit more than $4M\epsilon$ in the T-stage game by deviating from the stationary strategy profile x_0, provided $T \geq T_\epsilon / \epsilon$.

By Eq. (10.6) and since x_λ is a λ-discounted equilibrium, for every $\lambda \in (0, 1]$ and every mixed action profile $x'^i \in \Delta(A^i)$ we have

$$\begin{aligned} r_*^i(x'^i, x_0^{-i}) &= \lim_{\lambda \to 0} \gamma_\lambda^i(x'^i, x_\lambda^{-i}) \\ &\leq \lim_{\lambda \to 0} \gamma_\lambda^i(x_\lambda) \\ &= r_*^i(x_0) = w^i. \end{aligned} \quad (10.7)$$

Now fix $\epsilon > 0$, a player $i \in I$, and a strategy $\sigma^i \in \Sigma^i$. For every history $h_t \in H$, under the mixed action profile $(\sigma^i(h_t), x_0^{-i})$ absorption occurs with probability at least δ, and by Eq. (10.7) the expected absorbing payoff is at most w^i. Eq. (10.5) implies that

$$\gamma_T^i(\sigma^i, x_0^{-i}) \leq w^i + 2M\epsilon, \quad \forall T \geq T_\epsilon / \epsilon.$$

Therefore, x_0 is a $4M\epsilon$-equilibrium in the T-stage game, for every $T \geq T_\epsilon / \epsilon$, and the claim follows. $\square$

10.3 Uniform Equilibrium in Two-Player Absorbing Games

In this section, we restrict attention to two-player absorbing games, and we prove that such games have uniform ϵ-equilibrium for every $\epsilon > 0$, which need not be stationary. We note that to date it is not known whether every absorbing game that involves at least four players admits a uniform ϵ-equilibrium, for every $\epsilon > 0$.

Theorem 10.4   *Every two-player absorbing game admits a uniform ϵ-equilibrium, for every $\epsilon > 0$.*

To prove the theorem, we will construct for every $\epsilon > 0$ a uniform ϵ-equilibrium that consists of three parts: an equilibrium play, statistical tests, and punishment strategies. We now explain the role of each of these parts.

1. The *equilibrium play* is the main part of the uniform ϵ-equilibrium, and it is used to generate high payoff to each player. It will usually consist of a mixed action profile that the players adopt in every stage.
2. Players may be able to deviate from the equilibrium play and profit. To dissuade them from doing so, each player will check that the other player follows the equilibrium play. There will be two ways in which players check each other:

   • If under the equilibrium play the game is supposed to be absorbed, then there is a $T_* \in \mathbb{N}$ such that, up to stage T_*, the game should be absorbed with high probability. If the game is not absorbed by stage T_*, then one may deduce that some player deviated.
   • To ensure that no player deviates from the stationary strategy she should follow, each player will verify that the distribution of the realized actions of the other player is close to the mixed action she should play. By the strong law of large numbers, the difference between these two quantities is small with high probability. If it turns out that at some stage the difference is high for one of the players, the other player can conclude with high degree of confidence that the player did deviate.

     We will use two types of such *statistical tests*. In one, as described before, each player compares the distribution of the realized actions of the other player to the mixed action she should play. In the other, the players compare the expected average payoff to the one induced by the equilibrium play.

3. Once a deviation of a player is detected, the other player switches to a *punishment strategy*, which lowers the payoff of the deviator to her uniform min–max level.

If a player deviates by playing an action that leads to absorption, then the game is absorbed and the other player cannot punish the deviator. To ensure that the threat of punishment is effective, we will have to construct the equilibrium play in such a way that any deviation to an action that leads to absorption does not increase the payoff of the deviator relative to her payoff along the equilibrium play.

We now turn to the formal proof of the theorem.

Proof of Theorem 10.4   Let $\lambda \mapsto x_\lambda$ be a semi-algebraic mapping that assigns a λ-discounted stationary equilibrium x_λ to each discount factor $\lambda \in (0,1]$. Denote

$$x_0 := \lim_{\lambda \to 0} x_\lambda$$

and

$$w := \lim_{\lambda \to 0} \gamma_\lambda(x_\lambda) \in \mathbb{R}^2.$$

As in the proof of Theorem 10.3, for every action $a^1 \in A^1$ such that $p_*(a^1, x_0^2) > 0$ we have

$$u_*^1(a^1, x_0^2) \le w^1. \tag{10.8}$$

Indeed, since for every $\lambda \in (0,1]$ the stationary strategy profile x_λ is a λ-discounted equilibrium, and since $p_*(a^1, x_0^2) > 0$, we have

$$\begin{aligned} w^1 &= \lim_{\lambda \to 0} \gamma_\lambda^1(x_\lambda) \\ &\ge \lim_{\lambda \to 0} \gamma_\lambda^1(a^1, x_\lambda^2) \\ &= r_*^1(a^1, x_0^2), \end{aligned}$$

where the last inequality follows from Eq. (10.4) and Theorem 10.1(2).

The λ-discounted min–max values of the players are

$$v_\lambda^1 := \inf_{\sigma^2 \in \Sigma^2} \sup_{\sigma^1 \in \Sigma^1} \gamma_\lambda^1(\sigma^1, \sigma^2),$$

$$v_\lambda^2 := \inf_{\sigma^1 \in \Sigma^1} \sup_{\sigma^2 \in \Sigma^2} \gamma_\lambda^2(\sigma^1, \sigma^2).$$

The quantity v_λ^1 is the value of the two-player zero-sum absorbing game where the goal of Player 2 is to minimize Player 1's payoff. Similarly, the quantity v_λ^2 is the value of the two-player zero-sum absorbing game where the goal of Player 1 is to minimize Player 2's payoff. It follows that for each $i \in \{1,2\}$ the function $\lambda \mapsto v_\lambda^i$ is semi-algebraic, and therefore the two limits

$$v_0^i := \lim_{\lambda \to 0} v_\lambda^i, \quad i \in \{1,2\},$$

exist. By Theorem 9.13, for every $\epsilon > 0$, each player $i \in \{1, 2\}$ has a strategy that uniformly lowers player $(3 - i)$'s payoff to $v_0^{3-i} + \epsilon$. That is, there exists a strategy $\widehat{\sigma}_\epsilon^i \in \Sigma^i$ and a $T_0(\epsilon) \in \mathbb{N}$ such that

$$\gamma_T^{3-i}(\widehat{\sigma}_\epsilon^i, \sigma^{3-i}) \leq v_0^i + \epsilon, \quad \forall \sigma^{3-i} \in \Sigma^{3-i}, \forall T \geq T_0(\epsilon). \tag{10.9}$$

Every equilibrium payoff of a player is at least her min–max value (Exercise 3.5). Consequently,

$$\gamma_\lambda^i(x_\lambda) \geq v_\lambda^i, \quad i = 1, 2, \forall \lambda \in (0, 1].$$

Letting $\lambda \to 0$, we obtain

$$w^i = \lim_{\lambda \to 0} \gamma_\lambda^i(x_\lambda) \geq \lim_{\lambda \to 0} v_\lambda^i = v_0^i.$$

Fix $\epsilon > 0$. We will handle four cases separately, each with its own equilibrium play and own statistical test.

- Case 1: $p_*(x_0) > 0$. In this case under the equilibrium play the players adopt at every stage the mixed action profile x_0. Since $p_*(x_0) > 0$, if no player deviates the game will eventually be absorbed. We will show that no player can profit by deviating in a way that leads to absorption, so the only possible profitable deviations of a player may be adopting actions that cause the play to never absorb. To deter such deviations, if the game is not absorbed after some given large number of stages, each player starts punishing the other player at the uniform min–max value.

- Case 2: $p_*(x_0) = 0$ and $r^i(x_0) \geq w^i$ for each $i \in \{1, 2\}$. In this case under the equilibrium play the players adopt at every stage the mixed action profile x_0. As in Case 1, no player will be able to profit by deviating in a way that leads the game to absorption. To deter deviations under which the play never absorbs, each player will check whether the average payoff of the other player does not exceed the amount she should get, which is given by $r(x_0)$. If an increase in the average payoff of the other player is detected, that player is punished at her min–max level.

- Case 3: $p_*(x_0) = 0$ and $r^2(x_0) < w^2$. We will prove that there is $a^1 \in A^1$ such that (a) $p_*(a^1, x_0^2) > 0$, (b) $r_*^1(a^1, x_0^2) = w^1$, and (c) $r_*^2(a^1, x_0^2) = w^2$. That is, an action that yields high absorbing payoff to both players when adopted together with x_0^2.

  In this case, there is a uniform $4M\epsilon$-equilibrium where at every stage Player 2 selects the mixed action x_0^2, Player 1 selects the mixed action $(1 - \epsilon)x_0^1 + \epsilon \mathbf{1}_{a^1}$, and the players utilize statistical tests to ensure that none of them deviate. We will construct another uniform $(2 + 4M)\epsilon$-equilibrium in which the equilibrium play is not stationary, but rather uses the strategy we developed for the "Big Match," and was presented in Exercise 9.7. With this strategy profile, the test for detecting deviations will involve only deviations that lead the game to never absorb: if the play is not absorbed after a predetermined number of stages, both players will switch to a punishment strategy, each one against the other player.

- Case 4: $p_*(x_0) = 0$ and $r^1(x_0) < w^1$. This case is analogous to Case 3.

We turn to the formal proof, which will present in detail the construction of the equilibrium play and the statistical tests.

Case 1: $p_*(x_0) > 0$.

Step 1:   Choosing constants.

Let $T_* \in \mathbb{N}$ be sufficiently large so that $(1 - p_*(x_0))^{T_*} \leq \epsilon$. That is, when the players play the mixed action profile x_0, the game is absorbed before stage T_* with probability at least $1 - \epsilon$.

Step 2:   Defining a strategy profile σ_*.

Let σ_*^1 be the following strategy of Player 1:

a1) Play the mixed action x_0^1 in the first T_* stages.
a2) At stage T_* forget past play and start following the strategy $\widehat{\sigma}_\epsilon^1$; that is, if the game is not absorbed by stage T_*, punish Player 2.

Let σ_*^2 be the analogous strategy of Player 2:

b1) Play the mixed action x_0^2 in the first T_* stages.
b2) At stage T_* forget past play and start following the strategy $\widehat{\sigma}_\epsilon^2$.

Points (a1) and (b1) describe the equilibrium play. The tests for detecting deviation (which here are not statistical) are described in points (a2) and (b2), and they call for punishment if the game is not absorbed by stage T_*.

Step 3:   The payoff under σ_*.

If the players follow the strategy pair $\sigma_* := (\sigma_*^1, \sigma_*^2)$, then absorption occurs until stage T_* with probability at least $1 - \epsilon$ in which case the expected absorbing payoff is $r_*(x_0)$. Consequently, for every $T \geq T_*/\epsilon$, we have

$$\gamma_T^i(\sigma_*) \geq (1 - \epsilon)\frac{T_* \cdot (-M) + (T - T_*)r_*^i(x_0)}{T} - M\epsilon \qquad (10.10)$$

$$\geq r_*^i(x_0) - 2M\epsilon,$$

where the term $-M\epsilon$ bounds the contribution to the expected payoff of the event that the game is not absorbed by stage T_*, the term $(1 - \epsilon)\frac{T_* \cdot (-M)}{T}$

bounds the contribution to the expected payoff in the first T_* stages of the event that the game is absorbed by stage T_*, and the term $(1 - \epsilon)\frac{(T-T_*)r_*^i(x_0)}{T}$ is the contribution to the expected payoff between stages $T^* + 1$ and T of the event that the game is absorbed by stage T.

Step 4: σ_* **is a** $5M\epsilon$**-equilibrium in the** T**-stage game.**

We now argue that no player $i \in \{1, 2\}$ can profit more than $5M\epsilon$ in the T-stage game by deviating, provided T is sufficiently large. We prove this claim for Player 1. The proof for Player 2 is analogous. Let then σ^1 be any strategy of Player 1.

If at some stage until stage T_* Player 1 plays an action a^1 such that $p_*(a^1, x_0^2) > 0$, then, by Eq. (10.8), $r_*^1(a^1, x_0^2) \leq w^1 = r_*^1(x_0)$. It follows that if the game is absorbed by stage T_*, then the expected absorbing payoff is at most $r_*^1(x_0) = w^1$. If the game is not absorbed by stage T_*, then, since at stage T_* Player 2 starts following a punishment strategy against Player 1, for every $T' \geq T_0(\epsilon)$ (see Eq. (10.9)), the expected payoff of Player 1 between stages $T_* + 1$ and $T_* + T'$ is at most $v_0^1 + \epsilon \leq w^1 + \epsilon$. Thus, for every $T \geq T_1 := \max\left\{\frac{T_*}{\epsilon}, T_* + T_0(\epsilon)\right\}$ we have

$$\gamma_T^1(\sigma^1, \sigma_*^2)$$

$$\leq \frac{T_* \cdot (-M)}{T}$$

$$+ \frac{T - T_*}{T}\left(\mathbf{P}_{\sigma^1, \sigma_*^2}(t_* \leq T_*) \cdot r_*^1(x_0) + \mathbf{P}_{\sigma^1, \sigma_*^2}(t_* > T_*) \cdot (v_0^1 + \epsilon)\right)$$

$$\leq r_*^1(x_0) + 3M\epsilon. \tag{10.11}$$

Together with Eq. (10.10), Eq. (10.11) implies that σ_* is indeed a T-stage $5M\epsilon$-equilibrium, for every $T \geq T_1$, as desired.

Case 2: $p_*(x_0) = 0$ and $r^i(x_0) \geq w^i$ for each $i \in \{1, 2\}$.

Step 1: Choosing constants.

For $t \in \mathbb{N}$, let $\bar{r}_t$ be the average payoff until stage t:

$$\bar{r}_t := \frac{1}{T}\sum_{t=1}^{T} r(a_t^1, a_t^2) \in \mathbb{R}^2.$$

By the strong law of large numbers, if the players play the stationary strategy pair x_0, then for t sufficiently large $\bar{r}_t$ is close to $r(x_0)$; that is, there exists a $T_* \in \mathbb{N}$ such that

$$\mathbf{P}_{x_0}\left(\|\bar{r}_t - r(x_0)\|_\infty \leq \epsilon, \quad \forall t \geq T_*\right) \geq 1 - \epsilon$$

for all $t \geq T_*$. Let τ be the first stage after stage T_* in which the average payoff is far from $r(x_0)$, that is,

$$\tau := \min\{t \geq T_* : \|\bar{r}_t - r(x_0)\|_\infty > \epsilon\}.$$

Then

$$\mathbf{P}_{x_0}(\tau = \infty) \geq 1 - \epsilon.$$

Step 2:   Defining a strategy profile σ_*.

Let σ_*^1 be the following strategy of Player 1:

a1) Play the mixed action x_0^1 until stage τ.
a2) At stage $\tau + 1$ forget past play and start following the strategy $\hat{\sigma}_\epsilon^1$.

Let σ_*^2 be the analogous strategy of Player 2:

b1) Play the mixed action x_0^2 until stage τ.
b2) At stage $\tau + 1$ forget past play and start following the strategy $\hat{\sigma}_\epsilon^2$.

Step 3:   The payoff under σ_*.

Since $\mathbf{P}_{x_0}(\tau < \infty) < \epsilon$,

$$\|\gamma_T(\sigma_*^1, \sigma_*^2) - r(x_0)\|_\infty \leq (1 + 2M)\epsilon.$$

for all $T \in \mathbb{N}$. Here, in the right-hand side, one ϵ originates from the fact that the average payoff of each player i is allowed to be higher than $r^i(x_0)$ by at most ϵ, and the term $2M\epsilon$ arises because the probability of the event that some player fails the statistical test is at most 2ϵ, and on this event the payoff is bounded by M.

Step 4: σ_* is a $(1 + 4M)$-equilibrium in the T-stage game.

We will now show that Player 1 cannot profit more than $2M\epsilon$ by deviating. The proof that Player 2 cannot profit more than $2M\epsilon$ by deviating is analogous. Fix then a strategy σ^1 of Player 1 and $T \geq \frac{T_0(\epsilon)}{\epsilon} + T_*$. To calculate $\gamma_T^1(\sigma^1, \sigma_*^2)$, we define several disjoint events, and calculate the expected payoff in the first T stages separately in each event.

1. Consider first the event $E_1 := \{t_* \leq T_*\}$; that is, the play is absorbed before stage T_*. By Eq. (10.8) we have $r_*^1(a^1, x_0^2) \leq w^1 \leq r^1(x_0)$: the expected payoff after stage t_* is at most w^1. Since $T \geq T_*/\epsilon$, it follows that

$$\mathbf{E}_{s,\sigma^1,\sigma_*^2}\left[\frac{1}{T}\sum_{t=1}^{T} r^1(s_t, a_t) \mid E_1\right] \leq w^1 + \epsilon. \tag{10.12}$$

2. Consider next the event $E_2 := \{T_* < t_* \leq \tau\}$; that is, the play is absorbed after stage T_* and before stage τ. Since $\tau \geq t_* > T_*$, the average payoff up to stage $\min\{t_*, \tau - 1\}$ is at most $r^1(x_0) + \epsilon$, and if $t_* = \tau$, then by Eq. (10.8) the payoff in stage t_* is at most w^1. Therefore,

$$\mathbf{E}_{s,\sigma^1,\sigma_*^2}\left[\frac{1}{T}\sum_{t=1}^{T} r^1(s_t,a_t) \mid E_2\right] \leq r^1(x_0) + \epsilon + \frac{w^1}{T}. \qquad (10.13)$$

3. Consider now the event $E_3 := \{\tau + T_0(\epsilon) \leq T, \tau < t_*\}$; that is, one of the players fails the statistical test, and there is enough time to punish her. The average payoff up to time $\tau - 1$ is at most $r^1(x_0) + \epsilon$, the payoff in stage τ is bounded by M, and since $T - \tau \geq T_0(\epsilon)$, by Eq. (10.9) the expected average payoff from stage $\tau + 1$ to stage T is at most $v_0^1 + \epsilon \leq w^1 + \epsilon = r^1(x_0) + \epsilon$. Consequently,

$$\mathbf{E}_{s,\sigma^1,\sigma_*^2}\left[\frac{1}{T}\sum_{t=1}^{T} r^1(s_t,a_t) \mid E_3\right] \leq r^1(x_0) + \epsilon + \frac{M}{T}. \qquad (10.14)$$

4. Consider now the event $E_4 := \{\tau \leq T < \tau + T_0(\epsilon), \tau < t_*\}$; that is, one of the players fails the statistical test, and there is not enough time to punish her. The average payoff up to time $\tau - 1$ is at most $r^1(x_0) + \epsilon$, and the payoff between stages τ and T is bounded by $\frac{T-\tau+1}{T} \cdot M$. Since $\frac{T_0(\epsilon)}{\epsilon} \leq T < \tau + T_0(\epsilon)$, it follows that $\frac{T-\tau+1}{T} \geq \epsilon + \frac{1}{T}$, and therefore

$$\mathbf{E}_{s,\sigma^1,\sigma_*^2}\left[\frac{1}{T}\sum_{t=1}^{T} r^1(s_t,a_t) \mid E_3\right] \leq r^1(x_0) + \left(\epsilon + \frac{1}{T}\right)M. \qquad (10.15)$$

5. Finally, consider the event $E_5 := \{T < \tau, t_*\}$. By the definition of τ,

$$\mathbf{E}_{s,\sigma^1,\sigma_*^2}\left[\frac{1}{T}\sum_{t=1}^{T} r^1(s_t,a_t) \mid E_3\right] \leq r^1(x_0) + \epsilon. \qquad (10.16)$$

Since the events E_1, E_2, E_3, E_4, and E_5 are disjoint and their union is H_∞, the set of all plays, it follows from Eqs. (10.12)–(10.16) that $\gamma_T^1(\sigma^1,\sigma_*^2) \leq r^1(x_0) + 2M\epsilon$, as claimed.

Case 3: $p_*(x_0) = 0$ and $r^2(x_0) < w^2$.

Step 1: There is an action $a_*^1 \in A^1$ such that $p_*(a_*^1,x_0^2) > 0$, $r_*^2(a_*^1,x_0^2) \geq w^2$ and $r_*^1(a_*^1,x_0^2) = w^1$.

For each $i = 1,2$ the mapping $\lambda \mapsto x_\lambda^i$ is semi-algebraic. Exercise 6.10 implies that for every $\lambda > 0$ sufficiently small the support of x_λ^i is independent of λ: there is a sufficiently small $\lambda_0 > 0$ such that $\text{supp}(x_\lambda^i) = \text{supp}(x_{\lambda'}^i)$ for every $\lambda, \lambda' \in (0, \lambda_0)$. Since $x_0^2 = \lim_{\lambda \to 0} x_\lambda^2$, we have $\text{supp}(x_0^2) \subseteq \text{supp}(x_\lambda^2)$ for every $\lambda \in (0, \lambda_0)$. By Theorem 5.5, the stationary strategy x_0^2 is a best response against x_λ^1 in the λ-discounted game, for every $\lambda \in (0, \lambda_0)$, that is,

$$\gamma_\lambda^2(x_\lambda^1, x_0^2) = \gamma_\lambda^2(x_\lambda), \quad \forall \lambda \in (0, \lambda_0).$$

It follows that

$$
\begin{aligned}
w^2 &= \lim_{\lambda \to 0} \gamma_\lambda^2(x_\lambda) \\
&= \lim_{\lambda \to 0} \gamma_\lambda^2(x_\lambda^1, x_0^2) \\
&= \left(1 - \lim_{\lambda \to 0} \alpha_\lambda(x_\lambda^1, x_0^2)\right) \lim_{\lambda \to 0} r^2(x_\lambda^1, x_0^2) + \lim_{\lambda \to 0} \alpha_\lambda(x_\lambda^1, x_0^2) \cdot \lim_{\lambda \to 0} r_*^2(x_\lambda^1, x_0^2) \\
&= \left(1 - \lim_{\lambda \to 0} \alpha_\lambda(x_\lambda^1, x_0^2)\right) r^2(x_0^1, x_0^2) + \lim_{\lambda \to 0} \alpha_\lambda(x_\lambda^1, x_0^2) \cdot \lim_{\lambda \to 0} r_*^2(x_\lambda^1, x_0^2).
\end{aligned}
$$

Since $r^2(x_0) < w^2$, this sequence of equations implies that

$$\lim_{\lambda \to 0} \alpha_\lambda(x_\lambda^1, x_0^2) > 0 \quad \text{and} \quad \lim_{\lambda \to 0} r_*^2(x_\lambda^1, x_0^2) \geq w^2. \tag{10.17}$$

By Theorem 10.1, the former condition implies in particular that $p_*(x_\lambda^1, x_0^2) > 0$ for every λ sufficiently small. Since $p_*(x_0) = 0$, the condition $p_*(x_\lambda^1, x_0^2) > 0$ implies that there are actions $a^1 \in \text{supp}(x_\lambda^1) \setminus \text{supp}(x_0^1)$ such that $p_*(a^1, x_0^2) > 0$.

Recall that $\text{supp}(x_\lambda^1)$ is independent of λ, provided $\lambda \in (0, \lambda_0)$. Let $\widehat{A}^1$ be the set of all actions in this support that are absorbing when played against x_0^2:

$$\widehat{A}^1 := \left\{a^1 \in A^1 : p_*(a^1, x_0^2) > 0, a^1 \in \text{supp}(x_\lambda^1) \ \forall \lambda \in (0, \lambda_0)\right\}.$$

Since $p_*(x_0^1, x_0^2) = 0$, it follows that $\widehat{A}^1 \subseteq \text{supp}(x_\lambda^1) \setminus \text{supp}(x_0^1)$, for every $\lambda \in (0, \lambda_0)$. By Theorem 5.5, for every $a^1 \in \widehat{A}^1$ we have

$$
\begin{aligned}
r_*^1(a^1, x_0^2) &= \lim_{\lambda \to 0} \gamma_\lambda^1(a^1, x_\lambda^2) \\
&= \lim_{\lambda \to 0} \gamma_\lambda^1(a^1, x_\lambda^2) = w^1.
\end{aligned}
$$

Since the function $x \mapsto p_*(x) r_*^2(x)$ is multilinear,

$$p_*(x_\lambda^1, x_0^2) \cdot r_*^2(x_\lambda^1, x_0^2) = \sum_{a^1 \in \widehat{A}^1} x_\lambda^1(a^1) \cdot p_*(a^1, x_0^2) \cdot r_*^2(a^1, x_0^2),$$

and therefore

$$r_*^2(x_\lambda^1, x_0^2) = \sum_{a^1 \in \widehat{A}^1} \frac{x_\lambda^1(a^1) \cdot p_*(a^1, x_0^2)}{p_*(x_\lambda^1, x_0^2)} \cdot r_*^2(a^1, x_0^2).$$

Taking the limit as λ goes to 0, we obtain, by Eq. (10.17), that

$$w^2 \leq \lim_{\lambda \to 0} r_*^2(x_\lambda^1, x_0^2) = \sum_{a^1 \in \widehat{A}^1} \lim_{\lambda \to 0} \left(\frac{x_\lambda^1(a^1) \cdot p_*(a^1, x_0^2)}{p_*(x_\lambda^1, x_0^2)} \right) r_*^2(a^1, x_0^2).$$

Note that $\sum_{a^1 \in \widehat{A}^1} \frac{x_\lambda^1(a^1) \cdot p_*(a^1, x_0^2)}{p_*(x_\lambda^1, x_0^2)} = 1$. Therefore, w^2 is smaller than or equal to a weighted average of $(r_*^2(a^1, x_0^2))_{a^1 \in \widehat{A}^1}$. It follows that there is an action $a_*^1 \in \widehat{A}^1$ such that $r_*^2(a_*^1, x_0^2) \geq w^2$.

Step 2: Choosing constants.

Fix $\epsilon > 0$. Exercise 9.7 implies[1] that there is a strategy $\widetilde{\sigma}^1 \in \Sigma^1$ that plays only actions in $\mathrm{supp}(x_0^1) \cup \{a_*^1\}$ and satisfies the following properties:

1. For every strategy $\sigma^2 \in \Sigma^2$,

$$\mathbf{E}_{\widetilde{\sigma}^1, \sigma^2} \left[r_*^2(a_0^1, a_{t_*}^2) \mathbf{1}_{\{t_* < \infty\}} \right] \leq w^2 + \epsilon. \tag{10.18}$$

2. If Player 2 plays the stationary strategy x_0^2, the game is absorbed with probability 1:

$$\mathbf{P}_{\widetilde{\sigma}^1, x_0^2}(t_* < \infty) = 1. \tag{10.19}$$

By Eq. (10.19), there exists $T_1 \in \mathbb{N}$ sufficiently large such that

$$\mathbf{P}_{\widetilde{\sigma}^1, x_0^2}(t_* < T_1) \geq 1 - \epsilon. \tag{10.20}$$

Step 3: Defining a strategy profile σ_*.

Let σ_*^1 be the following strategy of Player 1:

a1) Play the strategy $\widetilde{\sigma}^1$ until stage T_1.
a2) In stage $T_1 + 1$, forget past play and start following the punishment strategy $\widehat{\sigma}_\epsilon^1$.

Let σ_*^2 be the following strategy of Player 2:

b1) Play the mixed action x_0^2 until stage T_1.
b2) In stage $T_1 + 1$, forget past play and start following the punishment strategy $\widehat{\sigma}_\epsilon^2$.

---

[1] To apply Exercise 9.7, consider an auxiliary game in which Player 1 has two actions, T and B; the action B corresponds to the action a_*^1 in the original game, and the action T corresponds to the mixed action x_0^1 in the original game; that is, whenever Player 1 plays the action T in the auxiliary game, it is as if she plays the mixed action x_0^1 in the original game.

Step 4:   The payoff under σ_*.

We argue that for every $T \geq T_1/\epsilon$, the expected payoff in the T-stage game under σ_* is close to $r_*(a_*^1, x_0^2)$:

$$\|\gamma_T(\sigma_*) - r_*(a_*^1, x_0^2)\|_\infty \leq 3M\epsilon, \quad \forall\, T \geq T_1/\epsilon. \tag{10.21}$$

Indeed, since σ_*^2 plays the mixed action x_0^2 in every stage until stage T_1, if the play is absorbed at stage t, then the expected absorbing payoff at that stage is

$$\sum_{a^2 \in A^2} \frac{x_0^2(a^2) p_*(a^2) r_*(a^2)}{x_0^2(a^2) p_*(a^2)} = r_*(x_0^2).$$

Eq. (10.21) follows now from Eq. (10.20).

Step 5:   Player 1 cannot profit more than $(2+4M)\epsilon$ by deviating from σ_*^1.

If the play is not absorbed by stage T_1, Player 1 is punished at her uniform min–max level. Since Player 1 is the one who leads the game to absorption (by playing the action a_*^1), the only way in which Player 1 can possibly profit by deviating is by leading the game to absorption by some other action. In view of Eq. (10.7), such a deviation is not profitable.

We now formalize these ideas. Set $T_* := \frac{T_1 + T_0(\epsilon)}{\epsilon}$, and let $\sigma^1 \in \Sigma^1$ be any strategy of Player 1. We claim that $\gamma_T^1(\sigma^1, \sigma_*^2) \leq r_*^1(a_*^1, x_0^2) + 2M\epsilon$ for all $T \geq T_*$.

Define the following events:

- $E_1 := \{t_* \leq T_1, a_{t_*}^1 = a_*^1\}$: the game is absorbed up to stage t_* when Player 1 played a_*^1.
- $E_2 := \{t_* \leq T_1, a_{t_*}^1 \neq a_*^1\}$: the game is absorbed up to stage t_* when Player 1 played an action different than a_*^1.
- $E_3 := \{t_* > T_1\}$: the game is not absorbed by stage t_*.

We note that $\mathbf{P}_{\sigma^1, \sigma_*^2}(E_1 \cup E_2 \cup E_3) = 1$. By the definition of σ_*^2, on the event E_1 the expected absorbing payoff is $r_*(a_*^1, x_0^2)$, hence

$$\gamma_T(\sigma^1, \sigma_*^2 \mid E_1) \leq r_*^1(a_*^1, x_0^2) + M\epsilon, \quad \forall\, T \geq T_*.$$

As in Eq. (10.7), on the event E_2 the expected absorbing payoff is bounded from above by $r_*^1(x_0)$, hence

$$\gamma_T(\sigma^1, \sigma_*^2 \mid E_2) \leq r_*^1(a_*^1, x_0^2) + M\epsilon, \quad \forall\, T \geq T_*.$$

On the event E_3 Player 1 is punished, hence

$$\gamma_T(\sigma^1, \sigma_*^2 \mid E_3) \leq v_0^1 + 2\epsilon \leq r_*^1(a_*^1, x_0^2) + 2\epsilon, \quad \forall\, T \geq T_*.$$

It follows that

$$\gamma_T^1(\sigma^1, \sigma_*^2) \le r_*^1(a_*^1, x_0^2) + (2 + M)\epsilon,$$

as claimed.

Step 6: **Player 2 cannot profit more than $(1 + 4M)\epsilon$ by deviating from σ_*^2.**

The analysis for Player 2 is similar to the one for Player 1.

Define the following events:

- $E_1 := \{t_* \le T_1, a_{t_*}^1 \in \text{supp}. a_{t_*}^2\}$: the game is absorbed up to stage t_* when Player 1 played a_*^1.
- $E_2 := \{t_* \le T_1, a_{t_*}^1 \ne a_*^1\}$: the game is absorbed up to stage t_* when Player 1 played an action different than a_*^1.
- $E_3 := \{t_* > T_1\}$: the game is not absorbed by stage t_*.

As in Step 5, for every $T \ge T_* := \frac{T_1 + T_0(\epsilon)}{\epsilon}$ we have

$$\gamma_T^2(\sigma_*^1, \sigma^2 \mid E_1) \le r_*^2(x_0) + M\epsilon,$$
$$\gamma_T^2(\sigma_*^1, \sigma^2 \mid E_2) \le r_*^2(x_0) + 2M\epsilon,$$
$$\mathbf{P}_{\sigma_*^1, \sigma^2}(E_3) \le \epsilon.$$

It follows that for every $T \ge T_*$ we have

$$\gamma_T^2(\sigma_*^1, \sigma^2) \le r_*^2(x_0) + (1 + 2M)\epsilon,$$

and the claim follows. $\qquad\qquad\square$

10.4 Comments and Extensions

The vanishing discount factor approach was introduced by Vrieze and Thuijsman (1989) in their study of two-player nonzero-sum absorbing games (where they proved Theorem 10.4), and was then used to study various questions, such as the existence of initial states at which a uniform equilibrium payoff exists (Vieille, 2000d; see also Thuijsman and Vrieze, 1991); the existence of uniform equilibrium payoff in three-player absorbing games (Solan, 1999); and the existence of normal-form correlated equilibrium in multiplayer absorbing games (Solan and Vieille, 2002).

Whether every stochastic game admits a uniform equilibrium payoff is still an open problem. In Chapters 11–13, we will see three techniques that were used to prove the existence of a uniform equilibrium payoff in classes of stochastic games. There are many results on this problem that we will not cover, for example, Vieille (2000a, 2000b), Altman et al. (2008), and Flesch et al. (2008, 2009). Results concerning the existence of equilibrium payoffs under the limsup evaluation can be found in, for example, Nowak (2003b) and Jaśkiewicz and Nowak (2005, 2006). Simon (2016) discusses the main challenges on the way to solve this problem.

Algorithms that calculate uniform equilibrium payoffs have been devised for some classes of stochastic games, see, for example, Raghavan and Syed (2002) and Bourque and Raghavan (2014).

The uniform value is the limit of the discounted value as the discount factor goes to 0. As we noted in Exercise 8.9, the set of stationary discounted equilibria converges to a limit set as the discount factor goes to 0. A natural question is whether this limit set coincides with the set of uniform equilibrium payoffs. Sorin (1986) provided an example of a two-player nonzero-sum absorbing game where the set of stationary discounted equilibrium payoff contains a single point for every discount factor, yet this point is not a uniform equilibrium payoff (see the game in Exercise 8.4).

The model we are studying is played in discrete time. To capture situations in which the actual time that elapses between two consecutive stages is small, one can study stochastic games in continuous time in which time is not indexed by the set of positive integers $\mathbb{N}$ but by the set of nonnegative real numbers $\mathbb{R}_+$. While the existence of a uniform equilibrium payoff in stochastic games in discrete time is open, the existence of a uniform equilibrium in stochastic games in continuous time was established by Neyman (2017), by adapting the extensive-form correlated equilibrium that was constructed in Section 9.5 to the continuous-time framework. Levy (2013b) studied continuous-time stochastic games when players are restricted to Markovian strategies.

10.5 Exercises

Exercise 10.9 is used in the solution of Exercise 11.9.

1. In strategic-form games, the payoff function, which is defined on the set $\prod_{i \in I} \Delta(A^i)$, is multilinear. In this exercise, we will see that this is not the case in absorbing games. Write down the payoff function $\gamma : \prod_{i \in I} \Delta(A^i) \to \mathbb{R}$ of the following two-player zero-sum absorbing game. Is this function multilinear?

	L	R
T	0	1 *
B	1 *	0 *

2. In this exercise, we present the discounted payoff $\gamma_\lambda^i(x)$ in an absorbing game as a convex combination of player i's discounted payoffs for the various actions, $(\gamma_\lambda^i(a^i, x^{-i}))_{a^i \in \text{supp}(x^i)}$. Let x be a stationary strategy profile in a multiplayer absorbing game and let $i \in I$ be a player. Prove that

$$\gamma_\lambda^i(x) = \sum_{a^i \in A^i} x^i(a^i) \frac{\lambda + (1 - \lambda) p_*(a^i, x^{-i})}{\lambda + (1 - \lambda) p_*(x)} \gamma_\lambda^i(a^i, x^{-i}).$$

3. Complete the proof of Theorem 10.3: for every $\epsilon > 0$ there is $T_\epsilon \in \mathbb{N}$ such that

$$\left| \gamma_T^i(x_0) - w^i \right| \le 2M\epsilon, \quad \forall T \ge T_\epsilon / \epsilon.$$

4. Consider the following two-player absorbing game where the probability of absorption in all entries is positive: the probability of absorption in the entries (T, L), (T, R), and (B, L) is 1, the probability of absorption in the entry (B, R) is $\frac{1}{2}$, and both absorbing and nonabsorbing payoff in this entry is $(2, 0)$. Find a stationary strategy profile that is a uniform ϵ-equilibrium for every $\epsilon > 0$.

	L	R
T	2, 0 *	0, 1 *
B	0, 1 *	2, 0 $(\frac{1}{2})$*

5. (a) For every $\rho \in \left[\frac{1}{3}, \frac{1}{2}\right]$ and every $\epsilon > 0$ describe a uniform ϵ-equilibrium σ in the game in Exercise 8.4 such that $\|\gamma_T(\sigma) - (1 - \rho, 2\rho)\|_\infty \le \epsilon$ for all sufficiently large T.

   (b) Prove that in this game the set of uniform equilibrium payoffs is $\left\{(1 - \rho, 2\rho) \in \mathbb{R}^2 : \rho \in \left[\frac{1}{3}, \frac{1}{2}\right]\right\}$.

6. Let $\Gamma = \langle I, S, (A^i(s))_{s \in S}^{i \in I}, q, (r^i)_{i \in I} \rangle$ be a stochastic game that satisfies the following property: For every two states $s, s' \in S$ and every strategy profile $\sigma \in \Sigma$,

$$\mathbf{P}_{s,\sigma}(s_t = s' \text{ for some } t \in \mathbb{N}) = 1.$$

Prove that the game admits a uniform ϵ-equilibrium for every $\epsilon > 0$.

7. Let $\Gamma = \langle I, (A^i)_{i \in I}, p_*, (r^i, r_*^i)_{i \in I} \rangle$ be a multiplayer absorbing game that satisfies the following properties: (a) $r^i(a) = 0$ and $r_*^i(a) \geq 0$ for every $a \in A$, and (b) there are an $i \in I$ and a non-empty set of actions $\widehat{A}^i \subseteq A^i$ such that

$$p_*(a^i, a^{-i}) > 0 \iff a^i \in \widehat{A}^i.$$

Prove that the game Γ admits a uniform equilibrium payoff.

8. In this exercise, we will provide another way to implement Case 3 in the proof of Theorem 10.4. Suppose that the assumption of Case 3 in the proof of Theorem 10.4 holds. Prove that for every $\epsilon > 0$ there exists a uniform ϵ-equilibrium in which along the equilibrium path, at every stage Player 1 plays the mixed action x_0^1 and Player 2 plays the mixed action $(1 - \delta)x_0 + \delta a_*^2$.

Hint: Use Exercise 2.5.

9. In this exercise, we prove that every multiplayer positive recursive quitting game admits a normal-form correlated uniform equilibrium payoff.

A *quitting game* is an absorbing game $\Gamma = \langle I, (\{C^i, Q^i\})_{i \in I}, p_*, (r^i, r_*^i)_{i \in I} \rangle$ where each player has two actions, $A^i = \{C^i, Q^i\}$, interpreted as a continue action and a quit action, and the transition rule p_* satisfies $p_*(\vec{C}) = 0$ and $p_*(a) = 1$ for every action profile $a \in A \setminus \{\vec{C}\}$, where $\vec{C} = (C^i)_{i \in I}$.

Let Γ be a positive recursive quitting game. Player i is called *punishable* if there exists $j \neq i$ such that $r_*^i(Q^j, C^{-j}) \leq r_*^i(Q^i, C^{-i})$, that is, the payoff to player i when player j quits alone is not higher than player i's payoff if she, player i, quits alone. Such a player j is called a *punisher* of player i. If player i is punishable, choose one of the punishers of player i and denote her by j_i.

Let $z \in \mathbb{R}^I$ be the vector that is defined as follows:

$$z^i := \begin{cases} r_*^i(Q^i, C^{-i}) - 1, & \text{if } i \text{ is a punishable player}, \\ r_*^i(Q^i, C^{-i}) + 1, & \text{if } i \text{ is not a punishable player}. \end{cases}$$

Consider the quitting game Γ' that is similar to Γ, except that the nonabsorbing payoff is z. Let $\lambda \mapsto x_\lambda \in [0, 1]^I$ be a semi-algebraic

mapping that assigns to every $\lambda \in (0, 1]$ a λ-discounted stationary
equilibrium in the game Γ', and set $x_0 := \lim_{\lambda \to 0} x_\lambda$. Do the following:

(a) Prove that if there is no punishable player in Γ, then for every $\epsilon > 0$
  sufficiently small and every $i \in I$ the stationary strategy profile
  $([(1 - \epsilon)(C^i), \epsilon(Q^i)], C^{-i})$ is a stationary uniform 0-equilibrium in Γ.
(b) Prove that if there is $i \in I$ such that $r_*^j(Q^i, C^{-i}) \geq r_*^j(Q^j, C^{-j})$, for
  every $j \in I$, then for every $\epsilon > 0$ the stationary strategy profile
  $([(1 - \epsilon)(C^i), \epsilon(Q^i)], C^{-i})$ is a stationary uniform
  $(2\|r_*\|_\infty \cdot \epsilon)$-equilibrium in Γ.

We will assume from now on that the conditions of the first two parts do
not hold.

(c) Prove that if $x_0 \neq \vec{C}$, then x_0 is a stationary uniform 0-equilibrium
  in Γ.
(d) Prove that for every $\lambda > 0$ sufficiently small we have $x_\lambda \neq \vec{C}$.

We will assume from now on that $x_0 \neq \vec{C}$. Define

$$\mu^i := \lim_{\lambda \to 0} \frac{x_\lambda^i}{\sum_{j \in I} x_\lambda^j}. \tag{10.22}$$

(e) Prove the following claims:

  • μ^i is well defined for every $i \in I$.
  • $\sum_{i \in I} \mu^i = 1$.
  • $\mu^i = 0$ for every nonpunishable player $i \in I$.

(f) Consider a variation Γ_{obs} of the original quitting game Γ that includes
  an impartial mediator, who can privately send to each player a
  message, which is a natural number, at the outset of the game. Each
  player can base her choice of action on the natural number she
  received from the mediator. Write down the space of strategies of each
  player $i \in I$ in the game Γ_{obs}.
(g) Fix $m \in \mathbb{N}$, and suppose that the mediator acts according to the
  following mechanism $\text{Mec}(m)$:

  • She chooses a player $\widehat{i} \in I$ according to the probability distribution
    $\mu := (\mu^i)_{i \in I}$ defined in Eq. (10.22), and two numbers
    $\widehat{t} \in \{1, 2, \ldots, m^2\}$ and $\widehat{d} \in \{1, 2, \ldots, m\}$ according to the uniform
    distribution.
  • She sends the number $\widehat{t}$ to player $\widehat{i}$, the number $\widehat{t} + \widehat{d}$ to the punisher
    $j_{\widehat{i}}$ of player $\widehat{i}$, and the number $\widehat{t} + \widehat{d} + 1$ to all other players $i \neq \widehat{i}, j_{\widehat{i}}$.

Let σ^i be the strategy profile of player i in the game Γ_{obs} in which she quits in the stage that is equal to the signal she received, and denote $\sigma := (\sigma^i)_{i \in I}$. Prove that for every $\epsilon > 0$, there is $T_0 \in \mathbb{N}$ such that for every $T \geq T_0$,

$$\left\| \gamma_T(\text{Mec}(m), \sigma) - \sum_{i \in I} \mu^i \cdot r_*(Q^i, C^{-i}) \right\|_\infty \leq \epsilon,$$

where $\gamma_T(\text{Mec}(m), \sigma)$ is the T-stage payoff in the game Γ_{obs} when the mediator uses the mechanism $\text{Mec}(m)$ and the players adopt the strategy profile σ.

(h) Suppose that player i received the signal t_i. Let
$\rho := \mathbf{P}_{\text{Mec}(m)}(\widehat{i = j} \mid t_i)$ be the probability that player i assigns to the event that player j is the player who is supposed to quit first, given the information that she, player i, has after she got the signal from the mediator. Prove that $\mathbf{P}_{\text{Mec}(m)}(\rho = \mu^j) \geq 1 - \frac{2}{m}$.

(i) Prove that for every $\epsilon > 0$ there is an $m \in \mathbb{N}$ sufficiently large such that the strategy profile σ is a uniform ϵ-equilibrium in the game Γ'' that includes a mediator who follows the mechanism $\text{Mec}(m)$.

(j) Where in the proof did we use the assumption that the game is recursive and positive?

11

Ramsey's Theorem and Two-Player Deterministic Stopping Games

In this chapter, we prove Ramsey's Theorem, which states that for every coloring of the complete infinite graph by finitely many colors there is an infinite complete monochromatic subgraph. We then define the notion of undiscounted ϵ-equilibrium, relate it to uniform ϵ-equilibrium, and show that every two-player deterministic stopping game admits an undiscounted ϵ-equilibrium.

11.1 Ramsey's Theorem

Ramsey's Theorem[1] states that for every coloring of the complete infinite graph by finitely many colors there exists a monochromatic infinite complete subgraph. We now formally state this result and prove it.

Definition 11.1 Let C be a finite set and let $G = (V, E)$ be a graph. A C-coloring of $G = (V, E)$ is a mapping $c \colon E \to C$ that assigns an element in C to every edge in G.

Ramsey (1930) proved the following result.

Theorem 11.2 Let $G = (V, E)$ be the complete infinite graph: the set of vertices is the set of natural numbers, $V = \mathbb{N}$, and the set of edges is $E = \{(i, j) \in \mathbb{N} \times \mathbb{N} \colon i < j\}$. Let C be a finite set. For every C-coloring of G there exist an infinite subset $A \subseteq \mathbb{N}$ and $c^* \in C$ such that $c(i, j) = c^*$ for every $i, j \in A$ with $i < j$.

---

[1] Frank Plumpton Ramsey (Cambridge, United Kingdom, February 22, 1903 – London, United Kingdom, January 19, 1930) was a British philosopher, mathematician, and economist. Some of his significant contributions were subjective probabilities, utilities, decidability in mathematical logic, a mathematical theory of saving, and Ramsey's Theorem in graph theory.

195

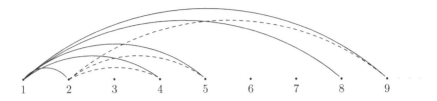

Figure 11.1 The construction in the proof of Theorem 11.2.

Proof   Set $i_1 := 1$ and $N_1 := \{1, 2, 3, \ldots\}$. For every color $c' \in C$ denote

$$B(i_1, c') := \{j \in N_1 : j > i_1, c(i_1, j) = c'\}.$$

These are all integers j in N_1 such that the color of the edge (i_1, j) is c'. Note that $\bigcup_{c' \in C} B(i_1, c') = N_1$, and, because the set C is finite, at least one of the sets $(B(i_1, c'))_{c' \in C}$ is infinite. Let c_1 be a color such that $B(i_1, c_1)$ is infinite, and set $N_2 := B(i_1, c_1)$.

In Figure 11.1, the vertices are the positive integers, and we drew solid arcs between $i_1 = 1$ and all the integers in $B(1, c_1)$.

Let i_2 be the minimal element in N_2. In Figure 11.1, $i_2 = 2$. For every color $c \in C$ denote

$$B(i_2, c') := \{j \in N_2 : j > i_2, c(\text{``}2, j) = c'\}.$$

These are all integers j in N_2 such that the color of the edge (i_2, j) is c'. Note that $\bigcup_{c' \in C} B(i_2, c') = N_2$. Since the set N_2 is infinite and the set C is finite, at least one of the sets $(B(i_2, c'))_{c' \in C}$ is infinite. Let c_2 be a color such that $B(i_2, c_2)$ is infinite, and set $N_3 := B(i_2, c_2)$. In Figure 11.1, we drew dotted arcs between $i_2 = 2$ and all the integers in $B(2, c_2)$.

Continue inductively to generate an increasing sequence $(i_n)_{n \in \mathbb{N}}$ of natural numbers, a sequence $(c_n)_{n \in \mathbb{N}}$ of colors, and a decreasing sequence $(N_n)_{n \in \mathbb{N}}$ of infinite subsets of $\mathbb{N}$, such that the following properties hold, where $B(i_n, c') := \{j \in N_n : j > i_n, c(i_n, j) = c'\}$:

- i_n is the first element in N_n.
- $N_{n+1} := B(i_n, c_n)$ is infinite.

The construction ensures that for every n and every $j \in N_{n+1}$, one has $c(i_n, j) = c_n$.

Since the set C is finite, there is a $c^* \in C$ that appears infinitely often in the sequence $(c_n)_{n \in \mathbb{N}}$, that is, there is a sequence $(n_k)_{k \in \mathbb{N}}$ such that $c_{n_k} = c^*$ for every $k \in \mathbb{N}$. Since the sequence of sets $(N_n)_{n \in \mathbb{N}}$ is nonincreasing, we have $i_{n_l} \in N_{n_k+1}$ for every $l > k$, and therefore $c(i_{n_k}, i_{n_l}) = c^*$ for every $l > k$. In particular, the infinite set $A := \{i_{n_k}, k \in \mathbb{N}\}$ and c^* satisfy the conclusion of the theorem. $\square$

11.2 Undiscounted Equilibrium

In Chapters 9 and 10, we considered the concept of uniform ϵ-equilibrium. In this chapter and in Chapters 12 and 13, it will be more convenient to study another equilibrium concept, called *undiscounted ϵ-equilibrium*, which applies to recursive games. As we will see later, the two concepts are equivalent in the class of games that we will study.

Definition 11.3   A *recursive game* is a stochastic game $\Gamma = \langle I, S, (A^i(s))_{s \in S}^{i \in I}, q, (r^i)_{i \in I} \rangle$ where the payoff in all nonabsorbing states is 0; that is, $r^i(s, a) = 0$ for every nonabsorbing state $s \in S$ and every action profile $a \in A(s)$.

When Γ is a recursive absorbing game, the payoff of all players in the nonabsorbing state is 0, and if we study existence of ϵ-equilibrium, we can assume that the payoff in the absorbing states is independent of the actions of the players, see Page 58. We can therefore denote the payoff at each state $s \in S$ by $r(s)$.

Denote by t_* the first stage in which the play reaches an absorbing state. Then t_* is a random variable whose distribution depends on the players' strategies. If $t_* < \infty$, then s_{t_*} is the absorbing state that the play reaches, and $r(s_{t_*})$ is the absorbing payoff.

Definition 11.4   Let $\Gamma = \langle I, S, (A^i(s))_{s \in S}^{i \in I}, q, (r^i)_{i \in I} \rangle$ be a recursive game and let $s \in S$. The *undiscounted payoff* of a strategy profile $\sigma \in \Sigma$ at the initial state s is

$$\gamma_\infty(s; \sigma) := \mathbf{E}_{s,\sigma} \left[\mathbf{1}_{\{t_* < \infty\}} r(s_{t_*}) \right].$$

Unlike the T-stage payoff, the undiscounted payoff takes into account absorption that occurs throughout the game, and not only in the first T stages. Unlike the λ-discounted payoff, the undiscounted payoff treats absorption in all stages equally, and does not discount payoffs in far away stages.

As the following theorem states, the undiscounted payoff is an approximation of the T-stage payoff for large Ts and of the λ-discounted payoff for small λ. The proof of the theorem is left to the reader as an exercise (Exercise 11.2).

Theorem 11.5   *Let* $\Gamma = \langle I, S, (A^i(s))_{s \in S}^{i \in I}, q, (r^i)_{i \in I} \rangle$ *be a recursive game, let* $s \in S$ *be a state, and let* $\sigma \in \Sigma$ *be a strategy profile. Then*

$$\gamma_\infty(s; \sigma) = \lim_{T \to \infty} \gamma_T(s; \sigma) = \lim_{\lambda \to 0} \gamma_\lambda(s; \sigma).$$

We next define an equilibrium notion that is based on the undiscounted payoff.

Definition 11.6   Let $\epsilon > 0$. A strategy profile σ_* is an *undiscounted* ϵ-*equilibrium* if for every state $s \in S$, every player $i \in I$, and every strategy $\sigma^i \in \Sigma^i$ of player i, we have

$$\gamma_\infty^i(s; \sigma^i, \sigma_*^{-i}) \leq \gamma_\infty^i(s; \sigma_*) + \epsilon.$$

In the next section, we will study one family of recursive games in which the two concepts of undiscounted ϵ-equilibrium and uniform ϵ-equilibrium agree.

11.3 Two-Player Recursive Absorbing Games with a Single Nonabsorbing Entry

In this section, we will be interested in a restricted family of recursive games, namely, recursive absorbing games in which all action profiles except one lead the play to absorption. Since in recursive absorbing games the stage payoff functions $(r^i)_{i \in I}$ are identically zero, there is no need to mention them, hence we denote a recursive absorbing game by $\Gamma = \langle I, (A^i)_{i \in I}, p_*, (r_*^i)_{i \in I} \rangle$.

Definition 11.7   A recursive absorbing game $\Gamma = \langle I, (A^i)_{i \in I}, p_*, (r_*^i)_{i \in I} \rangle$ *has a single nonabsorbing entry* if it satisfies the following property: there is an action profile $\tilde{a} \in A$ such that $p_*(\tilde{a}) = 0$ and $p_*(a) = 1$ for every $a \neq \tilde{a}$.

As the following theorem states, the concepts of uniform ϵ-equilibrium and undiscounted ϵ-equilibrium agree for absorbing games with a single nonabsorbing entry. The proof of the theorem is left to the reader as an exercise (Exercise 11.3).

Theorem 11.8   *Let $\Gamma = \langle I, (A^i)_{i \in I}, p_*, (r_*^i)_{i \in I} \rangle$ be a recursive absorbing game with a single nonabsorbing entry, and let $\epsilon > 0$. If a strategy profile $\sigma_* \in \Sigma$ is an undiscounted ϵ-equilibrium, then it is a uniform 2ϵ-equilibrium. Conversely, if a strategy profile $\sigma_* \in \Sigma$ is a uniform ϵ-equilibrium, then it is an undiscounted 2ϵ-equilibrium.*

Every action $a^i \neq \tilde{a}^i$ is an *absorbing action* of player i, because it leads the game to a sure absorption. We will call the action $\tilde{a}^i$ the *nonabsorbing* action of player i, and note that the game is not absorbed in a given stage if and only if all players play their nonabsorbing actions in that stage.

A recursive game is *positive* if $r_*^i(a) > 0$ for every $a \in A$ such that $p_*(a) > 0$. When the game is recursive and positive, $M := \|r_*\|_\infty = \max_{i \in I} \max_{a \in A_i} r_*^i(a)$ is the maximal payoff.

Theorem 11.9   *Let Γ be a two-player recursive absorbing game that has a single nonabsorbing entry and positive absorbing payoffs. Then for every $\epsilon \in \left(0, \frac{1}{2M}\right)$ the game admits a stationary undiscounted ϵ-equilibrium x such that $p_*(x) \geq \epsilon^2$.*

By Theorem 11.8, the stationary undiscounted ϵ-equilibrium that exists by Theorem 11.9 is a uniform 2ϵ-equilibrium.

Proof of Theorem 11.9 Since the game is recursive, $r^i(x) = 0$ for every $i \in I$ and every $x \in \prod_{i \in I} \Delta(A_i)$. In the proof of Theorem 10.4, we considered a semi-algebraic function $\lambda \mapsto x_\lambda$ of stationary λ-discounted equilibria that converges to a limit $x_0 = \lim_{\lambda \to 0} x_\lambda$, denoted by $w := \lim_{\lambda \to 0} \gamma_\lambda(x_\lambda)$ the limit of the λ-discounted equilibrium payoffs, and distinguished the following four cases:

(VT.1) $p_*(x_0) > 0$ (and then $r_*(x_0) = w$).

(VT.2) $p_*(x_0) = 0$ and $w^i \leq r^i(x_0) = 0$ for every player $i = \{1, 2\}$.

(VT.3) $p_*(x_0) = 0$ and $w^2 > r^2(x_0) = 0$.

(VT.4) $p_*(x_0) = 0$ and $w^1 > r^1(x_0) = 0$.

Note that for every $a^i \neq \widetilde{a}^i$,

$$w^i = \lim_{\lambda \to 0} \gamma_\lambda^i(x_\lambda) \geq \lim_{\lambda \to 0} \gamma_\lambda^i(a^i, x_\lambda^{-i}) = r_*^i(a^i, \widetilde{a}^{3-i}) > 0. \qquad (11.1)$$

In particular, $x_\lambda \neq \widetilde{a}$ for every $\lambda > 0$ sufficiently small. Eq. (11.1) also implies that Case (VT.2) cannot happen, and if one of Cases (VT.3) or (VT.4) holds, then the other holds as well.

If Case (VT.1) holds, then the proof of Theorem 10.4 implies that the mixed action profile x_0 is in fact an undiscounted 0-equilibrium (Exercise 11.5). If $p_*(x_0) \geq \epsilon^2$, then the theorem holds. We now handle the remaining cases.

If Case (VT.1) holds, then $x_0^i(\widetilde{a}^i) \geq 1 - \epsilon^2$ for each $i = 1, 2$, while if one of the Cases (VT.3) or (VT.4) holds, then $x_0 = \widetilde{a}$. For each $i \in \{1, 2\}$ the mapping $\lambda \mapsto x_\lambda^i$ is semi-algebraic, hence the set $\text{supp}(x_\lambda^i)$ is independent of λ, provided $\lambda > 0$ is sufficiently small. Denote by $\widehat{A}^i$ this common set of actions. If Cases (VT.1), (VT.3), or (VT.4) hold, then at least one of the sets $\widehat{A}^1$ and $\widehat{A}^2$ contains an element in addition to $\widetilde{a}^1$ or $\widetilde{a}^2$, respectively. Assume w.l.o.g. that $|\widehat{A}^1| \geq 2$.

As we have seen in Eq. (10.7),

$$w^1 = \lim_{\lambda \to 0} \gamma_\lambda^1(x_\lambda^1, x_\lambda^2) \geq \lim_{\lambda \to 0} \gamma_\lambda^1(a^1, x_\lambda^2)$$
$$\geq r_*^1(a^1, \widetilde{a}^2) - M\epsilon^2, \quad \forall a^1 \in A^1 \setminus \{\widetilde{a}^1\}, \qquad (11.2)$$

where the second inequality holds since $x_0^2(\widetilde{a}^2) \geq 1 - \epsilon^2$. Similarly $r_*^2(\widetilde{a}^1, a^2) \leq w^2 + M\epsilon^2$ for every $a^2 \in A^2 \setminus \{\widetilde{a}^2\}$. Moreover, for every $a^1 \in \widehat{A}^1 \setminus \{\widetilde{a}^1\}$ we have

$$w^1 = \lim_{\lambda \to 0} \gamma_\lambda^1(x_\lambda^1, x_\lambda^2) = \lim_{\lambda \to 0} \gamma_\lambda^1(a^1, x_\lambda^2) \leq r_*^1(a^1, \widetilde{a}^2) + M\epsilon^2, \quad \forall a^1 \in A^1,$$
$$(11.3)$$

	$\tilde{a}^2$	
$\tilde{a}^1$ ↑ $\hat{A}_1 = \text{supp}(x_\lambda^1)$ ↓	?, $\leq w^2 + M\epsilon^2$ *	?, $\leq w^2 + M\epsilon^2$ *
$[w^1 - M\epsilon^2, w^1 + M\epsilon^2], ?$ *	*	*
$[w^1 - M\epsilon^2, w^1 + M\epsilon^2], ?$ *	*	*
$\leq w^1 + M\epsilon, ?$ *	*	*

Figure 11.2 The situation in the proof of Theorem 11.9.

see Figure 11.2, where we provide the bounds on $r_*^i(a)$ given by Eqs. (11.2) and (11.3), and a question mark indicates that there are no contraints on the corresponding payoff.

We will define a mixed action x^1 for Player 1 that is derived from x_0^1 by increasing the probability assigned to each action $a^1 \neq \tilde{a}^1$, and prove that the pair $(x^1, \tilde{a}^2)$ is an undiscounted ϵ-equilibrium. Define a mixed action $x^1 \in \Delta(A^1)$ as follows:

$$
x^1(a^1) := \begin{cases} \epsilon^2 \cdot \dfrac{x_0(a^1)}{\sum_{a^1 \neq \tilde{a}^1} x_0(a^1)}, & \text{if } a^1 \neq \tilde{a}^1, \\ 1 - \epsilon^2, & \text{if } a^1 = \tilde{a}^1. \end{cases}
$$

Note that

$$
p_*(x^1, \tilde{a}^2) = \epsilon^2.
$$

Moreover, $\gamma_\infty^1(x^1, \tilde{a}^2)$ is a convex combination of $r_*^1(a^1, \tilde{a}^2)$ for $a^1 \in \hat{A}^1 \setminus \{\tilde{a}^1\}$, hence as shown by Figure 11.2,

$$
\gamma_\infty^1(x^1, \tilde{a}^2) = \sum_{a^1 \in \hat{A}^1 \setminus \{\tilde{a}^1\}} \frac{x^1(a^1)}{x^1(\hat{A}^1 \setminus \{\tilde{a}^1\})} r_*^1(a^1, \tilde{a}^2) \in \left[w^1 - M\epsilon^2, w^1 + M\epsilon^2 \right].
$$

(11.4)

Since payoffs are positive,

$$
\begin{aligned}
\gamma_\infty^2(x^1, \tilde{a}^2) &= r_*^2(x^1, \tilde{a}^2) \\
&= r_*^2(x_0^1, \tilde{a}^2) \\
&= \left(1 - \lim_{\lambda \to 0} \alpha_\lambda(x_\lambda^1, \tilde{a}^2) \right) \cdot 0 \\
&\quad + \left(\lim_{\lambda \to 0} \alpha_\lambda(x_\lambda^1, \tilde{a}^2) \right) \cdot \left(\lim_{\lambda \to 0} r_*^2(x_\lambda^1, \tilde{a}^2) \right) \\
&\leq \lim_{\lambda \to 0} r_*^2(x_\lambda^1, \tilde{a}^2) \leq w^2 + M\epsilon^2.
\end{aligned}
$$

(11.5)

By Eqs. (11.2) and (11.4), Player 1 cannot gain more than $2M\epsilon^2 \leq \epsilon$ by deviating. By the analogue of Eq. (11.2) for Player 2 and Eq. (11.5), Player 2 cannot profit more than $2M\epsilon^2 \leq \epsilon$ by deviating. We conclude that the stationary strategy profile $(x^1, \widetilde{a}^2)$ is an undiscounted ϵ-equilibrium. $\qquad \square$

11.4 Two-Player Deterministic Stopping Games

In this section, we define a new family of games, called deterministic stopping games. A deterministic stopping game is a sequential game in which each player has two actions, Continue and Quit. As long as both players choose Continue, they get no payoff and the play goes on. Once at least one player chooses Quit, the game terminates, and the terminal payoff depends on the stage in which the game is absorbed and on which players choose Quit at that stage. Thus, a deterministic stopping game is a two-player recursive game with countably many states. This is the only chapter in this book where we will consider stochastic games with an infinite set of states.

Definition 11.10   A *two-player deterministic stopping game* Γ_{DSG} is a stochastic game in which

- The set of players is $I = \{1, 2\}$.
- The set of states S is the union of a set of nonabsorbing states $S_1 := \mathbb{N}$ and a set of absorbing states $S_2 := \mathbb{N} \times \big\{\{1\}, \{2\}, \{1, 2\}\big\}$.
- In each state in S_1 each player $i \in I$ has two actions: $A^i(t) = \{C^i, Q^i\}$, where C^i stands for *Continue* and Q^i stands for *Quit*.
- The payoffs in states $t \in S_1$ are zero: $r^i(t, a) = 0$ for every $t \in S_1$ and every $a \in A(t) := A^1(t) \times A^2(t)$.
- Transitions in states $t \in S_1$ are as follows:

$$q(t + 1 \mid t, C^1, C^2) = 1,$$
$$q((t + 1, \{1\}) \mid t, Q^1, C^2) = 1,$$
$$q((t + 1, \{2\}) \mid t, C^1, Q^2) = 1,$$
$$q((t + 1, \{1, 2\}) \mid t, Q^1, Q^2) = 1.$$

We denote by $r(t, J)$ the payoff in the absorbing state (t, J), for every $(t, J) \in \mathbb{N} \times \big\{\{1\}, \{2\}, \{1, 2\}\big\}$. This is the payoff if in the first $t - 1$ stages both players continue, and in stage t the player(s) in J quit while the player not in J (if there is such a player) continue.

We will study the game when the initial state is $t = 1$. Hence, as long as the both players continue, the state s_t in stage t is t. Accordingly, we denote states in S_1 by the letter t and not by the letter s.

A pure strategy of a player is a mapping from histories to available actions. Since in deterministic stopping games the game is absorbed once some player quits, for the purpose of studying undiscounted ϵ-equilibria we can assume that a *pure strategy* is an element in $\mathbb{N} \cup \{\infty\}$: The element ∞ represents the strategy that always continues, and the element $t \in \mathbb{N}$ represents the strategy that quits at stage t.

Similarly, we can represent a *behavior strategy* for player i by a function $x^i : \mathbb{N} \rightarrow [0, 1]$ with the interpretation that $x^i(t)$ is the probability by which player i chooses Q^i at stage t if the game was not absorbed before that stage.

Thus, an alternative definition of a two-player deterministic stopping game is as a game in strategic form where both players' action sets are $\mathbb{N} \cup \{\infty\}$.

Definition 11.11   A *two-player deterministic stopping game* is a strategic-form game that is given by six sequences of real numbers $(r^i(t, S))_{i=1,2; \, S=\{1\}, \{2\}, \{1,2\}, \, t \in \mathbb{N}}$.

- The set of players is $I = \{1, 2\}$.
- The set of strategies of each player is $\mathbb{N} \cup \{\infty\}$.
- The payoff function is

$$u^i(t_1, t_2) := \mathbf{1}_{\{t_1 < t_2\}} r^i(t_1, \{1\}) + \mathbf{1}_{\{t_1 > t_2\}} r^i(t_2, \{2\}) + \mathbf{1}_{\{t_1 = t_2 < \infty\}} r^i(t_1, \{1, 2\}).$$

The main result of this chapter is the following.

Theorem 11.12   *In every two-player deterministic stopping game in which payoffs are bounded and positive, there exists an undiscounted ϵ-equilibrium, for every $\epsilon > 0$.*

The rest of the chapter is devoted to the proof of Theorem 11.12. We will first handle the case of periodic games, to which the next section is devoted.

11.5 Periodic Deterministic Stopping Games

Definition 11.13   Let $k \in \mathbb{N}$. A two-player deterministic stopping game is *periodic* with period k if the sequence $(r^i(t, J))_{t \in \mathbb{N}}$ has period k for each $i \in \{1, 2\}$ and each subset of players $J \in \{\{1\}, \{2\}, \{1, 2\}\}$. That is,

$$r^i(t + k, J) = r^i(t, J), \quad \forall i \in \{1, 2\}, \, \forall J \in \{\{1\}, \{2\}, \{1, 2\}\}, \, \forall t \in \mathbb{N}.$$

	0	1	2	$\cdots$	k
0	$0,0$	$r(1,\{2\})$ *	$r(2,\{2\})$ *		$r(k,\{2\})$ *
1	$r(1,\{1\})$ *	$r(1,\{1,2\})$ *	$r(1,\{1\})$ *		$r(1,\{1\})$ *
2	$r(2,\{1\})$ *	$r(1,\{2\})$ *	$r(2,\{1,2\})$ *		$r(2,\{1\})$ *
⋮	⋮	⋮	⋮		⋮
k	$r(k,\{1\})$ *	$r(1,\{2\})$ *	$r(2,\{2\})$ *		$r(k,\{1,2\})$ *

Figure 11.3 The game Γ_{AB}

A two-player periodic deterministic stopping game is equivalent to an absorbing game that has one nonabsorbing entry in which each player has $k+1$ actions. Indeed, consider a two-player periodic deterministic stopping game Γ_{DSG} with period k. A pure strategy of a player is a number in $\mathbb{N} \cup \{\infty\}$: the stage in which the player quits. Instead of deciding at the outset of the game when to quit, the player could decide, at the beginning of every period, whether or not to quit in that period, and if she decides to quit, in which stage of the period to quit. This provides a representation of the game as a recursive absorbing game Γ_{AB} as follows (see Figure 11.3):

- The set of actions of each player is $\{0, 1, 2, \ldots, k\}$. Action $\ell \in \{1, 2, \ldots, k\}$ corresponds to quitting in the ℓth stage of the coming period; action 0 corresponds to continuing throughout the coming period.
- The action pair $(0,0)$ is nonabsorbing (and the stage payoff is $(0,0)$).
- All other action pairs lead to absorption with probability 1. The absorbing payoff is as follows:

$$r_*^i(\ell, \ell) = r^i(\ell, \{1,2\}), \quad \forall \ell \in \{1, 2, \ldots, k\},$$
$$r_*^i(\ell_1, \ell_2) = r^i(\ell_1, \{1\}), \quad \forall \ell_1, \ell_2 \in \{1, 2, \ldots, k\}, \ \ell_1 < \ell_2,$$
$$r_*^i(\ell_1, \ell_2) = r^i(\ell_2, \{2\}), \quad \forall \ell_1, \ell_2 \in \{1, 2, \ldots, k\}, \ \ell_1 > \ell_2.$$

A stationary strategy of player i in the auxiliary game Γ_{AB} is described by a vector $x^i = (x^i(\ell))_{\ell=1}^k$, where for every $\ell \in \{1, 2, \ldots, k\}$ the quantity $x^i(\ell)$ is the probability that player i quits in stages $\ell \mod k$ of the game Γ_{DSG}. The stationary strategy profile (x^1, x^2) is absorbing if and only if $\sum_{\ell=1}^k (x^1(\ell) + x^2(\ell)) > 0$.

By Theorem 11.9, the game Γ_{AB} has a stationary undiscounted ϵ-equilibrium, which is absorbing with probability at least ϵ^2.

11.6 Proof of Theorem 11.12

Fix $\epsilon > 0$, and denote $M := \|r\|_\infty$. Let Z be an ϵ-dense set in the rectangle $[-M, M]^2$, that is, a finite set with the property that for every $z' \in [-M, M]^2$ there is a $z \in Z$ such that $\|z - z'\|_\infty \le \epsilon$.

For every two integers $k < l$ define a periodic deterministic stopping game $\Gamma_{k,l}$ with period $l - k$ as follows: the payoff at stage t, denoted $r_{k,l}(t, \cdot)$ is given by

$$r_{k,l}(t, J) := r(k + t \mod l - k, J), \quad \forall J \in \big\{\{1\}, \{2\}, \{1, 2\}\big\}, \forall t \in \mathbb{N}.$$

In words, in the auxiliary game $\Gamma_{k,l}$ the players repeat the portion of the game Γ_{DSG} between stages k and $l - 1$. Denote by $\gamma_{k,l,\infty}(\sigma) \in \mathbb{R}^2$ the undiscounted payoff in the recursive absorbing game $\Gamma_{k,l}$ under the strategy profile σ.

Let $x_{k,l}$ be a stationary undiscounted ϵ-equilibrium in the game $\Gamma_{k,l}$. Denote the expected payoff under $x_{k,l}$ by

$$g_{k,l} := \gamma_{k,l,\infty}(x_{k,l}).$$

Let $z_{k,l} \in Z$ be an element that is ϵ-close to $g_{k,l}$:

$$\|g_{k,l} - z_{k,l}\|_\infty \le \epsilon.$$

Let $C = Z \times \{\{1\}, \{2\}, \{1, 2\}\}$, and recall that $E = \{(i, j) \in \mathbb{N} \times \mathbb{N} : i < j\}$ is the set of edges in the complete infinite graph G. We are going to define a C-coloring of G, that is, a mapping $c \colon E \to C$. The coloring c is defined as follows: for every $(k, l) \in E$, $c_{k,l} = \{z_{k,l}\} \times I_{k,l}$, where $I_{k,l}$ is the set of players $i \in \{1, 2\}$ who quit with positive probability under $x_{k,l}$.

By Theorem 11.2, there are a $c^* \in C$ and an infinite sequence of positive integers $k_1 < k_2 < \cdots$ such that $c_{k_n, k_{n+1}} = c^*$ for every $n \in \mathbb{N}$.

Case 1: $k_1 = 1$.

Denote $c^* = (I_*, z_*)$. We will construct an undiscounted 3ϵ-equilibrium whose payoff is close to c^*. Define strategies σ_*^1 and σ_*^2 for the two players in Γ as follows

$$\sigma_*^1(k_n + t - 1) = x_{k_n, k_{n+1}}^1(t), \quad \forall n \in \mathbb{N}, \forall t \in \{1, \ldots, k_{n+1} - k_n\},$$

$$\sigma_*^2(k_n + t - 1) = x_{k_n, k_{n+1}}^2(t), \quad \forall n \in \mathbb{N}, \forall t \in \{1, \ldots, k_{n+1} - k_n\}.$$

In words, between the stages k_n and $k_{n+1} - 1$ the players follow the stationary undiscounted ϵ-equilibrium in the periodic deterministic stopping game $\Gamma_{k_n,k_{n+1}}$, which corresponds to these stages.

We will prove that $\sigma_* = (\sigma_*^1, \sigma_*^2)$ is an undiscounted 3ϵ-equilibrium. We first calculate $\gamma_\infty(\sigma_*)$. Denote by t_* the stopping time that indicates the first stage in which some players quit (and the game effectively terminates), and by I_* the set of players who quit at stage t_*. Since for every $n \in \mathbb{N}$ the strategy profile $x_{k_n,k_{n+1}}$ satisfies Theorem 11.9,

- Under the strategy profile $x_{k_n,k_{n+1}}$, the probability of absorption between the stages k_n and $k_{n+1} - 1$ is at least ϵ^2:

$$\mathbf{P}_{x_{k_n,k_{n+1}}}(t_* < k_{n+1} \mid t_* \geq k_n) \geq \epsilon^2. \tag{11.6}$$

- The expected absorbing payoff if absorption occurs between the stages k_n and $k_{n+1} - 1$ is close to z_*:

$$\left\| \mathbf{E}_{x_{k_n,k_{n+1}}}[\mathbf{1}_{\{t_* < k_{n+1}\}} r(t_*, I_*) \mid t_* \geq k_n] - z_* \right\|_\infty < \epsilon. \tag{11.7}$$

Since for each $i = 1,2$ the strategy σ_*^i is the concatenation of the strategies $(x_{k_n,k_{n+1}})_{n\in\mathbb{N}}$, we deduce from Eq. (11.6) that

$$\mathbf{P}_{\sigma_*}(t_* < \infty) = 1.$$

Indeed,

$$\mathbf{P}_{\sigma_*}(t_* = \infty) = \lim_{N \to \infty} \mathbf{P}_{\sigma_*}(t_* \geq k_N)$$

$$= \prod_{n=1}^{N-1} \mathbf{P}_{\sigma_*}(t_* \geq k_{n+1} \mid t_* \geq k_n)$$

$$\leq \lim_{N \to \infty} (1 - \epsilon^2)^N = 0.$$

Moreover, the expected absorbing payoff is close to z_*:

$$\left\| \mathbf{E}_{\sigma_*}[r(t_*, I_*)] - z_* \right\|_\infty < \epsilon. \tag{11.8}$$

Indeed,

$$\mathbf{E}_{\sigma_*}[r(t_*, I_*)] = \sum_{n=1}^{\infty} \mathbf{P}_{\sigma_*}(k_n \leq t_* < k_{n+1}) \cdot \mathbf{E}_{\sigma_*}[r(t_*, I_*) \mid k_n \leq t_* < k_{n+1}].$$

$$\tag{11.9}$$

Eq. (11.8) follows from Eqs. (11.7) and (11.9), since $\sum_{n=1}^{\infty} \mathbf{P}_{\sigma_*}(k_n \leq t_* < k_{n+1}) = 1$.

We now prove that Player 1 cannot gain much by deviating. For Player 2 the proof is analogous. By Kuhn's Theorem, it is sufficient to show that Player 1 cannot gain much by deviating to a pure strategy. There are two possible pure deviations for Player 1:

- She can deviate by never quitting.
- She can deviate by quitting at stage t with probability 1.

The first deviation is available to Player 1 in each of the auxiliary games $\Gamma_{k_n,k_{n+1}}$. Denote by $\emptyset^1$ the strategy of Player 1 in which she always continues. Since $x_{k_n,k_{n+1}}$ is an undiscounted ϵ-equilibrium in $\Gamma_{k_n,k_{n+1}}$, we deduce that by deviating from $x^1_{k_n,k_{n+1}}$ to $\emptyset^1$ Player 1 cannot gain more than ϵ:

- If $\mathbf{P}_{\emptyset^1,x^2_{k_n,k_{n+1}}}(t_* < \infty) = 0$, then $0 = \gamma^1_{k_n,k_{n+1}}(\emptyset^1, x^2_{k_n,k_{n+1}}) \leq z^1_* + \epsilon$.
- If $\mathbf{P}_{\emptyset^1,x^2_{k_n,k_{n+1}}}(t_* < \infty) > 0$, then

$$\mathbf{E}_{\emptyset^1,x^2_{k_n,k_{n+1}}}[r^1(t_*, I_*)] = \gamma^1_{k_n,k_{n+1}}(\emptyset^1, x^2_{k_n,k_{n+1}}) \leq z^1_* + \epsilon.$$

Consequently,

$$\gamma^1_\infty(\emptyset^1, \sigma^2_*) = \mathbf{E}_{\emptyset^1, \sigma^2_*}[r^1(t_*, I_*)]$$

$$= \sum_{n=1}^{\infty} \mathbf{P}_{\sigma_*}(k_n \leq t_* < k_{n+1}) \cdot \mathbf{E}_{\sigma_*}[r^1(t_*, I_*) \mid k_n \leq t_* < k_{n+1}]$$

$$\leq z^1_* + \epsilon$$

$$\leq \gamma^1_\infty(\sigma_*) + 2\epsilon.$$

We next show that Player 1 cannot gain by the second type of deviation, that is, by quitting in stage t. Let n be the unique integer such that $k_n \leq t < k_{n+1}$; that is, stage t is in the nth block. The argument is analogous to the previous one. Denote by $\widehat{x}^1$ the strategy of Player 1 in the auxiliary game $\Gamma_{k_n,k_{n+1}}$ in which she quits with probability 1 in stage $t - k_n + 1$. The deviation to $\widehat{x}^1$ is available to Player 1 in the auxiliary game $\Gamma_{k_n,k_{n+1}}$, and, since the strategy $x_{k_n,k_{n+1}}$ is an undiscounted ϵ-equilibrium, this deviation does not improve Player 1's payoff by more than ϵ in $\Gamma_{k_n,k_{n+1}}$. That is,

$$\mathbf{E}_{\widehat{x}^1, x^2_{k_n,k_{n+1}}}[r^1(t_*, I_*)] \leq g^1_{k_n,k_{n+1}} + \epsilon \leq z^1_* + 2\epsilon.$$

The argument now proceeds as above.

Case 2: $k_1 > 1$.

By the first two cases, when the initial state is k_1 both players have an undiscounted 3ϵ-equilibrium. Denote the corresponding payoff by g. States $\{1, 2, \ldots, k_1 - 1\}$ form a finite stage game, which terminates at stage k_1 with payoff g. This game has an equilibrium when the initial state is 1, which can be found by backward induction. The strategy pair that is constructed by the backward induction process is an undiscounted 3ϵ-equilibrium.

11.7 Comments and Extensions

Ramsey's Theorem was proven by Ramsey (1930), and initiated what is now called *Ramsey Theory*, which studies the occurrence of simple substructures in complex structures.

Theorem 11.12 was proven by Shmaya and Solan (2004). There are several extensions of the theorem that come to mind:

1. Extension to games where the payoff function is not necessarily positive.
2. Extension to games with more than two players.
3. Extension to games in which the transition is not deterministic; that is, if both players continue, the play moves to another state that is drawn randomly according to some transition probability.
4. Extension to games in which there is more than one nonabsorbing entry in each state.

To see which extension is covered by our technique of proof, we look closely at how we argued for Theorem 11.12. The proof is divided into three parts. First, we defined for every periodic game a color, by approximating an equilibrium payoff in the periodic game. Second, we applied Ramsey's Theorem to the complete infinite graph, which produced a sequence of periodic games. Third, we concatenated stationary ε-equilibria in these periodic games to form an undiscounted 3ε-equilibrium in the original infinite game.

The assumption that the payoffs are positive was used in the following argument. Suppose that $\mathbf{P}_{\emptyset^1, \sigma_*^2}(t_* < \infty) < 1$, where $\emptyset^1$ is the strategy of Player 1 in which she always continues. When the payoffs are positive, such a deviation lowers the payoff of Player 1. When the payoffs are negative and, say, $\gamma_\infty^1(\emptyset^1, \sigma_*^2) < 0$, such a deviation may be profitable. To overcome this difficulty, one can strengthen Theorem 11.9, to show that in two-player recursive absorbing games that have a single nonabsorbing entry, the game admits a stationary undiscounted ϵ-equilibrium x such that $p_*(x) \geq \epsilon^2$, and if $r^i(x) < 0$, then the per-stage probability that the other player plays a quitting action is at least ϵ^2. This additional property ensures that if $\gamma^1_\infty(\emptyset^1, \sigma^2_*) < 0$ then necessarily $\mathbf{P}_{\emptyset^1, \sigma^2_*}(t_* < \infty) = 1$, hence the above issue does not arise.

The fact that there are two players was used to ensure that the periodic game has a stationary undiscounted ϵ-equilibrium, for every $\epsilon > 0$. When there are more than two players, a stationary undiscounted ϵ-equilibrium need not exist. Nevertheless, by Solan (1999), three-player absorbing games with a single nonabsorbing entry admit a periodic ϵ-equilibrium for every $\epsilon > 0$. Even though these undiscounted ϵ-equilibria are not necessarily stationary, one can properly concatenate them to construct an undiscounted ϵ-equilibrium in the three-player stopping game.

When there are more than three players, we do not know whether the periodic game admits an undiscounted ϵ-equilibrium for every $\epsilon > 0$, hence in particular we do not know whether stopping games that involve more than three players admit an undiscounted ϵ-equilibrium for every $\epsilon > 0$. Nevertheless, Theorem 11.12 serves as a reduction: if every I-player absorbing game with a single nonabsorbing entry has an undiscounted ϵ-equilibrium, for every $\epsilon > 0$, then every I-player deterministic stopping game has an undiscounted ϵ-equilibrium, for every $\epsilon > 0$.

When the transition rule is not deterministic, the periodic game is defined by its starting point, and by a stopping time that indicates when it restarts. Theorem 10.4 can be applied to show that every such game admits an undiscounted ϵ-equilibrium, for every $\epsilon > 0$. Moreover, one can generalize Ramsey's Theorem to this more general setup. This extension was solved in Shmaya and Solan (2004).

When there is more than one nonabsorbing entry in each state, the technique used to prove Theorem 11.12 fails. Why is that? In deterministic stopping games, the state of the game in stage t is independent of the players' behavior. If there is more than one nonabsorbing entry in each state, this is no longer the case, as the evolution of the state variable depends on the players' actions. The extension of Ramsey's Theorem to this setup fails.

In Exercise 10.9, we proved that every quitting game admits a normal-form correlated uniform ϵ-equilibrium, for every $\epsilon > 0$. The approach that we developed in this chapter allows us to extend this result to every deterministic multiplayer stopping game (see Exercise 11.9). This result was further extended to every multiplayer stopping game (not necessarily deterministic) by Heller (2012). In a similar fashion, this approach was used by Mashiah-Yaakovi (2014) to prove the existence of a subgame-perfect uniform ϵ-equilibrium in stopping games with perfect information.

11.8 Exercises

1. In this exercise, we prove a finite version of Ramsey's Theorem (Theorem 11.2). Prove that for every two positive integers k and l there is a positive integer $R(k,l)$ such that for every $n \geq R(k,l)$ and every coloring of the complete graph with n vertices by two colors, Yellow and Green, either there is a complete subgraph with k vertices such that all the edges between these k vertices are colored Yellow, or there is a complete subgraph with l vertices such that all the edges between these l vertices colored Green.

2. Prove Theorem 11.5: Let $\Gamma = \langle I, S, (A^i(s))_{s \in S}^{i \in I}, q, (r^i)_{i \in I} \rangle$ be a recursive game, let $s \in S$ by a state, and let $\sigma \in \Sigma$ be a strategy profile. Then

$$\gamma_\infty(s;\sigma) = \lim_{T \to \infty} \gamma_T(s;\sigma) = \lim_{\lambda \to 0} \gamma_\lambda(s;\sigma).$$

3. In this exercise, we extend Theorem 11.8. Consider a recursive absorbing game with a single nonabsorbing entry, and let $\epsilon > 0$. Prove that if a strategy profile $\sigma_* \in \Sigma$ is an undiscounted ϵ-equilibrium, then it is a uniform 2ϵ-equilibrium, and if a strategy profile $\sigma_* \in \Sigma$ is a uniform ϵ-equilibrium, then it is an undiscounted 2ϵ-equilibrium.

4. In this exercise, we extend Theorem 11.8 to a multiplayer deterministic stopping game. Consider a multiplayer deterministic stopping game with uniformly bounded and positive payoffs, and let $\epsilon > 0$. Prove that if the strategy profile σ is an undiscounted ϵ-equilibrium, then it is a uniform 2ϵ-equilibrium, and if the strategy profile σ is a uniform ϵ-equilibrium, then it is an undiscounted 2ϵ-equilibrium.

5. Show that in Case (VT.1) in the proof of Theorem 11.9, the stationary strategy profile x_0 is an undiscounted 0-equilibrium.

6. Consider the two-player deterministic stopping game Γ that is displayed below, where the payoff function at each stage t depends on whether t is odd or even. Prove that for every $\epsilon > 0$, $(5,6)$ is an undiscounted ϵ-equilibrium payoff when the initial stage is 1, and $(8,3)$ is an undiscounted ϵ-equilibrium payoff when the initial stage is 2.

	C^2	Q^2		C^2	Q^2
C^1		2, 10	C^1		11, 1
Q^1	2, 9	8, 2	Q^1	9, 2	7, $\frac{11}{3}$

| Odd Stages | Even Stages |

7. In the following two-player recursive absorbing game, find four stationary undiscounted 0-equilibria that yield different payoffs.

	L	R
T		2, 1 *
B	1, 2 *	3, 3 *

8. Prove that the following two-player recursive absorbing game admits a stationary undiscounted 0-equilibrium in which Player 1 plays the action T with probability smaller than 1, and Player 2 plays the action L with probability smaller than 1. There is no need to calculate this equilibrium.

	L	C	R
T		2, 1 *	1, − 1 *
M	1, 0 *	0, 2 *	0, 1 *
B	0, 1 *	3, 0 *	−1, 2 *

9. Let Γ be a multiplayer positive deterministic stopping game with uniformly bounded payoffs. Prove that Γ admits a normal-form correlated uniform ϵ-equilibrium, for every $\epsilon > 0$.
   Hint: Use Exercise 10.9.

10. In Theorem 11.9, we proved that every two-player positive recursive absorbing game with a single nonabsorbing entry has a stationary undiscounted ϵ-equilibrium, for every $\epsilon > 0$. Prove the same result for two-player absorbing games with a single nonabsorbing entry, when the probability of absorption is not necessarily 1; that is, when $p_*(a)$ is required to be positive for every $a \neq (\tilde{a}^1, \tilde{a}^2)$ (and not necessarily 1).

12
Infinite Orbits and Quitting Games

In this chapter, we define infinite orbits and show how to approximate such orbits. We prove that for every mapping that has no fixed point there is an approximate infinite orbit with unbounded variation, and we use this result to show that a certain class of quitting games admits undiscounted ϵ-equilibria.

A *quitting game* is a deterministic stopping game in which the payoffs are independent of the stage of the game. Alternatively, this is an absorbing game in which each player has two actions, *Continue* and *Quit*; if all players continue, the game continues to the next stage; otherwise the game is absorbed with probability 1. We now define these games formally.

Definition 12.1 A *quitting game* is an absorbing game with a single nonabsorbing entry $\Gamma = \langle I, (A^i)_{i \in I}, p_*, (r_*^i)_{i \in I} \rangle$, where $A^i = \{C^i, Q^i\}$ for each player $i \in I$, $p_*(\vec{C}) = 0$, where $\vec{C} = (C^i)_{i \in I}$, and $p_*(a) = 1$ for every action profile $a \neq \vec{C}$.

Quitting games constitute a simple family of stochastic games, yet even for this family, the existence of an undiscounted ϵ-equilibrium is not known when $|I| \geq 4$. In this chapter and in Chapter 13, we will prove that for every $\epsilon > 0$ an undiscounted ϵ-equilibrium exists in certain families of quitting games.

12.1 Approximating Infinite Orbits

Let (X, d) be a metric space: X is a set and $d \colon X \times X \to \mathbb{R}$ is a metric on X. Readers who are not familiar with metric spaces should think of the Euclidean space, that is, $X = \mathbb{R}^n$ and d is some metric on X, which can be the Euclidean metric $d_2(x, y) := \sqrt{\sum_{i=1}^{n}(x_i - y_i)^2}$, the L_1-metric $d_1(x, y) := \sum_{i=1}^{n} |x_i - y_i|$, or the maximum metric $d_\infty(x, y) := \max_{1 \leq i \leq n} |x_i - y_i|$.

Definition 12.2   Let X be a set and let $f : X \to X$. An *infinite orbit* of f is a sequence $(x^k)_{k=0}^{\infty}$ such that $x^{k+1} = f(x^k)$ for every $k \geq 0$.

Every point $x \in X$ defines an infinite orbit $(x^k)_{k=1}^{\infty}$ that starts at x by

$$x^1 = x, \quad x^{k+1} = f(x^k), \quad \forall k \geq 1.$$

Definition 12.3   Let (X, d) be a metric space and let $(x^k)_{k=1}^{\infty}$ be a sequence of points in X. The *variation* of $(x^k)_{k=1}^{\infty}$ is

$$\operatorname{var}\left((x^k)_{k=1}^{\infty}\right) := \sum_{k=1}^{\infty} d(x^k, x^{k+1}).$$

The variation of a sequence is nonnegative, and it may be *bounded* (that is, $\operatorname{var}\left((x^k)_{k=1}^{\infty}\right) < \infty$) or *unbounded* (that is, $\operatorname{var}\left((x^k)_{k=1}^{\infty}\right) = \infty$). The variation is 0 if and only if x^1 is a fixed point of f, that is, $f(x^1) = x^1$. Even if the infinite orbit $(x^k)_{k=1}^{\infty}$ is such that the limit $\lim_{k \to \infty} x^k$ exists, the variation of the infinite orbit may be unbounded (see Example 12.5).

The following example shows that sometimes all infinite orbits have bounded variation, even if f has no fixed points.

Example 12.4   Let $X = [0, 1]$ and let d be the Euclidean distance. Let $f : X \to X$ be the following function (see Figure 12.1):

$$f(x) := \begin{cases} 1, & \text{if } x = 0, \\ x/2, & \text{if } x \neq 0. \end{cases}$$

If $x > 0$, then the orbit that starts at x is $x, \frac{x}{2}, \frac{x}{4}, \frac{x}{8}, \ldots$, and its variation is equal to x. If $x = 0$, then the orbit that starts at x is $0, 1, \frac{1}{2}, \frac{1}{4}, \frac{1}{8}, \ldots$, and its variation is equal to 2.   ♦

In the following example, all orbits converge and have unbounded variation.

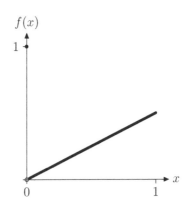

Figure 12.1 The function f in Example 12.4.

Example 12.5 Let $X = [-1, 1]$ and let d be the Euclidean distance. Let $f : X \to X$ be the following function:

$$f(1) = -\tfrac{1}{2}, \ f\left(-\tfrac{1}{2}\right) = \tfrac{1}{3}, \ f\left(\tfrac{1}{3}\right) = -\tfrac{1}{4}, \ f\left(-\tfrac{1}{4}\right) = \tfrac{1}{5}, \ \dots,$$

$$f\left((-1)^{m+1}\tfrac{1}{m}\right) = (-1)^m \tfrac{1}{m+1}, \dots,$$

and $f(x) = 1$ for every $x \notin \Omega := \left\{ 1, \ -\tfrac{1}{2}, \tfrac{1}{3}, \ -\tfrac{1}{4}, \dots, (-1)^{m+1}\tfrac{1}{m}, \dots \right\}$. For every $x \notin \Omega$ the infinite orbit that starts at x is

$$x, 1, \ -\tfrac{1}{2}, \tfrac{1}{3}, \ -\tfrac{1}{4}, \dots,$$

which converges to 0 and has unbounded variation. For every $x \in \Omega$, the infinite orbit that starts at x is a suffix of the infinite orbit above; hence, it converges to 0 and has unbounded variation as well. ♦

In this section, we are interested in infinite orbits with unbounded variation. Example 12.4 shows that even when f is not the identity function, all orbits may have bounded variation. As the following result shows, when f has no fixed points, there always exist finite sequences that (a) have high variation and (b) are approximate orbits.

Theorem 12.6 *Let (X, d) be a complete metric space, let $f : X \to X$ be a mapping with no fixed points, and let $x_* \in X$. For every $c, C > 0$ there exist a $K \in \mathbb{N}$ and a sequence $(x^k)_{k=1}^{K}$ of points in X such that the following properties hold:*

(A.0) $x^1 = x_*$: *the sequence starts at x_*.*

(A.1) $\sum_{k=1}^{K} d(x^k, f(x^k)) > C$.

(A.2) $\sum_{k=1}^{K-1} d(x^{k+1}, f(x^k)) < c$.

Figure 12.2 provides a graphical depiction of Theorem 12.6; the solid lines represent the distance between x^k and $f(x^k)$, and the dashed lines represent the distance between $f(x^k)$ and x^{k+1}. The theorem asserts that the total length of the solid lines is larger than C, while the total length of the dashed lines is smaller than c.

Figure 12.2 A graphical depiction of Theorem 12.6.

Example 12.7 (Example 12.4, continued) In this example, all orbits have bounded variation. We now construct for every $c, C > 0$ a sequence $(x^k)_{k=1}^K$ that has properties (A.1) and (A.2). Without loss of generality we can assume that C is an integer. Consider the following finite sequence $\vec{x}_{[l]} = (x^k)_{k=1}^{l+1}$ of $l + 1$ numbers:

$$1, \frac{1}{2}, \frac{1}{4}, \ldots, \frac{1}{2^l}, 0.$$

We have

$$\sum_{k=1}^l d(x^k, f(x^k)) = 1 - \frac{1}{2^l} \quad and \quad \sum_{k=1}^l d(f(x^k), x^{k+1}) = \frac{1}{2^l}.$$

Now let l be sufficiently large so that $\frac{2}{l-1} < c$, and consider the finite sequence $(x^k)_{k=1}^K$ that is constructed by concatenating the C blocks $(\vec{x}_{[l]}, \vec{x}_{[l+1]}, \ldots, \vec{x}_{[l+C]})$, whose length is $K = Cl + \frac{C(C+1)}{2}$. Along this sequence

$$\sum_{k=1}^K d(x^k, f(x^k)) = \sum_{l=1}^C \left(1 - \frac{1}{2^l} + 1\right) > 2C - c,$$

where the additional term 1 is due to the distance from the last element in the block $x_{[l]}$, which is 0, to its image under f. Moreover,

$$\sum_{k=1}^{K-1} d(f(x^k), x^{k+1}) = \sum_{l=1}^C \frac{1}{2^l} < \frac{2}{l-1} < c. \qquad \blacklozenge$$

Proof of Theorem 12.6 Assume without loss of generality that $c \leq C$. Denote by X' the closed ball centered at x_* with radius $C + 2c$. For each $x \in X'$, let $D(x)$ be the open ball centered at x with radius $\min\left\{\frac{c}{2}, \frac{c}{2C}d(x, f(x))\right\}$. Since $c \leq C$, $f(x)$ is not in $D(x)$, for every $x \in X'$.

The set X' is compact and the sets $(D(x))_{x \in X'}$ form an open cover of X'. Therefore, there are points $x_1, x_2, \ldots, x_L$ such that $\bigcup_{l=1}^L D(x_l) = X$. Assume w.l.o.g. that x_* is one of these points. Denote the minimum of the radii of the balls $D(x_1), D(x_2), \ldots, D(x_L)$ by ρ:

$$\rho := \min\left\{\frac{c}{2}, \frac{c}{2C}d(x_1, f(x_1)), \ldots, \frac{c}{2C}d(x_L, f(x_L))\right\} \in \left(0, \frac{c}{2}\right].$$

For each $y \in X$, let $l(y) \in \{1, 2, \ldots, L\}$ denote an index such that $y \in D(x_{l(y)})$. Define a sequence $(x^k)_k$ as follows:

$$x^1 := x_*,$$
$$x^{k+1} := x_{l(f(x^k))}, \quad \text{provided } f(x^k) \in X'.$$

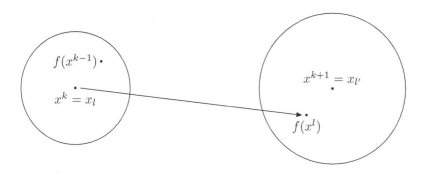

Figure 12.3 Construction of the sequence $(x^k)_{k=1}^\infty$.

In words, x^{k+1} is the center of one of the balls that contains $f(x^k)$ (see Figure 12.3). The sequence $(x^k)_k$ is infinite if it remains in X', and it is finite if $f(x^k) \notin X'$ for some k.

Since $f(x) \notin D(x)$ for every $x \in X'$,

$$d(x^k, f(x^k)) \ge \rho, \quad \forall k \ge 1. \tag{12.1}$$

Further, since $x^{k+1} \in D(f(x^k))$ for every $k \in \{1, \dots, K_* - 1\}$,

$$d(f(x^k), x^{k+1}) < \tfrac{c}{2C} d(x^{k+1}, f(x^{k+1})). \tag{12.2}$$

If the sequence $(x^k)_k$ is infinite, then by Eq. (12.1) there is $K \in \mathbb{N}$ such that

$$\sum_{k=1}^{K} d(x^k, f(x^k)) > C. \tag{12.3}$$

We argue that if the sequence $(x^k)_k$ is finite, say, if $f(x^K) \notin X'$, then Eq. (12.3) holds as well. Indeed, in that case,

$$C + 2c \le d(x^1, f(x^K)) \tag{12.4}$$

$$\le \sum_{k=1}^{K} d(x^k, f(x^k)) + \sum_{k=1}^{K-1} d(f(x^k), x^{k+1}) \tag{12.5}$$

$$< \sum_{k=1}^{K} d(x^k, f(x^k)) + \frac{c}{2C} \sum_{k=1}^{K-1} d(f(x^{k+1}), x^{k+1}) \tag{12.6}$$

$$< \left(1 + \frac{c}{2C}\right) \sum_{k=1}^{K} d(x^k, f(x^k)), \tag{12.7}$$

where Eq. (12.4) holds since x^1 is the center of X' and $f(x^K) \notin X'$, Eq. (12.5) holds by the triangle inequality, and Eq. (12.6) holds by Eq. (12.2). This implies that

$$\sum_{k=1}^{K} d(x^k, f(x^k)) > \frac{C + 2c}{\left(1 + \frac{c}{2C}\right)} = 2C \cdot \frac{C + 2c}{2C + c} > C,$$

as claimed.

Let K_* be the smallest natural number such that Eq. (12.3) holds. We will show that the sequence $(x^k)_{k=1}^{K_*}$ enjoys properties (A.0)–(A.2). By Definition (A.0) holds. By the choice of K_*, (A.1) also holds. Finally, by Eq. (12.2),

$$\sum_{k=1}^{K_*-1} d(f(x^k), x^{k+1}) = \sum_{k=1}^{K_*-2} d(f(x^k), x^{k+1}) + d(f(x^{K_*-1}, x^{K_*})$$

$$< \frac{c}{2C} \sum_{k=2}^{K_*-1} d(x^k, f(x^k)) + \frac{c}{2}$$

$$< \frac{c}{2C} \sum_{k=1}^{K_*-1} d(x^k, f(x^k)) + \frac{c}{2}$$

$$\leq \frac{c}{2C} \cdot C + \frac{c}{2} = c,$$

where the last inequality follows from the minimality of K_*. Consequently, the sequence $(x^k)_{k=1}^{K_*}$ has property (A.2). $\qquad\square$

12.2 An Example of a Three-Player Quitting Game

In this section, we present and analyze a specific three-player positive recursive quitting game that was studied in Flesch et al. (1997). As we will see, the game admits no undiscounted ϵ-equilibrium in stationary strategies, yet it does admit an undiscounted 0-equilibrium in nonstationary strategies. Thus the game exhibits the fundamental difference between two-player positive recursive absorbing games and positive recursive absorbing games with more than two players.

Consider the three-player quitting game that is displayed in Figure 12.4, where Player 1 chooses a row, Player 2 chooses a column, and Player 3 chooses a matrix. This game is symmetric, in the sense that the payoff if Player 1 quits alone $(1, 3, 0)$, is the cyclic shift of the payoff if Player 2 quits alone, which is $(0, 1, 3)$, and of the payoff if Player 3 quits alone, which is $(3, 0, 1)$. Similarly,

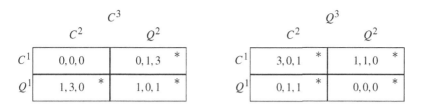

Figure 12.4 A three-player quitting game.

the payoff if Players 1 and 2 quit together, $(1, 0, 1)$, is the cyclic shift of the payoff if Players 2 and 3 quit together, which is $(1, 1, 0)$, and of the payoff if Players 3 and 1 quit together, which is $(0, 1, 1)$.

First, we will prove that this game has neither a stationary undiscounted 0-equilibrium, nor a stationary undiscounted ϵ-equilibrium.

Theorem 12.8 *The three-player quitting game in Figure 12.4 has no stationary undiscounted 0-equilibrium.*

Proof A stationary strategy can be represented by a single number, the probability to quit in each stage. We identify a stationary strategy x^i of player i with the per-stage probability with which player i chooses the action Q^i. Thus, in this chapter, a stationary strategy is identified with a real number in $[0, 1]$.

Suppose to the contrary that a stationary equilibrium $x = (x^1, x^2, x^3)$ exists. We first claim that $x^1 \in (0, 1)$. Indeed, if $x^1 = 0$ (Player 1 always continues), then Player 2's best response is $x^2 = 1$ (Player 2 always quits), and then Player 3's best response is $x^3 = 0$ (Player 3 always continues); but then Player 1's best response is $x^1 = 1$ (Player 1 always quits), a contradiction. If $x^1 = 1$ (Player 1 always quits), then Player 2's best response is $x^2 = 0$ (Player 2 always continues), and then Player 3's best response is $x^3 = 1$ (Player 3 always quits); but then Player 1's best response is $x^1 = 0$ (Player 1 always continues), a contradiction. The symmetry of the game implies that $x^2, x^3 \in (0, 1)$ as well.

Since $x^1 \in (0, 1)$, Player 1 is indifferent between continuing and quitting:

$$\frac{3(1 - x^2)x^3 + x^2 x^3}{1 - (1 - x^2)(1 - x^3)} = \gamma^1_\infty(C^1, x^2, x^3) = \gamma^1_\infty(Q^1, x^2, x^3) = 1 - x^3.$$

This equation solves to

$$x^2 = \frac{(x^3)^2 + 2x^3}{(x^3)^2 + 1}.$$

Since $x^3 \in (0,1)$, we see that $(x^3)^2 + 2x^3 > (x^3)^3 + x^3$. Therefore,

$$x^2 = \frac{(x^3)^2 + 2x^3}{(x^3)^2 + 1} > \frac{(x^3)^3 + x^3}{(x^3)^2 + 1} = x^3.$$

Thus, $x^2 > x^3$, and by symmetry we deduce that $x^3 > x^1$ and $x^1 > x^2$, a contradiction. $\square$

Theorem 12.9   *The three-player quitting game in Figure 12.4 has no stationary undiscounted ϵ-equilibrium, for every $\epsilon > 0$ sufficiently small.*

Proof   Assume to the contrary that there is a sequence $(\epsilon_n)_{n\in\mathbb{N}}$ that converges to 0 such that for every $n \in \mathbb{N}$, the game has a stationary undiscounted ϵ_n-equilibrium. The set of vectors $(\epsilon, x^1_\epsilon, x^2_\epsilon, x^3_\epsilon) \in [0,1]^4$, where $(x^1_\epsilon, x^2_\epsilon, x^3_\epsilon)$ is a stationary undiscounted ϵ-equilibrium, is semi-algebraic (see Exercise 12.3), hence if the contrapositive assumption holds, then there is an interval $(0, \epsilon_0)$ such that a stationary undiscounted ϵ-equilibrium exists for every $\epsilon \in (0, \epsilon_0)$. By Theorem 6.11, there is a mapping $\epsilon \mapsto (x^1_\epsilon, x^2_\epsilon, x^3_\epsilon)$ such that $(x^1_\epsilon, x^2_\epsilon, x^3_\epsilon)$ is a stationary ϵ-equilibrium for every $\epsilon \in (0, \epsilon_0)$. In particular, the following three limits exist:[1]

$$x^1_0 := \lim_{\epsilon\to 0} x^1_\epsilon,$$
$$x^2_0 := \lim_{\epsilon\to 0} x^2_\epsilon,$$
$$x^3_0 := \lim_{\epsilon\to 0} x^3_\epsilon.$$

We distinguish three cases, and derive a contradiction in each case.

Case 1:   Under (x^1_0, x^2_0, x^3_0) at least two players quit with positive probability.
    Assume without loss of generality that $x^1_0, x^2_0 > 0$. Then, in particular, $\lim_{\epsilon\to 0} x^2_\epsilon > 0$. Since $x^2_0 > 0$, we have $p_*(x^1, x^2_0, x^3_0) > 0$ for every mixed action x^1 of Player 1. This implies that

$$\lim_{\epsilon\to 0} r_*(x^1, x^2_\epsilon, x^3_\epsilon) = r_*(x^1, x^2_0, x^3_0), \quad \forall x^1 \in [0,1]. \tag{12.8}$$

Since $(x^1_\epsilon, x^2_\epsilon, x^3_\epsilon)$ is a stationary undiscounted ϵ-equilibrium,

$$\begin{aligned} r^1_*(x^1, x^2_\epsilon, x^3_\epsilon) &= \gamma^1_\infty(x^1, x^2_\epsilon, x^3_\epsilon) \\ &\le \gamma^1_\infty(x^1_\epsilon, x^2_\epsilon, x^3_\epsilon) + \epsilon \\ &= r^1_*(x^1_\epsilon, x^2_\epsilon, x^3_\epsilon) + \epsilon, \quad \forall x^1 \in [0,1]. \end{aligned}$$

[1] An alternative argument that does not use the theory of semi-algebraic sets is that since the set of stationary strategies is compact, we can let (x^1_0, x^2_0, x^3_0) be an accumulation point of the sequence $(x^1_\epsilon, x^2_\epsilon, x^3_\epsilon)$ as ϵ goes to 0.

Letting $\epsilon \to 0$ and using Eq. (12.8), we deduce that for every $x^1 \in [0, 1]$,

$$\gamma_\infty^1(x^1, x_0^2, x_0^3) = r_*(x^1, x_0^2, x_0^3)$$
$$= \lim_{\epsilon \to 0} r_*(x^1, x_\epsilon^2, x_\epsilon^3)$$
$$\leq \lim_{\epsilon \to 0} \left(r_*^1(x^1, x_\epsilon^2, x_\epsilon^3) + \epsilon \right)$$
$$= r_*^1(x_0^1, x_0^2, x_0^3)).$$

Since this inequality holds for every $x^1 \in [0, 1]$, Player 1 cannot profit by deviating from the stationary strategy profile (x_0^1, x_0^2, x_0^3). Analogous arguments show that Players 2 and 3 cannot profit by deviating from the stationary strategy profile (x_0^1, x_0^2, x_0^3) either, hence the strategy profile (x_0^1, x_0^2, x_0^3) is an undiscounted 0-equilibrium. This contradicts Theorem 12.8.

Case 2: Under (x_0^1, x_0^2, x_0^3) exactly one player, say Player 1, quits with positive probability.

Since $x_0^1 > 0$, while $x_0^2 = x_0^3 = 0$, we have $\lim_{\epsilon \to 0} x_\epsilon^1 > 0$, while $\lim_{\epsilon \to 0} x_\epsilon^2 = \lim_{\epsilon \to 0} x_\epsilon^3 = 0$. This implies that

$$\lim_{\epsilon \to 0} r_*(x_\epsilon^1, x_\epsilon^2, x_\epsilon^3) = (1, 3, 0),$$

and

$$\lim_{\epsilon \to 0} r_*(x_\epsilon^1, x_\epsilon^2, Q^3) = x_0(1, 0, 1) + (1 - x_0)(3, 0, 1).$$

Therefore, for ϵ sufficiently small,

$$\gamma_\infty^3(x_\epsilon^1, x_\epsilon^2, x_\epsilon^3) = r_*^3(x_\epsilon^1, x_\epsilon^2, x_\epsilon^3) < \frac{1}{2} < 1 = r_*^3(x_\epsilon^1, x_\epsilon^2, Q^3) = \gamma_\infty^3(x_\epsilon^1, x_\epsilon^2, Q^3),$$

which contradicts the assumption that $(x_\epsilon^1, x_\epsilon^2, x_\epsilon^3)$ is an ϵ-equilibrium.

Case 3: Under (x_0^1, x_0^2, x_0^3) no player quits: $(x_0^1, x_0^2, x_0^3) = (0, 0, 0)$.

The per-stage probability of absorption of the mixed action profile (x^1, x^2, x^3) is

$$p(x^1, x^2, x^3) = 1 - (1 - x^1)(1 - x^2)(1 - x^3).$$

For each action profile $a \neq (C^1, C^2, C^3)$ denote by $p(a \mid x^1, x^2, x^3)$ the conditional probability that the play is absorbed by the action profile a under the strategy profile (x^1, x^2, x^3). The table in Figure 12.5 provides the formula of $p(a \mid x^1, x^2, x^3)$ for each action profile $a \in A \setminus \vec{C}$.

Since $(x_0^1, x_0^2, x_0^3) = (0, 0, 0)$, we have

$$\lim_{\epsilon \to 0} x_\epsilon^1 = \lim_{\epsilon \to 0} x_\epsilon^2 = \lim_{\epsilon \to 0} x_\epsilon^3 = 0. \tag{12.9}$$

a	$p(a \mid x^1, x^2, x^3)$	$r_*(a)$
(Q^1, C^2, C^3)	$\frac{x^1(1-x^2)(1-x^3)}{1-(1-x^1)(1-x^2)(1-x^3)}$	$(1,3,0)$
(C^1, Q^2, C^3)	$\frac{(1-x^1)x^2(1-x^3)}{1-(1-x^1)(1-x^2)(1-x^3)}$	$(0,1,3)$
(C^1, C^2, Q^3)	$\frac{(1-x^1)(1-x^2)x^3}{1-(1-x^1)(1-x^2)(1-x^3)}$	$(3,0,1)$
(Q^1, Q^2, C^3)	$\frac{x^1 x^2(1-x^3)}{1-(1-x^1)(1-x^2)(1-x^3)}$	$(1,0,1)$
(Q^1, C^2, Q^3)	$\frac{x^1(1-x^2)x^3}{1-(1-x^1)(1-x^2)(1-x^3)}$	$(0,1,1)$
(C^1, Q^2, Q^3)	$\frac{(1-x^1)x^2 x^3}{1-(1-x^1)(1-x^2)(1-x^3)}$	$(1,1,0)$
(Q^1, Q^2, Q^3)	$\frac{x^1 x^2 x^3}{1-(1-x^1)(1-x^2)(1-x^3)}$	$(0,0,0)$

Figure 12.5 Absorption probability for each action profile under (x, y, z).

Since (C^1, C^2, C^3) is not an undiscounted ϵ-equilibrium for $\epsilon \in (0, \epsilon_0)$, we must have $x_\epsilon^1 + x_\epsilon^2 + x_\epsilon^3 > 0$ for every $\epsilon > 0$ sufficiently small, so that $p(x_\epsilon^1, x_\epsilon^2, x_\epsilon^3) > 0$ for every $\epsilon > 0$ sufficiently small.

For every $a \in A \setminus \{\vec{C}\}$ the function $\epsilon \mapsto p(a \mid x_\epsilon^1, x_\epsilon^2, x_\epsilon^3)$ is semi-algebraic, hence the limit $\lim_{\epsilon \to 0} p(a \mid x_\epsilon^1, x_\epsilon^2, x_\epsilon^3)$ exists. Moreover, $\sum_{a \in A \setminus \{\vec{C}\}} \lim_{\epsilon \to 0} p(a \mid x_\epsilon^1, x_\epsilon^2, x_\epsilon^3) = 1$. From the table in Figure 12.5 and Eq. (12.9), we deduce that

$$\lim_{\epsilon \to 0} p(C^1, Q^2, Q^3 \mid x_\epsilon^1, x_\epsilon^2, x_\epsilon^3) = 0,$$

$$\lim_{\epsilon \to 0} p(Q^1, C^2, Q^3 \mid x_\epsilon^1, x_\epsilon^2, x_\epsilon^3) = 0,$$

$$\lim_{\epsilon \to 0} p(Q^1, Q^2, C^3 \mid x_\epsilon^1, x_\epsilon^2, x_\epsilon^3) = 0,$$

$$\lim_{\epsilon \to 0} p(Q^1, Q^2, Q^3 \mid x_\epsilon^1, x_\epsilon^2, x_\epsilon^3) = 0.$$

It follows that

$$\lim_{\epsilon \to 0} \gamma_\infty(x_\epsilon^1, x_\epsilon^2, x_\epsilon^3) = \lim_{\epsilon \to 0} r_*(x_\epsilon^1, x_\epsilon^2, x_\epsilon^3)$$

$$= \left(\lim_{\epsilon \to 0} p_{x_\epsilon^1, x_\epsilon^2, x_\epsilon^3}(Q^1, C^2, C^3) \right) \cdot (1, 3, 0) \qquad (12.10)$$

$$+ \left(\lim_{\epsilon \to 0} p_{x_\epsilon^1, x_\epsilon^2, x_\epsilon^3}(C^1, Q^2, C^3) \right) \cdot (0, 1, 3)$$

$$+ \left(\lim_{\epsilon \to 0} p_{x_\epsilon^1, x_\epsilon^2, x_\epsilon^3}(C^1, C^2, Q^3) \right) \cdot (3, 0, 1).$$

Exercise 6.10 implies that in a neighborhood of 0, each of the three functions $\epsilon \mapsto p_{x_\epsilon^1, x_\epsilon^2, x_\epsilon^3}(C^1, C^2, Q^3)$, $\epsilon \mapsto p_{x_\epsilon^1, x_\epsilon^2, x_\epsilon^3}(C^1, Q^2, C^3)$, and $\epsilon \mapsto p_{x_\epsilon^1, x_\epsilon^2, x_\epsilon^3}(Q^1, C^2, C^3)$ is either the constant 0 or positive.

Since $(x_\epsilon^1, x_\epsilon^2, x_\epsilon^3)$ is an undiscounted ϵ-equilibrium, we deduce that

$$\lim_{\epsilon \to 0} \gamma^1(x_\epsilon^1, x_\epsilon^2, x_\epsilon^3) \geq \lim_{\epsilon \to 0}\left(\gamma^i(Q^1, x_\epsilon^2, x_\epsilon^3) - \epsilon\right) = 1.$$

Moreover, if $x_\epsilon^1 > 0$ for every $\epsilon > 0$ sufficiently small, then

$$\lim_{\epsilon \to 0} \gamma^1(x_\epsilon^1, x_\epsilon^2, x_\epsilon^3) \leq \lim_{\epsilon \to 0}\left(\gamma^i(Q^1, x_\epsilon^2, x_\epsilon^3) + \epsilon\right) = 1.$$

Since the condition $\lim_{\epsilon \to 0} p_{x_\epsilon^1, x_\epsilon^2, x_\epsilon^3}(Q^1, C^2, C^3) > 0$ implies that $x_\epsilon^1 > 0$ for every $\epsilon > 0$ sufficiently small, we deduce that $\lim_{\epsilon \to 0} \gamma^1(x_\epsilon^1, x_\epsilon^2, x_\epsilon^3) \geq 1$, with equality if $\lim_{\epsilon \to 0} p_{x_\epsilon^1, x_\epsilon^2, x_\epsilon^3}(Q^1, C^2, C^3) > 0$. Analogous conclusions hold for Players 2 and 3.

We claim that there is no convex combination

$$u = \alpha_1 \cdot (1, 3, 0) + \alpha_2 \cdot (0, 1, 3) + \alpha_3 \cdot (3, 0, 1)$$

such that

- $u^i \geq 1$ for every $i = 1, 2, 3$.
- If $\alpha_i > 0$ then $u^i = 1$.

In view of Eq. (12.10) and the previous paragraph, this constitutes a contradiction. Why can't such a convex combination exist? It cannot be that $\alpha_i > 0$ for a single player $i \in \{1, 2, 3\}$, because then $u^{i+2} = 0 < 1$. It cannot be that $\alpha_i > 0$ for two players in $\{1, 2, 3\}$, say, Players 1 and 2, because then $u^1 < 1$. And it cannot be that $\alpha_i > 0$ for all players $i \in \{1, 2, 3\}$, because then we must have $u = (1, 1, 1)$; but the sum of coordinates of each of the vectors $(1, 3, 0)$, $(0, 1, 3)$, and $(3, 0, 1)$ is 4. $\qquad\square$

We will prove that the game has a periodic undiscounted 0-equilibrium with period 3. In this equilibrium,

- in stages $1, 4, 7, \dots$ Player 1 quits with probability $\frac{1}{2}$ while Players 2 and 3 continue;
- in stages $2, 5, 8, \dots$ Player 2 quits with probability $\frac{1}{2}$ while Players 1 and 3 continue;
- and in stages $3, 6, 9, \dots$ Player 3 quits with probability $\frac{1}{2}$ while Players 1 and 2 continue.

For $i \in \{1, 2, 3\}$, we will denote a strategy of player i by $x^i = (x^i(t))_{t \in \mathbb{N}}$, where $x^i(t)$ is the probability with which player i quits at stage t, provided no player quits before stage t.

Theorem 12.10   *The following strategy profile* $x_* = (x_*^i)_{i=1}^3$ *is an equilibrium in the three-player quitting game in Figure 12.4:*

$$x_*^1(1) = \frac{1}{2}, \ x_*^1(2) = 0, \ x_*^1(3) = 0, \ x_*^1(4) = \frac{1}{2}, \ x_*^1(5) = 0, \ x_*^1(6) = 0, \dots,$$

$$x_*^2(1) = 0, \ x_*^2(2) = \frac{1}{2}, \ x_*^2(3) = 0, \ x_*^2(4) = 0, \ x_*^2(5) = \frac{1}{2}, \ x_*^2(6) = 0, \dots,$$

$$x_*^3(1) = 0, \ x_*^3(2) = 0, \ x_*^3(3) = \frac{1}{2}, \ x_*^3(4) = 0, \ x_*^3(5) = 0, \ x_*^3(6) = \frac{1}{2}, \dots.$$

Proof   **Step 1:** $\gamma_\infty(x_*) = (1, 2, 1).$

Under the strategy profile x_*, with probability $\frac{1}{2}$ Player 1 quits in the first stage; with probability $\frac{1}{4}$ Player 2 quits in the second stage; with probability $\frac{1}{8}$ Player 3 quits in the third stage; and with probability $\frac{1}{8}$ the game continues to stage 4. Since the strategy profile x_* is periodic with period 3, we have

$$\gamma_\infty(x_*) = \frac{1}{2}(1, 3, 0) + \frac{1}{4}(0, 1, 3) + \frac{1}{8}(3, 0, 1) + \frac{1}{8}\gamma_\infty(x_*).$$

The solution of this equation is

$$\gamma_\infty(x_*) = (1, 2, 1).$$

Step 2:   No player can profit by deviating from x_*.

By Kuhn's Theorem, it is sufficient to prove that no player can profit by deviating to a pure strategy. By the symmetry of the game and of the strategies x_* it is sufficient to show that the following deviations are not profitable:

• No player can profit by quitting in the first stage.
• Player 1 cannot profit by never quitting.

Player 1 receives 1 whether or not she quits in the first stage, hence she cannot profit by quitting in the first stage. If Player 2 deviates and quits in the first stage, her payoff is $\frac{1}{2}$, and therefore this deviation decreases her payoff. If Player 3 deviates and quits in the first stage, her payoff is 1, and therefore again such a deviation is not profitable.

It remains to show that Player 1 cannot profit by always continuing. Let then $\vec{C}^1$ be the strategy of Player 1 in which she always continues. Recall that t_* is the stage in which absorption occurs, and denote by $\gamma_\infty(\vec{C}^1, x_*^2, x_*^3 \mid t_* \le t)$ the undiscounted payoff conditional on the event that the game is absorbed before or at stage t:

$$\gamma_\infty(\vec{C}^1, x_*^2, x_*^3 \mid t_* \leq t) = \mathbf{E}_{\vec{C}^1, x_*^2, x_*^3}[r(a_{t_*}) \mid t_* \leq t].$$

Under the strategy profile $(\vec{C}^1, x_*^2, x_*^3)$, at the second stage the game is absorbed by Player 2 with probability $\frac{1}{2}$, and at the third stage the game is absorbed by Player 3 with probability $\frac{1}{4}$. Consequently,

$$\gamma_\infty^1(\vec{C}^1, x_*^2, x_*^3) = \frac{1}{2} \cdot 0 + \frac{1}{4} \cdot 3 + \frac{1}{4} \cdot \gamma_\infty^1(\vec{C}^1, x_*^2, x_*^3),$$

which solves to $\gamma_\infty^1(\vec{C}^1, x_*^2, x_*^3) = 1$, and therefore this deviation is not profitable to Player 1. □

12.3 Multiplayer Quitting Games in Which Players Do Not Want Others to Join Them in Quitting

The three-player quitting game that we studied in Section 12.2 enjoys the following properties:

- It is recursive: the payoff if no player ever quits is 0.
- The payoff of a player who quits alone is 1.
- The payoff of a player who quits with others is at most 1.

In this section, we will prove that every multiplayer quitting game with these four properties admits an undiscounted ϵ-equilibrium, for every $\epsilon > 0$. We will thus study the class of recursive quitting games that satisfy the following condition (P).

(P) For every player $i \in I$ and every subset of players $J \subseteq I$ we have

$$r_*^i(Q^i, C^{-i}) = 1 \geq r_*^i\big(Q^{J \cup \{i\}}, C^{I \setminus (J \cup \{i\})}\big).$$

An implication of Property (P) is that if some player quits, her payoff is at most 1:

$$r_*^i(Q^i, x^{-i}) \leq 1 = r_*^i(Q^i, C^{-i}), \quad \forall i \in I, \forall x^{-i} \in [0,1]^{|I|-1}. \tag{12.11}$$

Theorem 12.11 *Every multiplayer recursive quitting game that satisfies Condition* (P) *has an undiscounted ϵ-equilibrium, for every $\epsilon > 0$.*

From Theorem 11.8, we deduce that every multiplayer recursive quitting game that satisfies Condition (P) has a uniform ϵ-equilibrium, for every $\epsilon > 0$.

Comment 12.12　　The assumptions that the game is recursive and $r_*^i(Q^i, C^{-i}) = 1$ for all players $i \in I$ are normalization assumptions. Theorem 12.11 and its proof are valid whenever $r_*^i(Q^i, C^{-i})$ is at least player i's payoff when the play never absorbs, for every $i \in I$.

Proof　　Fix a recursive quitting game $\Gamma = \langle I, (A^i)_{i \in I}, p_*, (r_*^i)_{i \in I} \rangle$ that satisfies Property (P) and an $\epsilon > 0$. Denote $M := \|r_\infty\|_\infty$.

Step 1:　　Definition of a family of auxiliary strategic-form games.

For each $w \in [-M, M]^{|I|}$ define a strategic-form game $\widehat{G}(w) = \langle I, (A^i)_{i \in I}, (\widehat{u}_w^i)_{i \in I} \rangle$ as follows:

- The set of players is I, the set of players in Γ.
- The set of actions of each player $i \in I$ is $A^i = \{Q^i, C^i\}$, her set of actions in Γ.
- The payoff function of each player $i \in I$, denoted $\widehat{u}_w^i$, is given by

$$\widehat{u}_w^i(a) := \begin{cases} w^i, & \text{if } a = \vec{C}, \\ r_*^i(a), & \text{if } a \neq \vec{C}. \end{cases}$$

In words, the auxiliary game $\widehat{G}(w)$ captures one stage of the quitting game Γ, where, if the quitting game is not absorbed, the continuation payoff is w.

A mixed action profile in $\widehat{G}(w)$ is a vector $x \in [0,1]^I$, with the interpretation that x^i is the probability that player i chooses the action Q^i. The probability of absorption under the mixed action profile x is

$$p_*(x) := 1 - \prod_{i \in I}(1 - x^i).$$

The expected payoff in the auxiliary game $\widehat{G}(w)$ under the mixed action profile x is

$$\widehat{u}_w(x) = \left(\prod_{i \in I}(1 - x^i)\right) w + \sum_{\emptyset \subset J \subseteq I}\left(\prod_{i \in J} x^i\right)\left(\prod_{i \notin J}(1 - x^i)\right) r_*(Q^J, C^{I \setminus J})$$

$$= (1 - p_*(x))w + p_*(x)r_*(x). \tag{12.12}$$

Step 2:　　Equilibria of the auxiliary game $\widehat{G}(w)$.

Recall that for every $\epsilon > 0$, a mixed action profile $x = (x^i)_{i \in I}$ is an ϵ-equilibrium in the auxiliary game $\widehat{G}(w)$ if

$$\widehat{u}_w^i(x) \geq \max_{x'^i \in \Delta(A^i)} \widehat{u}_w^i(x'^i, x^{-i}) - \epsilon.$$

Since the auxiliary game $\widehat{G}(w)$ is a strategic-form game with finitely many players and actions, it has a 0-equilibrium.

We here make the following observations.

Claim 12.13   1. *If $\vec{C}$ is an equilibrium of $\widehat{G}(w)$, then $w^i \geq 1$ for every $i \in I$.*

2. *Let $i_0 \in I$ and let $w \in \left[-M, M\right]^{|I|}$ satisfy $w^{i_0} = 1$ and $w^i \geq 1$ for every $i \in I$. Then the strategy profile $x_* := ([(1 - \epsilon)(C^{i_0}), \epsilon(Q^{i_0})], C^{-i_0})$ is a $2M\epsilon$-equilibrium of $\widehat{G}(w)$. Moreover, player i_0 cannot profit by deviating from x_* in the auxiliary game $\widehat{G}(w)$.*

3. *Let x be a stationary strategy that satisfies (a) $p_*(x) > 0$, and (b) $\widehat{u}_w(x) = w$. Then $r_*(x) = w$.*

Proof   To see that item 1 of the claim holds, note that, by the definition of $\widehat{u}_w$, by Property (P), and since $\vec{C}$ is an equilibrium of $\widehat{G}(w)$,

$$w^i = \widehat{u}_w^i(\vec{C}) \geq \widehat{u}_w^i(Q^i, C^{-i}) = 1, \quad \forall i \in I.$$

Let us prove item 2 of the claim. Since $w^{i_0} = 1$, $r_*^{i_0}(Q^{i_0}, C^{-i_0}) = 1$, and $x_*^{-i_0} = C^{-i_0}$,

$$\widehat{u}_w^{i_0}(C^{i_0}, x_*^{-i_0}) = w^{i_0} = 1,$$

and

$$\widehat{u}_w^{i_0}(Q^{i_0}, x_*^{-i_0}) = 1.$$

Thus, player i_0 cannot profit by deviating from $[(1 - \epsilon)(C^{i_0}), \epsilon(Q^{i_0})]$ in the game $\widehat{G}(w)$. We now argue that no other player $i \neq i_0$ can profit more than $2M\epsilon$ by deviating from x_* in the game $\widehat{G}(w)$. This follows from the following inequalities:

$$\widehat{u}_w^i(x_*) = (1 - \epsilon)w^i + \epsilon r_*^i(Q^{i_0}, C^{-i_0}) \geq 1 - 2M\epsilon,$$

$$\widehat{u}_w^i(Q^i, x_*^{-i}) = r_*^i(Q^i, [(1 - \epsilon)(Q^{i_0}), \epsilon(C^{i_0})], C^{-i, i_0}) \leq 1,$$

where the last inequality holds by Eq. (12.11).

We finally prove item 3 of the claim. By condition (b) and (12.12),

$$w = \widehat{u}_w(x) = (1 - p_*(x)w + p_*(x)r_*(x).$$

By condition (a), we have $p_*(x) > 0$, hence $r_*(x) = w$, as claimed.   ▲

Step 3: Definition of a set W and mappings $x \colon W \to \mathbb{R}^{|I|}$ and $f \colon W \to W$.
Define the set W by [2]

$$W := \left\{ w \in \left[-M, M\right]^{|I|} : \exists i \in I \text{ with } w^i \leq 1 \right\}.$$

We turn to define the mapping $x \colon W \to [0, 1]^I$.

(R1) If $\vec{C}$ is an equilibrium of the auxiliary game $\widehat{G}(w)$, then
$w^i \geq r_*^i(Q^i, C^{-i})$ for every $i \in I$. By the definition of the set W, there
exists a player $i_0 \in I$ such that $w^{i_0} = r_*^{i_0}(Q^{i_0}, C^{-i_0})$. Set

$$x(w) := \left(\left[(1 - \epsilon)(C^{i_0}), \epsilon(Q^{i_0})\right], C^{-i_0} \right).$$

By Claim 12.13(2), in this case $x(w)$ is a $2M\epsilon$-equilibrium of the
auxiliary game $\widehat{G}(w)$.

(R2) Otherwise, all equilibria x of the game $\widehat{G}(w)$ are absorbing; that is, they
satisfy $p_*(x) > 0$. Set $x(w)$ to be one of the equilibria of $\widehat{G}(w)$.

Define a mapping $f \colon W \to \mathbb{R}^I$ by

$$f(w) := \widehat{u}_w(x(w)). \tag{12.13}$$

Thus, $x(w)$ is an absorbing equilibrium of the auxiliary game $\widehat{G}(w)$ or an
absorbing $2M\epsilon$-equilibrium of this game in which a single player quits with
positive probability, and $f(w)$ is the corresponding equilibrium payoff or
$2M\epsilon$-equilibrium payoff.

Step 4: For every $w \in W$, we have $f(w) \in W$.

Fix $w \in W$. Suppose first that $x(w) = \vec{C}$, and let i_0 be the player from the
definition of $f(w)$ in this case. Then

$$
\begin{aligned}
f^{i_0}(w) &= \widehat{u}_w^{i_0}(x(w)) \\
&= (1 - \epsilon)\widehat{u}_w^{i_0}(\vec{C}) + \epsilon \widehat{u}_w^{i_0}(Q^{i_0}, C^{-i_0}) \\
&= (1 - \epsilon)w^{i_0} + \epsilon r_*^{i_0}(Q^{i_0}, C^{-i_0}) = 1,
\end{aligned}
$$

where the last equality holds by the definition of i_0. We conclude that in this
case $f(w) \in W$.

Suppose now that $x(w) \neq \vec{C}$, so in particular $p_*(x(w)) > 0$. Hence, there is
a player i_0 such that $x_w^{i_0} > 0$. and therefore under $x(w)$ player i_0 is indifferent

---

[2] The threshold of 1 in the definition of W arises because of the assumption that $r_*^i(Q^i, C^{-i}) = 1$
for every player i. If we had dropped this condition we would have defined

$$W := \left\{ w \in \left[-M, M\right]^{|I|} : \exists i \in I \text{ with } w^i \leq r_*^i(Q^i, C^{-i}) \right\}.$$

between Q^i and C^i (if $x_w^{i_0} \in (0,1)$) or weakly prefers Q^i to C^i (if $x_w^{i_0} = 1$). In both cases,

$$f^{i_0}(w) = \widehat{u}_w^{i_0}(x(w)) \le \widehat{u}_w^{i_0}(Q^{i_0}, x_w^{-i_0}) = r_*^{i_0}(Q^{i_0}, x^{-i_0}(w)) \le 1,$$

where the first inequality holds because player i_0 weakly prefers Q^i to C^i, and the last inequality holds by Eq. (12.11). Once again $f(w) \in W$.

Step 5: If $w = f(w)$ and $x(w)$ is defined by (R1), then $x(w)$ is a stationary undiscounted 0-equilibrium.

Since $x(w)$ is defined by (R1), $\vec{C}$ is an equilibrium of $\widehat{G}(w)$, hence by Claim 12.13(1),

$$w^i \ge 1, \quad \forall i \in I. \tag{12.14}$$

Moreover, by (R1) there is $i_0 \in I$ such that

$$x(w) = \left([(1-\epsilon)(C^{i_0}), \epsilon(Q^{i_0})], C^{-i_0} \right).$$

Then

$$f(w) = (1-\epsilon)w + \epsilon r_*(Q^{i_0}, C^{-i_0}),$$

and since $f(w) = w$, we deduce that

$$w = r_*(Q^{i_0}, C^{-i_0}). \tag{12.15}$$

It follows that $r_*^i(Q^{i_0}, C^{-i_0}) \ge 1$ for every $i \in I$.

Under the stationary strategy profile $x(w)$ the game is bound to be absorbed, and the undiscounted payoff is

$$\gamma_\infty(x(w)) = r_*(Q^{i_0}, C^{-i_0}) = w. \tag{12.16}$$

To prove that $x(w)$ is a stationary undiscounted 0-equilibrium we need to show that no player can profit in Γ by deviating from $x(w)$. We start with player i_0, and suppose that the other players adopt the stationary strategy $x^{-i_0}(w) = C^{-i_0}$. If player i_0 ever quits, her payoff is 1, while since the game is recursive, if she continues forever her payoff is 0. Since $\gamma_\infty^{i_0}(x(w)) = w^{i_0} = 1$, it follows that player i_0 cannot profit by deviating from $x(w)$.

Fix now $i \ne i_0$. By Eqs. (12.11), (12.14), and (12.16),

$$r_*^i(Q^i, x^{-i}(w)) \le 1 \le w^i = \gamma_\infty^i(x(w)). \tag{12.17}$$

A deviation of player i must involve quitting at some stage, which, by Eq. (12.17), leads to a payoff at most w^i. This implies that player i cannot profit by deviating from $x(w)$.

Step 6: If $w = f(w)$ and $x(w)$ is defined by (R2), then $x(w)$ is a stationary undiscounted 0-equilibrium.

If there is a single player $i \in I$ such that $x^i(w) > 0$, then the argument is similar to that in Step 5. We therefore assume that there are at least two players $i \in I$ with $x^i(w) > 0$.

Since $w = f(w) = \widehat{u}_w(x(w))$, and since $p_*(x(w)) > 0$, Claim 12.13(3) implies that

$$\gamma_\infty(x(w)) = r_*(x(w)) = w. \tag{12.18}$$

We now argue that no player can profit by deviating from the stationary strategy profile $x(w)$ in the game Γ. Fix then a player $i \in I$. Since w is a fixed point of f, and since $x(w)$ is an equilibrium of $\widehat{G}(w)$,

$$w^i = f^i(w) = \widehat{u}_w^i(x(w)) \geq \max\{\widehat{u}_w^i(Q^i, x^{-i}(w)), \widehat{u}_w^i(C^i, x^{-i}(w))\}.$$

Since at least two players quit with positive probability under $x(w)$, $p_*(C^i, x^{-i}(w)) > 0$, and since

$$\begin{aligned} w^i \geq u_w^i(C^i, x^{-i}(w)) &= (1 - p_*(C^i, x^{-i}(w))) \cdot w^i \\ &\quad + p_*(C^i, x^{-i}(w)) r_i^*(C^i, x^{-i}(w)) \end{aligned}$$

it follows that

$$r_i^*(C^i, x^{-i}(w)) \leq w^i.$$

Moreover,

$$r_i^*(Q^i, x^{-i}(w)) = \widehat{u}_w^i(Q^i, x^{-i}(w)) \leq w^i.$$

Thus, whatever player i plays, the game is bounded to be absorbed, and the expected absorbing payoff of player i at stage t_* is at most w^i, whether she plays Q^i or C^i at that stage. This implies that player i cannot profit by deviating from the stationary strategy profile $x(w)$ in Γ.

By Steps 5 and 6, if the mapping f has a fixed point, then the quitting game Γ admits a stationary undiscounted 0-equilibrium. We therefore assume from now on that f has no fixed points.

Step 7: $p_*(x(w)) \geq \frac{1}{2M} \|w - \widehat{u}_w(x(w))\|_1$.

We now relate the probability of absorption under $x(w)$ to the difference between the continuation payoff w and the equilibrium payoff $\widehat{u}_w(x(w))$.

By Eq. (12.12),

$$\widehat{u}_w(x(w)) = (1 - p_*(x(w))) \cdot w + p_*(x(w)) \cdot r_*(x(w)),$$

Stage	Mixed action profile
1	x_K
2	x_{K-1}
3	x_{K-2}
⋮	⋮
t	x_{K-t+1}
$t+1$	x_{K-t}
⋮	⋮
K	x_1
$K+1$	$\vec{C}$
$K+2$	$\vec{C}$
⋮	⋮

Figure 12.6 The strategy profile σ_*.

and therefore

$$\|\widehat{u}_w(x(w)) - w\|_\infty \le 2M p_*(x(w)).$$

The result follows.

Step 8: Definition of a strategy profile σ_*.

The mapping f is defined on the metric space $(W, d_\infty(\cdot, \cdot))$, where $d_\infty(\cdot, \cdot)$ is the maximum distance. Since f has no fixed points, we can apply Theorem 12.6 with $C = \frac{2M}{\epsilon}$, $c = \frac{\epsilon}{M}$, and $(1, 1, \ldots, 1) \in W$ as the initial point of the sequence, and conclude that there are $K \in \mathbb{N}$ and a sequence $(w_k)_{k=1}^K$ of points in W such that the following properties hold:

(A.0) $w_1 = (1, 1, \ldots, 1)$.
(A.1) $\sum_{k=1}^K d_\infty(w_k, f(w_k)) > 2M/\epsilon$.
(A.2) $\sum_{k=1}^{K-1} d_\infty(w_{k+1}, f(w_k)) < \epsilon/M$.

Recall that $f(w_k)$ is a $2M\epsilon$-equilibrium payoff in the auxiliary game $\widehat{G}(w_k)$ with payoff w_{k+1}. Denote by x_k the equilibrium strategy profile by which $f(w_k)$ is defined, so that $f(w_k) = \widehat{u}_{w_k}(x_k)$.

Define the following strategy profile σ_* (see Figure 12.6): For $t \in \{1, 2, \ldots, K\}$, at stage t each player i plays the mixed action x^i_{K-t+1}. From stage $K+1$ and on, all players continue.

Step 9: Under σ_*, the game Γ is absorbed with high probability until stage K, that is, $\mathbf{P}_{\sigma_*}(t_* \le K) \ge 1 - \epsilon$.

Under σ_* the probability of absorption at stage t, given that the game was not absorbed before stage t, is $p_*(x_{K-t+1})$. By Step 7, this probability is at least $\frac{1}{2M}d_\infty(w_{K-t+1}, f(w_{K-t+1}))$. It follows that the probability that the game is not absorbed in the first K stages is at most $\prod_{k=1}^{K}\left(1 - \frac{1}{2M}d_\infty(w_k, f(w_k))\right)$. Since $\ln(1-z) \le -z$ for every $z \in (0,1)$, we have

$$\ln\left(\prod_{k=1}^{K}\left(1 - \tfrac{1}{M}d_\infty(w_k, f(w_k))\right)\right) = \sum_{k=1}^{K}\ln\left(1 - \tfrac{1}{M}d_\infty(w_k, f(w_k))\right)$$

$$\le -\tfrac{1}{2M}\sum_{k=1}^{K}d_\infty(w_k, f(w_k)) < -\tfrac{1}{\epsilon}.$$

Consequently,

$$\mathbf{P}_{\sigma_*}(t_* > K) = \prod_{k=1}^{K}\left(1 - \tfrac{1}{2M}d_\infty(w_k, f(w_k))\right) < \exp(-1/\epsilon) < \epsilon.$$

Step 10: $\gamma_\infty^i(\sigma_*) \ge f^i(w_K) - 2M\epsilon$, for every $i \in I$.

We next prove that the expected payoff under σ_* is high. To this end, define a stochastic process $(\eta_t)_{t=0}^{K-1}$ as follows. For each $t = 0, 1, \ldots, K-1$,

- If the game was not absorbed up to stage t, set

$$\eta_t := f^i(w_{K-t}) + \sum_{j=0}^{t-1}|w_{K-j}^i - f^i(w_{K-j-1})|.$$

- If the absorption stage t_* satisfies $t_* \le t$, set

$$\eta_t := r_*^i(Q^{I_*}, C^{I \setminus I_*}) + \sum_{j=0}^{t_*-1}|w_{K-j}^i - f^i(w_{K-j-1})|,$$

where I_* is the set of players who quit at stage t_*.

Denote by $\mathcal{H}_t$ the sigma-algebra on the set H_∞ of plays that is spanned by the cylinder sets that are defined by histories of length t, see Page 144. We claim that the process $(\eta_t)_{t=0}^{K-1}$ is a submartingale under the strategy profile σ_*, that is,

$$\eta_t \le \mathbf{E}_{\sigma_*}[\eta_{t+1} \mid \mathcal{H}_t]. \tag{12.19}$$

Indeed, on the event $\{t_* \leq t\}$ we have $\eta_{t+1} = \eta_t$, and on the event $\{t_* > t\}$ we have

$$\mathbf{E}_{\sigma_*}[\eta_{t+1} \mid \mathcal{H}_t]$$

$$= \mathbf{E}_{\sigma_*}\left[\mathbf{1}_{\{t_*=t+1\}} r_*^i(Q^{I_*}, C^{I \setminus I_*}) + \mathbf{1}_{\{t_*>t+1\}} f^i(w_{K-t-1}) \mid \mathcal{H}_t\right] \quad (12.20)$$

$$+ \sum_{j=0}^{t} |w_{K-j}^i - f^i(w_{K-j-1})|$$

$$= \mathbf{E}_{\sigma_*}\left[\mathbf{1}_{\{t_*=t+1\}} r_*^i(Q^{I_*}, C^{I \setminus I_*}) + \mathbf{1}_{\{t_*>t+1\}} w_{K-t-1}^i \mid \mathcal{H}_t\right] \quad (12.21)$$

$$+ \sum_{j=0}^{t} |w_{K-j}^i - f^i(w_{K-j-1})|$$

$$+ \mathbf{P}_{\sigma_*}(t_* > t + 1 \mid \mathcal{H}_t) \cdot \left(f^i(w_{K-t-1}) - w_{K-t-1}^i\right)$$

$$\geq \mathbf{E}_{\sigma_*}\left[\mathbf{1}_{\{t_*=t+1\}} r_*^i(Q^{I_*}, C^{I \setminus I_*}) + \mathbf{1}_{\{t_*>t+1\}} w_{K-t-1}^i \mid \mathcal{H}_t\right] \quad (12.22)$$

$$+ \sum_{j=0}^{t-1} |w_{K-j}^i - f^i(w_{K-j-1})|$$

$$= f^i(w_{K-t}) + \sum_{j=0}^{t-1} |w_{K-j}^i - f^i(w_{K-j-1})| \quad (12.23)$$

$$= \eta_t, \quad (12.24)$$

where Eq. (12.20) holds by the definition of η_{t+1}, to derive Eq. (12.21) we added and subtracted the quantity $\mathbf{E}_{\sigma_*}[\mathbf{1}_{\{t_*>t\}} w_{K-t-1}^i]$, Eq. (12.22) holds since $|z| + \lambda z \geq 0$ for every $z \in \mathbb{R}$ (applied to $z = f^i(w_{K-t-1}) - w_{K-t-1}^i$), Eq. (12.23) holds by the definition of f (Eq. (12.13)), and Eq. (12.24) holds by the definition of η_t and since $t_* > t$. Therefore, as we claimed,

$$f^i(w_K) = \eta_0$$

$$\leq \mathbf{E}_{\sigma_*}[\eta_{K-1}] \quad (12.25)$$

$$\leq \mathbf{P}_{\sigma_*}(t_* \leq K) \cdot \mathbf{E}_{\sigma_*}[r_*^i(Q^{I_*}, C^{I \setminus I_*}) \mid t_* \leq K] + M \cdot \mathbf{P}_{\sigma_*}(t_* > K)$$

$$+ \sum_{j=0}^{K-1} |w_{K-j}^i - f^i(w_{K-j-1})| \quad (12.26)$$

$$\leq \gamma_\infty^i(\sigma_*) + (M + 2)\epsilon, \quad (12.27)$$

where Eq. (12.25) follows since the process $(\eta_t)_{t=0}^{K-1}$ is a submartingale under the strategy profile σ_*, Eq. (12.26) holds by the definition of η_{K-1}, and Eq. (12.27) holds by Step 9 and Condition (A.2).

Comment 12.14   The inequality $\eta_0 \leq \mathbf{E}_{\sigma_*}[\eta_{K-1}]$ holds because the sequence $(\eta_t)_{t=0}^{K-1}$ is a submartingale. A reader who is not familiar with martingale theory will note that this inequality holds by summing Eq. (12.19) over $t = 0, 1, \ldots, K - 2$, taking the expectation, and deleting common terms on both sides.

Step 11:   The strategy profile σ_* is an undiscounted $2(M + 2)\epsilon$-equilibrium.

In view of Step 10, we need to show that, for every player $i \in I$ and every strategy $\sigma^i \in \Sigma^i$ in Γ,

$$\gamma_\infty^i(\sigma^i, \sigma_*^{-i}) \leq f^i(w_K) + (M + 2)\epsilon.$$

The proof is analogous to the proof of Step 10. Define a stochastic process $(\eta_t)_{t=0}^{K-1}$ as follows. For every $t = 0, 1, \ldots, K - 1$,

- If the game was not absorbed before stage t, set

$$\eta_t := f^i(w_{K-t}) - \sum_{j=0}^{t-1} |w_{K-j}^i - f^i(w_{K-j-1})|.$$

- If the game was absorbed before stage t, say, at stage $t_* < t$, set

$$\eta_t := r_*^i(Q^{I_*}, C^{I \setminus I_*}) - \sum_{j=0}^{t_*-1} |w_{K-j}^i - f^i(w_{K-j-1})|.$$

Now fix a strategy σ^i of player i. We claim that the process $(\eta_t)_{t=1}^K$ is a supermartingale under the strategy profile $(\sigma^i, \sigma_*^{-i})$, that is,

$$\eta_t \geq \mathbf{E}_{\sigma^i, \sigma_*^{-i}}[\eta_{t+1} \mid \mathcal{H}_t]. \tag{12.28}$$

Indeed, if the game was absorbed up to stage t, then $\eta_{t+1} = \eta_t$ and Eq. (12.28) holds. On the event $\{t_* > t\}$,

$$\mathbf{E}_{\sigma_*}[\eta_{t+1} \mid \mathcal{H}_t]$$
$$= \mathbf{E}_{\sigma_*}\left[\mathbf{1}_{\{t_*=t+1\}} r_*^i(Q^{I_*}, C^{I \setminus I_*}) + \mathbf{1}_{\{t_*>t+1\}} f^i(w_{K-t-1}) \mid \mathcal{H}_t \right]$$
$$- \sum_{j=0}^{t} |w_{K-j}^i - f^i(w_{K-j-1})|$$

$$= \mathbf{E}_{\sigma_*}\left[\mathbf{1}_{\{t_*=t+1\}}r_*^i(Q^{I_*},C^{I\setminus I_*}) + \mathbf{1}_{\{t_*>t+1\}}w_{K-t-1}^i \mid \mathcal{H}_t\right]$$

$$- \sum_{j=0}^{t}|w_{K-j}^i - f^i(w_{K-j-1})|$$

$$+ \mathbf{P}_{\sigma_*}(t_* > t+1 \mid \mathcal{H}_t)\cdot\left(f^i(w_{K-t-1}) - w_{K-t-1}^i\right)$$

$$\leq \mathbf{E}_{\sigma_*}\left[\mathbf{1}_{\{t_*=t+1\}}r_*^i(Q^{I_*},C^{I\setminus I_*}) + \mathbf{1}_{\{t_*>t+1\}}w_{K-t-1}^i \mid \mathcal{H}_t\right]$$

$$- \sum_{j=0}^{t-1}|w_{K-j}^i - f^i(w_{K-j-1})|$$

$$= f^i(w_{K-t}) - \sum_{j=0}^{t-1}|w_{K-j}^i - f^i(w_{K-j-1})|$$

$$= \eta_t,$$

and Eq. (12.28) holds as well. Since $w_1 = (1,1,\ldots,1)$, $\vec{C}$ is an equilibrium of $\widehat{G}(w_1)$, hence x_1 is defined by (R1). Consequently,

$$f^i(w_1) \geq 1 - M\epsilon. \tag{12.29}$$

We therefore have, as we claimed,

$$f^i(w_K) = \eta_0 \tag{12.30}$$

$$\geq \mathbf{E}_{\sigma^i,\sigma_*^{-i}}[\eta_{K-1}] \tag{12.31}$$

$$\geq \mathbf{P}(t_* \leq K)\cdot \mathbf{E}_{\sigma^i,\sigma_*^{-i}}\left[r_*^i(Q^{I_*},C^{I\setminus I_*}) \mid t_* \leq K\right]$$

$$+ \mathbf{P}(t_* > K)\cdot f^i(w_1) - \sum_{j=0}^{K-1}|w_{K-j}^i - f^i(w_{K-j-1})| \tag{12.32}$$

$$\geq \mathbf{P}(t_* \leq K)\cdot \gamma_\infty^i(\sigma^i,\sigma_*^{-i} \mid t_* \leq K)$$

$$+ \mathbf{P}(t_* > K)\cdot(1 + M\epsilon) - \epsilon. \tag{12.33}$$

where Eq. (12.31) holds since the sequence $(\eta_t)_{t=1}^{K}$ is a supermartingale under (σ^i,σ_*^{-i}), Eq. (12.32) holds by the definition of η_{K-1}, and Eq. (12.33) holds by Eq. (12.29) and Condition (A.2).

After stage K, under σ_* all players continue, which, by Property (P), implies that $\gamma_\infty^i(\sigma^i,\sigma_*^{-i} \mid t_* > K) \leq 1$. Therefore,

$$f^i(w_K) \geq \mathbf{P}(t_* \leq K)\cdot \gamma_\infty^i(\sigma^i,\sigma_*^{-i} \mid t_* \leq K) - \mathbf{P}(t_* > K)\cdot(1 + M\epsilon) - \epsilon$$

$$\geq \gamma_\infty^i(\sigma^i,\sigma_*^{-i}) - (M+2)\epsilon.$$

The claim follows. $\qquad\square$

12.4 Comments and Extensions

Every stochastic game admits a stationary λ-discounted equilibrium x_λ. Given such an equilibrium, the expected payoff to the players from state t and on depends on the current state s_t and not on the stage t, and it is given by $\gamma_\lambda(s; x_\lambda)$. Theorem 10.4 asserts a similar phenomenon w.r.t. uniform equilibria in two-player absorbing games: there is a vector $w \in \mathbb{R}^I$, and for every $\epsilon > 0$ there is a uniform ϵ-equilibrium σ_ϵ, such that as long as the game is not absorbed and punishment was not triggered, and conditional on the event that punishment will not be triggered in the future, the expected payoff from stage t and on is close to w, irrespective of the current stage. The three-player example presented in Section 12.2 shows that this property does not hold for undiscounted ϵ-equilibrium in three-player quitting games: given an undiscounted ϵ-equilibrium, one cannot find a payoff vector $w \in \mathbb{R}^I$, such that, as long as the game is not absorbed and punishment was not triggered, and conditional on the event that punishment will not be triggered in the future, the undiscounted payoff from stage t and on is close to w. It turns out that there are two-player nonzero-sum stochastic games where this property does not hold as well, see Simon (2006).

Theorem 12.11 was proven by Solan and Vieille (2001), who introduced infinite orbits to the study of quitting games. To prove this theorem, Solan and Vieille (2001) used sophisticated probabilistic estimates for the probability of absorption under the strategy profile σ_*, rather than Theorem 12.6. Theorem 12.6 was proved by Solan and Solan (2020), and the proof of Theorem 12.11 that we provided here follows ideas in this paper. Infinite orbits were extended and used for studying equilibria in stochastic games by Simon (2007, 2012).

12.5 Exercises

1. Provide an example of a complete metric space (X, d) and a mapping $f: X \to X$ with a unique fixed point such that all infinite orbits of f have bounded variation.

2. In this exercise, we prove a stronger version of Theorem 12.6. Let (X, d) be a complete metric space and let $f: X \to X$ be a mapping with no fixed points. Prove that for every $c > 0$ and every $x_* \in X$ there is a sequence $(x^k)_{k=1}^\infty$ of points in X such that

   (A.0) $x^1 = x_*$.
   (A.1) $\sum_{k=1}^\infty d(x^k, f(x^k)) = \infty$.
   (A.2) $\sum_{k=1}^\infty d(x^{k+1}, f(x^k)) < c$.

3. Prove that the correspondence that assigns to each $\epsilon > 0$ the set of all stationary undiscounted ϵ-equilibria in a multiplayer quitting game is semi-algebraic.

4. Where in the proof of Theorem 12.11 did we use the assumption that $r_*^i(Q^i, C^{-i}) = 1$ for each $i \in I$? How would you prove the result in the absence of this assumption?

5. Consider the following four-player quitting game, where Player 1 chooses a row, Player 2 chooses a column, Player 3 chooses either the top two matrices or the bottom two matrices, and Player 4 chooses either the two left matrices or the two right matrices.

C^4

C^3	C^2	Q^2
C^1		$4,1,0,0$ *
Q^1	$1,4,0,0$ *	$1,1,1,1$ *

Q^4

C^3	C^2	Q^2
C^1	$0,0,4,1$ *	$1,1,0,1$ *
Q^1	$1,0,1,1$ *	$0,1,0,1$ *

C^4

Q^3	C^2	Q^2
C^1	$0,0,1,4$ *	$0,1,1,1$ *
Q^1	$1,1,1,0$ *	$1,0,1,0$ *

Q^4

Q^3	C^2	Q^2
C^1	$1,1,1,1$ *	$1,0,1,0$ *
Q^1	$0,1,0,1$ *	$0,0,0,0$ *

(a) Determine whether the game satisfies Property (P) or not.

(b) Prove that there is no pure stationary undiscounted 0-equilibrium.

(c) Prove that there is no periodic undiscounted 0-equilibrium of period 4 in which in each stage only one player quits with positive probability.

(d) Find a periodic undiscounted 0-equilibrium of period 2 in which in odd stages Player 1 quits with probability $x \in (0, 1)$, Player 3 quits with positive probability $z \in (0, 1)$, and Players 2 and 4 continue with probability 1, and in even stages Player 2 quits with probability x, Player 4 quits with probability z, and Players 1 and 3 continue with probability 1. To find the equilibrium, you may use a freeware that solves a system of polynomial equalities, like *wolfram alpha*.

13

Linear Complementarity Problems
and Quitting Games

In this chapter, we present linear complementarity problems and use them to provide sufficient conditions that guarantee the existence of an undiscounted ϵ-equilibrium in quitting games.

13.1 Linear Complementarity Problems

In this section, we present linear complementarity problems. When using matrix notations, all vectors will be column vectors. We will denote by $\vec{0} = (0, 0, \ldots, 0)$ the vector all of whose coordinates are 0. When $x = (x^i)_{i=1}^n$ and $y = (y^i)_{i=1}^n$ are vectors in $\mathbb{R}^n$, we will write $x \geq y$ if $x^i \geq y^i$ for every $i \in \{1, 2, \ldots, n\}$. When R is an $n \times n$ matrix, we denote its columns by $R_{[1]}, R_{[2]}, \ldots, R_{[n]}$, and its (i, j)-entry by $R_{[j]}^i$.

Definition 13.1   Let R be an $n \times n$ matrix and let $q \in \mathbb{R}^n$. The *linear complementarity problem* $\mathrm{LCP}(R, q)$ is the following problem that consists of linear equalities and inequalities:

$$\text{Find} \quad w \in \mathbb{R}^n \text{ and } z = (z^0, z^1, \ldots, z^n) \in \Delta(\{0, 1, \ldots, n\}),$$

$$\text{such that} \quad w = z^0 q + \sum_{i=1}^n z^i R_{[i]}, \tag{13.1}$$

$$w^i \geq R_{[i]}^i, \quad \forall i = 1, 2, \ldots, n,$$

$$z^i = 0 \text{ or } w^i = R_{[i]}^i, \quad \forall i = 1, 2, \ldots, n.$$

The last condition in Problem (13.1) is the *complementarity condition*.

We note that if $q \in \mathbb{R}^n$ satisfies $q^i \geq R_{[i]}^i$ for every $i = 1, 2, \ldots n$, then problem (13.1) admits at least one solution, namely, $z = (1, 0, \ldots, 0)$ and $w = q$. This solution is called the *trivial solution*.

Example 13.2   Let $R = \begin{pmatrix} 0 & 1 \\ 1 & 0 \end{pmatrix}$. For $q = (1, -1)$, the linear complementarity problem LCP(R, q) has two solutions:

- $w = (0, 1)$ and $z = (0, 1, 0)$.
- $w = (1, 0)$ and $z = (0, 0, 1)$.

For $q' = (1, 1)$, in addition to the previous two solutions, the problem LCP(R, q') has the trivial solution:

- $w = (1, 1)$ and $z = (1, 0, 0)$.

For $q'' = (-1, -2)$, in addition to the two solutions of LCP(R, q), the problem LCP(R, q'') has a third solution:

- $w = (0, 0)$ and $z = \left(\frac{1}{4}, \frac{1}{2}, \frac{1}{4} \right)$.

Example 13.3   Let $R = \begin{pmatrix} 0 & -1 \\ -1 & 0 \end{pmatrix}$. For $q = (-1, -1)$, the linear complementarity problem LCP(R, q) has no solution. In fact, the problem LCP(R, q) has a solution if and only if q lies in the nonnegative orthant. If both coordinates of q are positive, then there are two solutions: the trivial solution and the solution $w = (0, 0)$ and $z = \left(\frac{1}{1+q^1+q^2}, \frac{q^2}{1+q^1+q^2}, \frac{q^1}{1+q^1+q^2} \right)$. If one of the coordinates of q is equal to 0, say, $q^1 = 0$ and $q^2 > 0$, then the problem LCP(R, q) has a continuum of solutions:

- $z = (z^0, 1 - z^0, 0)$, $w = (0, z^0(1 + q^2) - 1)$, for every $z^0 \in [\frac{1}{1+q^2}, 1]$.

Definition 13.4   An $n \times n$ matrix R is called a *Q-matrix* if for every $q \in \mathbb{R}^n$ the linear complementarity problem LCP(R, q) has at least one solution.

Example 13.5   The matrix $R = \begin{pmatrix} 0 & -1 & -1 \\ 1 & 0 & 1 \\ -1 & 0 & 0 \end{pmatrix}$ is not a Q-matrix, because for $q = \begin{pmatrix} -1 \\ -1 \\ -1 \end{pmatrix}$ the linear complementarity problem LCP(R, q) has no solution. Indeed, suppose (w, z) is a solution of LCP(R, q). Then

$$z^0 q + z^1 R_{[1]} + z^2 R_{[2]} + z^3 R_{[3]} = w \geq \vec{0}.$$

Since $q^1 < 0$, $R_{[2]}^1 < 0$, $R_{[3]}^1 < 0$, and $w^1 \geq 0$, we must have $z^1 = 1$, but then

$$z^0 q + z^1 R_{[1]} + z^2 R_{[2]} + z^3 R_{[3]} = R_{[1]} = \begin{pmatrix} 0 \\ 1 \\ -1 \end{pmatrix},$$

which is not a nonnegative vector.   ◆

$$\begin{pmatrix} 1 & 0 & 0 \\ 3 & 1 & 0 \\ 2 & 0 & 1 \end{pmatrix} \qquad \begin{pmatrix} 1 & 0 & 3 \\ 3 & 1 & 0 \\ 0 & 3 & 1 \end{pmatrix} \qquad \begin{pmatrix} 1 & 4 & 0 & 0 \\ 0 & 1 & 4 & 0 \\ 0 & 0 & 1 & 4 \\ 4 & 0 & 0 & 1 \end{pmatrix}$$

Figure 13.1 Three Q-matrices.

$$\begin{pmatrix} 0 & 2 & -3 \\ -3 & 0 & 1 \\ 5 & -3 & 0 \end{pmatrix} \qquad \begin{pmatrix} 1 & 4 & 0 & 0 \\ 4 & 1 & 0 & 0 \\ 0 & 0 & 1 & 4 \\ 0 & 0 & 0 & 1 \end{pmatrix}$$

Figure 13.2 Two matrices that are not Q-matrices.

The matrices that appear in Figure 13.1 below are Q-matrices. The matrices that appear in Figure 13.2 are not Q-matrices. We are not aware of a characterization of Q-matrices that allows one to identify when a given matrix is a Q-matrix, and verifying, for example, that the middle matrix in Figure 13.1 is a Q-matrix requires a nonnegligible amount of calculations.

The following result provides a simple way to generate new Q-matrices from known Q-matrices.

Theorem 13.6   *Let R be an $n \times n$ matrix, let $c \in \mathbb{R}$, and let R' be the matrix that is obtained from R by adding c to all elements in the first row. Then R is a Q-matrix if and only if R' is a Q-matrix.*

Proof   Let $\vec{c} = (c, 0, 0, \dots, 0) \in \mathbb{R}^n$ be the vector whose first coordinate is c and all the remaining coordinates are 0. To prove the lemma, we will show that (w, z) is a solution of the linear complementarity problem LCP(R, q) if and only if (w', z) is a solution of the linear complementarity problem LCP(R', q'), where $w' = w + \vec{c}$ and $q' = q + \vec{c}$. Indeed, (w, z) is a solution of the problem LCP(R, q) if and only if

$$w = z^0 q + \sum_{i=1}^{n} z^i R r_{[i]},$$

$$w^i \geq R_{[i]}^i, \quad \forall i = 1, 2, \dots, n, \qquad (13.2)$$

$$z^i = 0 \text{ or } w^i = R_{[i]}^i, \quad \forall i = 1, 2, \dots, n.$$

These equalities and inequalities hold if and only if

$$w' = z^0 (q + \vec{c}) + \sum_{i=1}^{n} z^i (R_{[i]} + \vec{c}),$$

$$w'^i \geq R_{[i]}^i + \vec{c}^{\,i}, \quad \forall i = 1, 2, \dots, n, \qquad (13.3)$$

$$z^i = 0 \text{ or } w'^i = R_{[i]}^i + \vec{c}^{\,i}, \quad \forall i = 1, 2, \dots, n.$$

Indeed, to obtain the system (13.3) we added c to the first coordinate in all equalities and inequalities in (13.2). It remains to note that the system (13.3) is equivalent to the property that (w', z) is a solution of $LCP(R', q')$. $\qquad\Box$

13.2 Stationary Equilibria in Quitting Games

The significance of linear complementarity problems for stochastic games is exhibited by the following result, which states that for a given positive recursive quitting game, if a certain matrix that is derived from the payoff function is not a Q-matrix, then the game admits a stationary undiscounted 0-equilibrium. In the rest of this chapter we will denote the number of players by n.

Theorem 13.7 *Let* $\Gamma = \langle I, (\{C^i, Q^i\})_{i \in I}, p_*, (r_*^i)_{i \in I}\rangle$ *be a positive recursive quitting game. Denote by* $n := |I|$ *the number of players and by R the $n \times n$ matrix whose ith column is the vector* $r_*(Q^i, C^{-i})$, *that is, the payoff if player i quits alone. If R is not a Q-matrix, then the game admits a stationary undiscounted 0-equilibrium.*

Proof   Since the matrix R is not a Q-matrix, there is a vector $q \in \mathbb{R}^n$ for which the linear complementarity problem $LCP(R, q)$ has no solution. Consider the auxiliary quitting game $\widehat{\Gamma}$ that is identical to Γ, except that the nonabsorbing payoff in $\widehat{\Gamma}$ is q, and not $\vec{0}$. Thus, the quitting game $\widehat{\Gamma}$ is not recursive.

To distinguish the payoff (resp., the λ-discounted payoff) in the original game Γ from the payoff (resp., the λ-discounted payoff) in the auxiliary game $\widehat{\Gamma}$, we denote the former by $\gamma(x)$ (resp., $\gamma_\lambda(x)$), and the latter by $\widehat{\gamma}(x)$ (resp., $\widehat{\gamma}_\lambda(x)$). Recall that we identify a mixed action of player i with the probability to select Q^i.

Let $\lambda \mapsto x_\lambda$ be a semi-algebraic mapping that assigns a stationary strategy profile in the auxiliary game $\widehat{\Gamma}$ to each discount factor $\lambda \in (0, 1]$, and denote $x_0 := \lim_{\lambda \to 0} x_\lambda$. We argue that if the stationary strategy profile x_0 is absorbing, that is, if $p_*(x_0) > 0$, then

$$\lim_{\lambda \to 0} \gamma_\lambda(x_\lambda) = \gamma_\infty(x_0) = r_*(x_0) = \widehat{\gamma}_\infty(x_0) = \lim_{\lambda \to 0} \widehat{\gamma}_\lambda(x_\lambda). \tag{13.4}$$

Indeed, since x_0 is absorbing, by the definition of the undiscounted payoff we have $\gamma_\infty(x_0) = r_*(x_0) = \widehat{\gamma}_\infty(x_0)$. Moreover, by Theorem 10.1, $\lim_{\lambda \to 0} \gamma_\lambda(x_\lambda) = r_*(x_0)$ and $\lim_{\lambda \to 0} \widehat{\gamma}_\lambda(x_\lambda) = r_*(x_0)$.

By Corollary 8.14, there is a semi-algebraic mapping $\lambda \mapsto \widehat{x}_\lambda$ that assigns a stationary λ-discounted equilibrium of the game $\widehat{\Gamma}$ to each discount factor $\lambda \in (0, 1]$. Denote $\widehat{x}_0 := \lim_{\lambda \to 0} \widehat{x}_\lambda$.

We will show that if $\widehat{x}_0$ is nonabsorbing, then the problem $\mathrm{LCP}(R,q)$ has a solution, which contradicts the choice of q. We will then use earlier results to show that $\widehat{x}_0$ is a stationary undiscounted 0-equilibrium.

Step 1: $p_*(\widehat{x}_\lambda) > 0$ for every $\lambda > 0$ sufficiently small.

Since the mapping $\lambda \mapsto \widehat{x}_\lambda$ is semi-algebraic, so is the function $\lambda \mapsto p_*(\widehat{x}_\lambda)$. By Exercise 6.10, either (a) $p_*(\widehat{x}_\lambda) = 0$ for every $\lambda > 0$ sufficiently small, or (b) $p_*(\widehat{x}_\lambda) > 0$ for every $\lambda > 0$ sufficiently small. Assume, by contradiction, that (a) holds. This implies that $\widehat{x}_\lambda = \vec{C}$ for every $\lambda > 0$ sufficiently small. Consequently, for every $\lambda > 0$ sufficiently small and each player $i \in I$,

$$q^i = \gamma_\lambda^i(\vec{C}) \geq \gamma_\lambda^i(Q^i, C^{-i}).$$

By Theorem 10.1,

$$q^i \geq \lim_{\lambda \to 0} \gamma_\lambda^i(\vec{C}) \geq r_*^i(Q^i, C^{-i}).$$

It follows that the problem $\mathrm{LCP}(R,q)$ admits the trivial solution, which contradicts the choice of q.

Step 2:   If $\widehat{x}_0 = \vec{C}$, then the linear complementarity problem $\mathrm{LCP}(R,q)$ has a solution.

For each player $i \in I$ and every $\lambda \in (0,1)$, set

$$z_\lambda^i := \frac{\widehat{x}_\lambda^i}{\lambda + \sum_{j \in I} \widehat{x}_\lambda^j},$$

and

$$z_\lambda^0 := \frac{\lambda}{\lambda + \sum_{j \in I} \widehat{x}_\lambda^j}.$$

The denominators in the definitions of $(z_\lambda^i)_{i \in I}$ and z_λ^0 are positive, and therefore $(z_\lambda^i)_{i=0,1,\ldots,n}$ are well defined. Since the mapping $\lambda \mapsto \widehat{x}_\lambda$ is semi-algebraic, it follows that the function $\lambda \mapsto z_\lambda^i$ is semi-algebraic for every $i \in \{0,1,\ldots,n\}$. Set

$$z_0^i := \lim_{\lambda \to 0} z_\lambda^i, \quad \forall i \in \{0,1,\ldots,n\}.$$

The vector $(z_\lambda^i)_{i=0}^n$ is a probability distribution for every $\lambda \in (0,1]$, hence so is $z_0 := (z_0^i)_{i=0}^n$.

Set

$$w := \lim_{\lambda \to 0} \widehat{\gamma}_\lambda(\widehat{x}_\lambda). \tag{13.5}$$

We claim that

$$w = z_0^0 q + \sum_{i \in I_*} z_0^i R_{[i]}. \tag{13.6}$$

Indeed, by the definition of the discounted payoff,

$$\widehat{\gamma}_\lambda(\widehat{x}_\lambda) = \frac{\lambda q \prod_{i \in I}(1 - \widehat{x}_\lambda^i) + \sum_{\emptyset \neq J \subseteq I}\left(r_*(Q^J, C^{-J}) \prod_{i \in J} \widehat{x}_\lambda^i \prod_{i \notin J}(1 - \widehat{x}_\lambda^i)\right)}{\lambda \prod_{i \in I}(1 - \widehat{x}_\lambda^i) + \sum_{\emptyset \neq J \subseteq I}\left(\prod_{i \in J} \widehat{x}_\lambda^i \prod_{i \notin J}(1 - \widehat{x}_\lambda^i)\right)}. \tag{13.7}$$

Since $\widehat{x}_0 = \vec{C}$, we have $\lim_{\lambda \to 0} \widehat{x}_\lambda^i = 0$ for every $i \in I$, hence letting $\lambda \to 0$ in Eq. (13.7), we obtain

$$\lim_{\lambda \to 0} \widehat{\gamma}_\lambda(\widehat{x}_\lambda) = \lim_{\lambda \to 0} \frac{\lambda q + \sum_{\emptyset \neq J \subseteq I}\left(r_*(Q^J, C^{-J}) \prod_{i \in J} \widehat{x}_\lambda^i\right)}{\lambda + \sum_{\emptyset \neq J \subseteq I}\left(\prod_{i \in J} \widehat{x}_\lambda^i\right)}. \tag{13.8}$$

When $|J| > 1$ and $i_0 \in J$,

$$\lim_{\lambda \to 0} \frac{\prod_{i \in J} \widehat{x}_\lambda^i}{\widehat{x}_\lambda^{i_0}} = \lim_{\lambda \to 0} \prod_{i \in J \setminus \{i_0\}} \widehat{x}_\lambda^i = 0.$$

That is, in the sums in the numerator and denominator of Eq. (13.8), the coefficient that corresponds to the set of players J vanishes relative to the coefficient that corresponds to the set $J = \{i_0\}$. It follows that Eq. (13.8) further simplifies to

$$\lim_{\lambda \to 0} \widehat{\gamma}_\lambda(\widehat{x}_\lambda) = \lim_{\lambda \to 0} \frac{\lambda q + \sum_{i \in I} \widehat{x}_\lambda^i \cdot r_*(Q^i, C^{-i})}{\lambda + \sum_{i \in I} \widehat{x}_\lambda^i},$$

and Eq. (13.6) follows.

We verify that (w, z_0) is a solution of the linear complementarity problem LCP(R, q), contradicting the assumption that this problem has no solution. In view of Eq. (13.6), it remains to show that $w^i \geq R_{[i]}^i$ for every $i \in I$ and that the complementarity condition holds.

The fact that $\widehat{x}_\lambda$ is a λ-discounted equilibrium of the auxiliary game $\widehat{\Gamma}$ and Eq. (13.4) imply that

$$w^i = \lim_{\lambda \to 0} \widehat{\gamma}_\lambda^i(\widehat{x}_\lambda) \geq \lim_{\lambda \to 0} \widehat{\gamma}_\lambda^i(Q^i, \widehat{x}_\lambda^{-i}) = \widehat{\gamma}_\infty^i(Q^i, C^{-i})$$
$$= R_{[i]}^i = r_*^i(Q^i, C^{-i}), \quad \forall i \in I. \tag{13.9}$$

If $z_0^i > 0$ for some player $i \in I$, then $z_\lambda^i > 0$ for every λ sufficiently close to 0, hence $\widehat{x}_\lambda^i > 0$ for every λ sufficiently close to 0, which implies that we have equality in Eq. (13.9). It follows that the complementarity condition holds, hence (w, z_0) is indeed a solution to the linear complementarity problem LCP(R, q), as claimed.

Step 3:   If $\widehat{x}_0$ is absorbing, then $\widehat{x}_0$ is a stationary 0-equilibrium in Γ.

The argument in Case 1 of the proof of Theorem 10.4 shows that if $\widehat{x}_0$ is absorbing, then it is a stationary uniform 0-equilibrium in the original game Γ. Indeed, by Case 1 of the proof of Theorem 10.4 we know that no player $i \in I$ can profit by deviating from $\widehat{x}_0$ to an action $a^i \in \{C^i, Q^i\}$ when $p_*(a^i, \widehat{x}_0^{-i}) > 0$, while, since the game is recursive and positive, no player can profit by deviating to C^i when $p_*(C^i, \widehat{x}_0^{-i}) = 0$.   □

We now present another relation between the linear complementarity problem and equilibria in quitting games: if there is a probability distribution $z \in \Delta(0, 1, \ldots, n)$ that is part of a solution of the linear complementarity problem LCP(R, q) for every $q \in \mathbb{R}^n$, then the quitting game admits a stationary undiscounted ϵ-equilibrium, for every $\epsilon > 0$.

Lemma 13.8   *Let R be an $n \times n$ matrix. If the probability distribution $z \in \Delta(\{0, 1, \ldots, n\})$ is part of a solution of the linear complementarity problem LCP(R, q), for every $q \in \mathbb{R}^n$, then $z^0 = 0$.*

The proof of the lemma is left to the reader (Exercise 13.8).

Theorem 13.9   *Let $\Gamma = \langle I, (\{C^i, Q^i\})_{i \in I}, (r_*^i)_{i \in I} \rangle$ be a positive recursive quitting game. Denote by $n := |I|$ the number of players and by R the $n \times n$ matrix whose ith column is the vector $r_*(Q^i, C^{-i})$. If there is a vector $(w, z) \in \mathbb{R}^n \times \Delta(\{0, 1, \ldots, n\})$ that is a solution of the linear complementarity problem LCP(R, q) for every $q \in \mathbb{R}^n$, then the game Γ admits a stationary undiscounted ϵ-equilibrium, for every $\epsilon > 0$.*

Proof   Let (w, z) be a solution of the problem LCP(R, q), for every $q \in \mathbb{R}^n$. By Lemma 13.8, $z^0 = 0$, and therefore $\sum_{i \in I} z^i = 1$. By the definition of a solution to LCP(R, q),

$$w = \sum_{i \in I} z^i r_*(Q^i, C^{-i}), \tag{13.10}$$

$$w^i \geq r_*^i(Q^i, C^{-i}), \quad \forall i \in I, \tag{13.11}$$

$$z^i > 0 \implies w^i = r_*^i(Q^i, C^{-i}), \quad \forall i \in I. \tag{13.12}$$

Fix $\epsilon > 0$ and let $x_\epsilon^i := \epsilon z^i$ be the stationary strategy of player i in which she quits at every stage with probability ϵz^i. Let $x_\epsilon := (x_\epsilon^i)_{i \in I}$. Under x_ϵ, in every stage, the probability that a single player quits is $\sum_{i \in I} x_\epsilon^i \prod_{j \neq i}(1 - x_\epsilon^j)$, which is of the order of ϵ. Similarly, in every stage the probability that exactly L players quit together is of the order of ϵ^L.

Denote by T_1 the event that the game is terminated by a single player; that is, at the first stage in which at least one player quits, exactly one player quits and the others continue. Denote by $T_{\geq 2}$ the event that the game is terminated by at least two players. Since x_ϵ is a stationary strategy profile under which the per-stage probability of absorption is positive, $\mathbf{P}_{x_\epsilon}(T_1 \cup T_{\geq 2}) = 1$. The discussion in the previous paragraph implies that $\lim_{\epsilon \to 0} \frac{\mathbf{P}_{x_\epsilon}(T_{\geq 2})}{\mathbf{P}_{x_\epsilon}(T_1)} = 0$. This implies that $\lim_{\epsilon \to 0} \mathbf{P}_{x_\epsilon}(T_1) = 1$.

Consequently, and since $\sum_{i \in I} z^i = 1$,

$$
\begin{aligned}
\lim_{\epsilon \to 0} \gamma_\infty(x_\epsilon) &= \lim_{\epsilon \to 0} r_*(x_\epsilon) \\
&= \lim_{\epsilon \to 0} \sum_{i \in I} \frac{\epsilon z^i}{\sum_{j \in I} \epsilon z^j} r_*(Q^i, C^{-i}) \\
&= \sum_{i \in I} \frac{z^i}{\sum_{j \in I} z^j} r_*(Q^i, C^{-i}) \qquad (13.13) \\
&= \sum_{i \in I} z^i r_*^i(Q^i, C^{-i}) \\
&= w.
\end{aligned}
$$

By Eqs. (13.13) and (13.11), for every $\delta > 0$ there is an $\epsilon_0 > 0$ such that

$$
\gamma_\infty^i(x_\epsilon) \geq w^i - \delta \geq r_*^i(Q^i, C^{-i}) - \delta, \quad \forall \epsilon \in (0, \epsilon_0). \qquad (13.14)
$$

We argue that x_ϵ is an undiscounted $(M\epsilon + 3\delta)$-equilibrium, provided $\epsilon > 0$ is sufficiently small. To this end, we fix $i \in I$ and show that player i cannot profit by deviating more than $(M\epsilon + 2\delta)$.

Suppose player i quits at some stage. Her expected absorbing payoff will be $r_*^i(Q^i, x_\epsilon^{-i})$. Since $x_\epsilon^i = \epsilon z^i$ for each $i \in I$ and since $\sum_{i \in I} z^i = 1$, we have

$$
r_*^i(Q^i, x_\epsilon^{-i}) \leq r_*^i(Q^i, C^{-i}) + M\epsilon \leq \gamma_\infty^i(x_\epsilon) + M\epsilon. \qquad (13.15)
$$

If $z^i = 0$, then under x_ϵ player i always continues, hence the only deviation she has is to quit at some stage, which, as we have seen in the previous paragraph, does not increase her payoff by more than $M\epsilon$.

If $z^i = 1$, that is, under x_ϵ all other players continue, then $\gamma_\infty(x_\epsilon) = r_*(Q^i, C^{-i})$. Since the game is positive and recursive, player i cannot profit by deviating: if she ever plays Q^i her payoff is $r_*^i(Q^i, C^{-i}) = \gamma_\infty^i(x_\epsilon)$, while if she always plays C^i her payoff is $0 < r_*^i(Q^i, C^{-i}) = \gamma_\infty^i(x_\epsilon)$.

It remains to treat the case $z^i \in (0,1)$. In that case, there is $j \in I \setminus \{i\}$ with $z^j > 0$, hence even if player i always continues, the play will be eventually absorbed by some other player. We will show that if the game is absorbed while player i continues, then her expected absorbing payoff is close to $r_i^*(Q^i, C^{-i})$ as well, and thereby complete the proof.

By Eq. (13.10), w^i is a convex combination of $(r_*^i(Q^j, C^{-j}))_{j \in I}$, and can be expressed as

$$w^i = \sum_{j \in I} \frac{z^j}{\sum_{j' \in I} z^{j'}} r_*^i(Q^j, C^{-j}) \tag{13.16}$$

$$= \frac{z^i}{\sum_{j' \in I} z^{j'}} r_*^i(Q^i, C^{-i}) + \frac{1 - z^i}{\sum_{j' \in I} z^{j'}} \sum_{j \neq i} \frac{z^j}{1 - z^i} r_*^i(Q^j, C^{-j}),$$

where the denominator is positive because $z^i < 1$. The fact that $z^i > 0$ and Eq. (13.12) imply that $w^i = r_*^i(Q^i, C^{-i})$; hence, we deduce from Eq. (13.16) that

$$w^i = \sum_{j \neq i} \frac{z^j}{\sum_{j' \neq i} z^{j'}} r_*^i(Q^j, C^{-j}). \tag{13.17}$$

As in the derivation of Eq. (13.13), as ϵ goes to 0 the probability that under (C^i, x_ϵ^{-i}) two players quit simultaneously goes to 0, hence by Eq. (13.17)

$$\lim_{\epsilon \to 0} \gamma_\infty^i(C^i, x_\epsilon^{-i}) = \lim_{\epsilon \to 0} r_*^i(C^i, x_\epsilon^{-i})$$

$$= \sum_{j \neq i} \frac{\epsilon z^j}{\sum_{j \neq i} \epsilon z^j} r_*^i(Q^j, C^{-j})$$

$$= \sum_{j \neq i} \frac{z^j}{\sum_{j \neq i} z^j} r_*^i(Q^j, C^{-j})$$

$$= w^i = r_*^i(Q^i, C^{-i}), \tag{13.18}$$

where the last equality holds by the complementarity condition (Eq. (13.12)). This shows that for every $\delta > 0$ there is $\epsilon_1 > 0$ such that

$$\gamma_\infty^i(C^i, x_\epsilon^{-i}) \leq w^i + \delta, \quad \forall \epsilon \in (0, \epsilon_1).$$

It follows from Eqs. (13.14) and (13.18) that for every $\epsilon \in (0, \min\{\epsilon_0, \epsilon_1\})$, we have

$$\gamma_\infty^i(C^i, x_\epsilon^{-i}) \leq w^i + \delta \leq \gamma_\infty^i(x_\epsilon) + 2\delta. \tag{13.19}$$

Eqs. (13.15) and (13.19) imply that player i cannot profit more than $(M\epsilon + 2\delta)$ by deviating from x_ϵ. $\qquad\square$

13.3 Cyclic Equilibrium in Three-Player Positive Quitting Games

In Section 12.2, we analyzed a specific three-player quitting game, and showed that it has a periodic undiscounted ϵ-equilibrium. In this section, we extend the analysis to all three-player positive recursive quitting games.

Theorem 13.10 *Every three-player positive recursive quitting game admits a periodic undiscounted ϵ-equilibrium, for every $\epsilon > 0$.*

In view of Theorem 13.7, it remains to show the following.

Theorem 13.11 *Let Γ be a three-player positive recursive quitting game. Denote by R the 3×3 matrix whose ith column is the vector $r_*(Q^i, C^{-i})$. If R is a Q-matrix, then the game admits a periodic undiscounted ϵ-equilibrium, for every $\epsilon > 0$.*

Proof By Theorem 13.6, we can assume without loss of generality that the three diagonal entries of R are 1; that is, $r_*^i(Q^i, C^{-i}) = 1$ for each $i \in I$.

Step 1: If there is $i \in \{1, 2, 3\}$ such that $R_{[i]} \geq (1, 1, 1)$, then the game admits a stationary undiscounted ϵ-equilibrium, for every $\epsilon > 0$.

In this case, the following pair (q, z) is a solution of the linear complementarity problem $\mathrm{LCP}(R, q)$, for every $q \in \mathbb{R}^n$:

$$z^0 = 0, \quad z^i = 1, \quad z^j = 0 \quad \forall j \in I \setminus \{i\}.$$

By Theorem 13.9, the game has a stationary undiscounted ϵ-equilibrium.

Step 2: There is no row in R that contains two entries smaller than 1 (and the diagonal entry in that row is 1).

If one of the rows in R contains two entries smaller than 1 and the condition in Step 1 does not hold, then, as in Example 13.5, the matrix R is not a Q-matrix, which contradicts the assumption.

Step 3: If there is a convex combination of the columns of R that is equal to $(1, 1, 1)$, then the game admits a stationary undiscounted ϵ-equilibrium, for every $\epsilon > 0$.

Suppose that $(z^i)_{i=1}^3 \in \Delta\{1, 2, 3\}$ satisfies

$$\sum_{i=1}^3 z^i R_{[i]} = (1, 1, 1).$$

$$\begin{pmatrix} 1 & <1 & >1 \\ >1 & 1 & <1 \\ <1 & >1 & 1 \end{pmatrix}$$

Figure 13.3 The matrix R.

Set $z^0 := 0$, $z := (z^i)_{i=0}^3$, and $w = (1,1,1)$. The reader can verify that (w,z) is a solution of the problem LCP(R,q) for every $q \in \mathbb{R}^3$; hence, by Theorem 13.9 the game admits a stationary undiscounted ϵ-equilibrium, for every $\epsilon > 0$.

By Step 1, we can assume that each column of R contains an entry smaller than 1. Step 2 implies that in fact every row and every column of R contains exactly one entry larger than 1 and exactly one entry smaller than 1. By Step 3, we can assume that there is no convex combination of the columns of R that is equal to $(1,1,1)$.

Assume without loss of generality that the entries of R look as in Figure 13.3.

Step 4: There is a convex combination of the three columns of R that lies in $[1,\infty)^3$.

Set $q = (0,0,0)$ and consider the linear complementarity problem LCP(R,q). Since R is a Q-matrix, this problem has a solution $(w,z) \in \mathbb{R}^3 \times \Delta(\{0,1,2,3\})$. In particular, $w^i \geq R^i_{[i]} = 1$ for each $i = 1,2,3$, and

$$w = z^0 q + \sum_{i=1}^3 z^i R_{[i]}.$$

Since $q = (0,0,0)$, it follows that at least one of z^1, z^2, or z^3 is positive, and $\sum_{i=1}^3 z^i R_{[i]} \geq (1,1,1)$. The vector (z^1,z^2,z^3) is not necessarily a probability distribution, because z^0 may be positive. However, the vector $\widehat{z} = (\widehat{z}^1,\widehat{z}^2,\widehat{z}^3)$ defined by

$$\widehat{z}^i := \frac{z^i}{z^1 + z^2 + z^3}, \quad i \in \{1,2,3\}$$

is a probability distribution and, because $z^1 + z^2 + z^3 \leq 1$, it satisfies $\sum_{i=1}^3 \widehat{z}^i R_{[i]} \geq (1,1,1)$. In particular, we identified a convex combination of the three columns of R that lies in $[1,\infty)^3$.

Step 5: Drawing the columns of R in the payoff space.

Figure 13.4 displays the three columns of R.

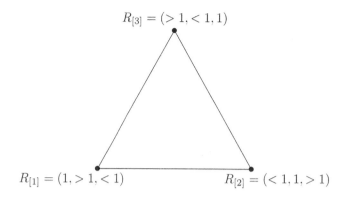

Figure 13.4 The vectors $(R_{[i]})_{i=1,2,3}$.

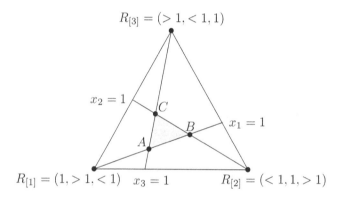

Figure 13.5 The vectors $(R_{[i]})_{i=1,2,3}$ and the lines $x^1 = 0$, $x^2 = 0$ and $x^3 = 0$.

Let us add to the figure the lines $x^1 = 1$, $x^2 = 1$, and $x^3 = 1$ (see Figure 13.5). Note that the line $x^1 = 1$ passes through $R_{[1]}$ (since $R^i_{[i]} = 1$) and, since $R^1_{[3]} > 1$ and $R^1_{[2]} < 1$, it passes between $R_{[2]}$ and $R_{[3]}$. Similarly, the line $x^2 = 1$ passes through $R_{[2]}$ and between $R_{[1]}$ and $R_{[3]}$, and the line $x^3 = 1$ passes through $R_{[3]}$ and between $R_{[1]}$ and $R_{[2]}$. The darker region is Figure 13.5 is the set of all vectors in the convex hull of $R_{[1]}$, $R_{[2]}$, and $R_{[3]}$ that lie in $[1, \infty)^3$. Since there is a convex combination of the columns of R that lies in $[1, \infty)^3$, the dark region is nonempty. Denote the three extreme points of the dark region by A, B, and C (see Figure 13.5). If $A = B = C$, then there is a convex combination of the columns of the matrix that is equal to $(1, 1, 1)$, and then, by Step 3, a stationary undiscounted ϵ-equilibrium exists for every $\epsilon > 0$. Assume that such a combination does not exist. Hence, the dark region in Figure 13.5 has a nonempty interior. As a result, there are numbers $\alpha, \beta, \gamma \in (0, 1)$ such that

$$
\begin{aligned}
A &= \alpha R_{[1]} + (1 - \alpha)B, \\
B &= \beta R_{[2]} + (1 - \beta)C, \\
C &= \gamma R_{[3]} + (1 - \gamma)A.
\end{aligned}
\tag{13.20}
$$

Step 6: Defining a periodic strategy profile σ_*.

Fix $\epsilon > 0$ and a positive integer $m = m(\epsilon)$ such that

$$
(1 - \alpha)^{1/m} > 1 - \epsilon, \quad (1 - \beta)^{1/m} > 1 - \epsilon, \quad (1 - \gamma)^{1/m} > 1 - \epsilon.
\tag{13.21}
$$

Consider the following strategy profile σ_* in Γ, which is periodic with period $3m$. In each period, the players proceed as follows:

- In each of the first m stages of the period, Player 1 quits with probability $1 - (1 - \alpha)^{1/m}$, and Players 2 and 3 play C^2 and C^3, respectively.
- In each of the following m stages of the period, Player 2 quits with probability $1 - (1 - \beta)^{1/m}$, and Players 1 and 3 play C^1 and C^3, respectively.
- In each of the following m stages of the period, Player 3 quits with probability $1 - (1 - \gamma)^{1/m}$, and Players 1 and 2 play C^1 and C^2, respectively.

Under σ_*, the probability that Player 1 continues in each stage of the first m stages of a period is $(1 - \alpha)^{1/m}$, hence the probability that she continues in these stages is $1 - \alpha$, and therefore the probability that she quits in these stages is α. Similarly, the probability that Player 2 quits in the next m stages of the period, conditioned that the play was not absorbed before these stages, is β, and the probability that Player 3 quits in the last m stages of the period, conditioned that the play was not absorbed before these stages, is γ.

For every $t \in \mathbb{N}$, we let

$$
\gamma_\infty(\sigma_* \mid t_* \geq t) := \mathbf{E}_{\sigma_*}[\mathbf{1}_{\{t_* < \infty\}} r_*(t_*) \mid t_* \geq t)
$$

denote the undiscounted payoff under σ_* conditioned that the play was not absorbed before stage t.

Step 7: Calculating $\gamma_\infty(\sigma_* \mid t_* \geq t)$.

Under σ_*, the probability of absorption in each period is bounded away from 0, hence under σ_* the play is absorbed with probability 1, that is, $\mathbf{P}_{\sigma_*}(t_* < \infty) = 1$.

Set $a := \gamma_\infty(\sigma_*)$, $b := \gamma_\infty(\sigma_* \mid t_* > m)$, and $c := \gamma_\infty(\sigma_* \mid t_* > 2m)$. That is, a is the undiscounted payoff under σ_* from the beginning of the game, b is the undiscounted payoff under σ_* if Player 1 did not quit in the first m stages, and c is the undiscounted payoff under σ_* if Players 1 and 2 did not quit in the first $2m$ stages. Since the strategy profile σ_* is periodic, we have $a = \gamma_\infty(\sigma_* \mid t_* > 3m)$.

Since the probabilities that the three players quit in each period are α, β, and γ, respectively, the quantities a, b, and c satisfy

$$a = \alpha R_{[1]} + (1 - \alpha)b,$$
$$b = \beta R_{[2]} + (1 - \beta)c,$$
$$c = \gamma R_{[3]} + (1 - \gamma)a.$$

Since $\alpha, \beta, \gamma \in (0, 1)$, this system has a unique solution. However, (A, B, C) is also a solution of this system, see the system of equations (13.20). We conclude that $\gamma_\infty(\sigma_*) = a = A$, and similarly $\gamma_\infty(\sigma_* \mid t_* > m) = b = B$ and $\gamma_\infty(\sigma_* \mid t_* > 2m) = c = C$.

We will now study $\gamma_\infty(\sigma \mid t_* \geq t)$ for $t \in \{1, 2, \ldots, m + 1\}$. Since $t \in \{1, 2, \ldots, m + 1\}$, between stages 1 and $t - 1$ the only player who may quit under σ_* is Player 1. Therefore,

$$A = \gamma_\infty(\sigma_*) = \mathbf{P}_{\sigma_*}(t_* < t) \cdot R_{[1]} + \mathbf{P}_{\sigma_*}(t_* \geq t) \cdot \gamma_\infty(\sigma \mid t_* \geq t),$$

and hence

$$\gamma_\infty(\sigma \mid t_* \geq t) = \frac{A - \mathbf{P}_{\sigma_*}(t_* < t) \cdot R_{[1]}}{1 - \mathbf{P}_{\sigma_*}(t_* < t)}.$$

Since $A^1 = R_{[1]}^1 = 1$ (see Figure 13.5), it follows that $\gamma_\infty^1(\sigma \mid t_* \geq t) = 1$, and the point $\gamma_\infty(\sigma \mid t_* \geq t)$ lies on the line that connects A and B. For $t = 1$ we have $\gamma_\infty(\sigma \mid t_* \geq 1) = A$, for $t = m+1$ we have $\gamma_\infty(\sigma \mid t_* \geq m+1) = B$, and as t increases from 1 to m the point $\gamma_\infty(\sigma \mid t_* \geq t)$ moves from A toward B.

Similarly, for every $t \in \{m+1, m+2, \ldots, 2m+1\}$, the point $\gamma_\infty(\sigma \mid t_* \geq t)$ lies on the line that connects B and C, and for every $t \in \{2m + 1, 2m + 2, \ldots, 3m + 1\}$, the point $\gamma_\infty(\sigma \mid t_* \geq t)$ lies on the line that connects C and A. In particular,

$$\gamma_\infty^i(\sigma \mid t_* \geq t) \geq 1, \quad \forall i \in I, \forall t \in \mathbb{N}. \tag{13.22}$$

That is, if the game was not absorbed in the first $t - 1$ stages, then the expected payoff to all players from stage t and on is at least 1.

Step 6: σ_* is an undiscounted $M\epsilon$-equilibrium.

We will show that Player 1 cannot profit more than $M\epsilon$ by deviating. If Player 1 is supposed to continue at stage t and she deviates and quits at that stage, then, by Eq. (13.22), her payoff if she does not deviate is at least 1, while since the player who is supposed to quit at stage t does so with probability smaller than ϵ (see Eq. (13.21)), Player 1's payoff by quitting is bounded by $1 + M\epsilon$. It follows that such a deviation cannot increase Player 1's payoff by more than ϵ.

Suppose that Player 1 continues throughout the period. Player 2 (resp. Player 3) still quits with probability β (resp. γ) in stages $\{m + 1, m + 2, \ldots, 2m\}$ (resp. $\{2m + 1, 2m + 2, \ldots, 3m\}$). We argue that Player 1's expected absorbing payoff along the period conditional on the event that absorption occurs in the period remains 1. Indeed, this payoff is $\frac{\beta R_{[2]}^1 + (1-\beta)\gamma R_{[3]}^1}{\beta + (1-\beta)\gamma}$. The system of equations (13.20) implies that

$$
\begin{aligned}
A &= \frac{\alpha R_{[1]} + (1 - \alpha)\beta R_{[2]} + (1 - \alpha)(1 - \beta)\gamma R_{[3]}}{\alpha + (1 - \alpha)\beta + (1 - \alpha)(1 - \beta)\gamma} \\
&= \frac{\alpha}{\alpha + (1 - \alpha)\beta + (1 - \alpha)(1 - \beta)\gamma} \cdot R_{[1]} \\
&\quad + \frac{(1 - \alpha)\beta + (1 - \alpha)(1 - \beta)\gamma}{\alpha + (1 - \alpha)\beta + (1 - \alpha)(1 - \beta)\gamma} \cdot \frac{\beta R_{[2]} + (1 - \beta)\gamma R_{[3]}}{\beta + (1 - \beta)\gamma},
\end{aligned}
$$

Since $A^1 = R_{[1]}^1 = 1$, we deduce that

$$
\frac{\beta R_{[2]}^1 + (1 - \beta)\gamma R_{[3]}^1}{\beta + (1 - \beta)\gamma} = 1,
$$

as claimed.

Thus, Player 1's profit if she quits at a given stage is at most $M\epsilon$, and she cannot profit by continuing throughout a period. It follows that she has no deviation that yields a profit higher than $M\epsilon$. Similarly, Players 2 and 3 cannot profit more than $M\epsilon$ by deviating from σ_*.   □

13.4 Comments and Extensions

Linear complementarity problems were studied extensively starting in the 1960s, as they generalize several optimization problems, like Nash equilibrium in two-player nonzero-sum games (see Exercise 13.1). Readers who are interested in this topic are referred to Cottle et al. (1992). In this literature, a linear complementarity problem is defined differently than the way we did. Denote the nonnegative orthant by

$$\mathbb{R}^n_+ := \left\{ x \in \mathbb{R}^n : x^i \geq 0 \ \forall i \in \{1, 2, \ldots, n\} \right\}.$$

Let R be an $n \times n$ matrix, and let $q \in \mathbb{R}^n$. The linear complementarity problem $\mathrm{LCP}(R, q)$ is the following problem that consists of linear equalities and inequalities:

$$\text{Find} \quad z, w \in \mathbb{R}^n_+, \tag{13.23}$$
$$\text{such that} \quad w = q + Rz,$$
$$z_i = 0 \text{ or } w_i = 0, \quad \forall i \in \{1, 2, \ldots, n\}.$$

Problem (13.23) differs from Problem (13.1) in two ways:

- Whereas in Problem (13.23) we require that w lies in the nonnegative orthant, in Problem (13.1) the vector w is compared to the diagonal of the matrix R.
- Whereas in Problem (13.1) the vector z is required to be a probability distribution on $\{0, 1, \ldots, n\}$, in Problem (13.23) the vector z is in the nonnegative orthant in $\mathbb{R}^n$ and the weight of q is 1. Alternatively, we could have required the vector z in Problem (13.23) to be a probability distribution on $\{0, 1, \ldots, n\}$ such that $z^0 > 0$.

In our application, it is more convenient to use the form (13.1) than the form (13.23).

In this chapter, we provided conditions that involve linear complementarity problems and ensure the existence of an undiscounted 0-equilibrium: if the matrix R whose ith column is the payoff vector $r_*(Q^i, C^{-i})$ is not a Q-matrix, then the game admits a stationary undiscounted 0-equilibrium. It follows from Ashkenazi-Golan et al. (2020) that if the matrix R and all its principal minors are Q-matrices, then for every $\epsilon > 0$ the quitting game admits an ϵ-equilibrium in which in every stage at most one player quits with positive probability. It is not known whether an undiscounted ϵ-equilibrium exists in the intermediate cases, where R is a Q-matrix, but some of its principal minors are not.

Theorem 13.7 appeared in Solan and Solan (2020), who proved that if the matrix R is a Q-matrix, then there exists a sunspot undiscounted ϵ-equilibrium for every $\epsilon > 0$. This is an undiscounted ϵ-equilibrium in an extended game in which the players observe at every stage a public signal in $[0, 1]$ that is drawn from the uniform distribution, independently of past signals. This result was extended by Solan and Solan (2021) to positive recursive quitting games in which players may have more than one continue action, and the absorbing payoff depends on the set of players who quit at the terminal stage, as well as on the continue actions that the other players adopt at that stage. This latter result was used by Solan et al. (2020) to prove that every positive recursive quitting game in which at least two players have at least two continue actions has an undiscounted ϵ-equilibrium, for every $\epsilon > 0$.

Solan (1999) proved that every three-player absorbing game admits a uniform ϵ-equilibrium, for every $\epsilon > 0$. Theorem 13.11 shows that in certain cases this uniform ϵ-equilibrium is periodic.

Exercise 13.1 is adapted from Cottle and Dantzig (1968).

13.5 Exercises

1. In this exercise, we show a relation between the linear complementarity problem and equilibrium in strategic-form games. We will use the following notations. For every $k \in \mathbb{N}$ denote by
   $e_k := (0, \ldots, 0, 1, 0, \ldots, 0)$ the vector with all coordinates equal 0, except for coordinate k, which is equal to 1, by $\vec{1}_n := (1, 1, \ldots, 1) \in \mathbb{R}^n$ the vector with all coordinates equal to 1, and by 0_k the $k \times k$ matrix with all entries equal to 0. All vectors are row vectors, and the column vector that corresponds to the row vector x is denoted by x^t. Similarly, when A is an $(n \times m)$ matrix, the transpose of A is denoted by A^t.

   Let A and B be two $n \times m$ positive matrices; that is, the entries of A and B are positive real numbers. Let G be the two-player strategic-form game where Player 1 has n actions, Player 2 has m actions, and the *losses* of the two players are given by the matrices A and B, respectively. Thus, the pair of vectors $(x, y) \in \Delta_n \times \Delta_m$ is an *equilibrium* of G if and only if

   $$x_* A y_*^t \leq x A y_*^t \quad \forall x \in \Delta_n, \quad x_* A y_*^t \leq x_* A y^t \quad \forall y \in \Delta_m.$$

   (a) Show that $(x_*, y_*) \in \Delta_n \times \Delta_m$ is an equilibrium of the game G if and only if

   $$(x_* A y_*^t)\vec{1}_n \leq A y_* \quad \text{and} \quad (x_* B y_*^t)\vec{1}_m \leq B^t x_*.$$

   (b) Let $R = \begin{pmatrix} 0_n & A \\ B^t & 0_m \end{pmatrix}$ and $q = -\vec{1}_{m+n}$. Show that if (x_*, y_*) is an equilibrium of the game G, then there is a vector $w \in \mathbb{R}^{n+m}$ such that (w, z) is a solution of the problem LCP(R, q), for a proper vector $z \in \mathbb{R}^{1+n+m}$.

   (c) Show that if the problem LCP(R, q) has a solution (w, z), with $z = (z_0, \zeta, \widehat{\zeta}) \in \mathbb{R}^{1+n+m}$, then the pair $\left(\frac{\zeta}{\sum_{i=1}^n \zeta^i}, \frac{\widehat{\zeta}}{\sum_{j=1}^m \widehat{\zeta}^j} \right)$ is an equilibrium of G.

2. Find a 3×3 matrix R and a vector $q \in \mathbb{R}^3$ such that the problem LCP(R, q) has exactly four solutions.

3. Find a 3×3 matrix R and a vector $q \in \mathbb{R}^3$ such that the problem LCP(R, q) has exactly seven solutions.

4. Let R be an $n \times n$ matrix that satisfies the following condition: there exists $i \in \{1, 2, \ldots, n\}$ such that $R^j_{[i]} \geq R^j_{[j]}$ for every $j \in \{1, 2, \ldots, n\}$. Prove that the matrix R is a Q-matrix.

5. Let R be an $n \times n$ matrix that satisfies the following condition: there exist $i, j \in \{1, 2, \ldots, n\}$ such that (a) $R^j_{[i]} < R^j_{[j]}$ and $R^i_{[k]} < R^i_{[i]}$ for every $k \neq i$. Prove that the matrix R is not a Q-matrix.

6. (a) Prove that the two 3×3 matrices that appear in Figure 13.1 are Q-matrices.

   (b) Prove that the matrices that appear in Figure 13.2 are not Q-matrices.

7. Let $\Gamma = \langle I, (\{C^i, Q^i\})_{i \in I}, p_*, (r^i_*)_{i \in I} \rangle$ be a positive recursive quitting game such that the matrix R whose ith column is the vector $r_*(Q^i, C^{-i})$ is a Q-matrix. Providing an example to show that the game Γ may have a stationary undiscounted 0-equilibrium. Does this contradict Theorem 13.7?

8. Prove Theorem 13.8: Let R be an $n \times n$ matrix. Show that if the probability distribution $z \in \Delta(\{0, 1, \ldots, n\}$ is part of a solution of the linear complementarity problem LCP(R, q), for every $q \in \mathbb{R}^n$, then $z^0 = 0$.

9. In this exercise, we extend Theorem 13.11 to the situation in which the absorption probability when some players quit is not necessarily 1. Let $\Gamma = \langle I, (\{Q^i, C^i\})_{i \in I}, p_*, (r^i_*)_{i \in I} \rangle$ be a three-player positive recursive absorbing game where each player $i \in I$ has two actions, $A^i = \{C^i, Q^i\}$, and the transition rule satisfies $p_*(C^1, C^2, \ldots, C^n) = 0$ and $p_*(Q^J, C^{-J}) > 0$ for every non-empty subset $\emptyset \neq J \subseteq N$. Denote by R the matrix whose ith column is the vector $r_*(Q^i, C^{-i})$.

   (a) Prove that if R is not a Q-matrix, then the game admits a stationary undiscounted 0-equilibrium.

   (b) Prove that if R is a Q-matrix, then the game admits a periodic undiscounted ϵ-equilibrium, for every $\epsilon > 0$.

10. Is it true that in the conditions of Step 4 of the proof of Theorem 13.11, there is a convex combination of the columns of R that lies in $(1, \infty)^3$? Justify your answer.

11. Let $\Gamma = \langle I, (\{Q^i, C^i\})_{i \in I}, (r^i_*)_{i \in I} \rangle$ be a four-player positive recursive quitting game where

$$r_*(Q^1, C^2, C^3, C^4) = (1,4,0,0),$$
$$r_*(C^1, Q^2, C^3, C^4) = (0,1,4,0),$$
$$r_*(C^1, C^2, Q^3, C^4) = (0,0,1,4),$$
$$r_*(C^1, C^2, C^3, Q^4) = (4,0,0,1).$$

For every $d \in (0,1)$ and every $m \in \mathbb{N}$ let $\sigma^i_{d,m} \in \Sigma^i$ be the following strategy of player i, that is periodic with period $4m$:

- Quit with probability d in each of the stages $(i-1)m$, $(i-1)m + 1, \ldots, im - 1$ of the period.
- Continue in all other stages of the period.

Show that for every $\epsilon > 0$ there are $d \in (0,1)$ and $m \in \mathbb{N}$ such that the strategy profile $\sigma_{d,m} := (\sigma^i_{d,m})^4_{i=1}$ is an undiscounted ϵ-equilibrium.

References

Altman, E., Avrachenkov, K., Bonneau, N., Debbah, M., El-Azouzi, R., and Sadoc Menasche, D. (2008) Constrained Cost-Coupled Stochastic Games with Independent State Processes, *Operations Research Letters*, **36**, 160–164.

Amir, R. (1996) Continuous Stochastic Games of Capital Accumulation with Convex Transitions, *Games and Economic Behavior*, **15**(2), 111–131.

Andersson, D. and Miltersen, P. B. (2009) The Complexity of Solving Stochastic Games on Graphs. In Dong, Y., Du, D. Z., and Ibarra, O. (eds.), *Algorithms and Computation*, ISAAC 2009. Lecture Notes in Computer Science, vol. 5878, Springer, pp. 112–121.

Ashkenazi-Golan, G., Krasikov, I., Rainer, C., and Solan, E. (2020) Absorption Paths and Equilibria in Quitting Games. https://arxiv.org/pdf/2012.04369.pdf.

Attia, L. and Oliu-Barton, M. (2020) A Formula for the Value of a Stochastic Game, *Proceedings of the National Academy of Sciences of the United States of America*, **116**(52), 26435–26443.

Başar, T. and Olsder, G. J. (1998) *Dynamic Noncooperative Game Theory*, 2nd ed., Academic Press.

Başar, T. and Zaccour, G. (2017) *Handbook of Dynamic Game Theory*, Springer.

Benedetti, R. and Risler, J. J. (1990) *Real Algebraic and Semi-Algebraic Sets*, Hermann.

Bewley, T. and Kohlberg, E. (1976) The Asymptotic Theory of Stochastic Games, *Mathematics of Operations Research*, **1**, 197–208.

Bewley, T. and Kohlberg, E. (1978) On Stochastic Games with Stationary Optimal Strategies, *Mathematics of Operations Research*, **3**, 104–125.

Billingsley, P. (1995) *Probability and Measure*, John Wiley & Sons.

Blackwell, D. (1962) Discrete Dynamic Programming, *The Annals of Mathematical Statistics*, **33**(2), 719–726.

Blackwell, D. (1965) Discounted Dynamic Programming, *The Annals of Mathematical Statistics*, **36**(1), 226–235.

Blackwell, D. and Ferguson, T. S. (1968) The Big Match, *The Annals of Mathematical Statistics*, **39**, 159–163.

Bochnak, J., Coste, M., and Roy, M. F. (2013) *Real Algebraic Geometry*, Springer Science & Business Media.

Bolte, J., Gaubert, S., and Vigeral, G. (2015) Definable Zero-Sum Stochastic Games, *Mathematics of Operations Research*, **40**(1), 171–191.

Border, K. C. (1985) *Fixed Point Theorems with Applications to Economics and Game Theory*, Cambridge University Press.

Bourque, M. and Raghavan, T. E. S. (2014) Policy Improvement for Perfect Information Additive Reward and Additive Transition Stochastic Games with Discounted and Average Payoffs, *Journal of Dynamics and Games*, **1**(3), 347–361.

Breton, M. (1991) Algorithms for Stochastic Games. In Raghavan T. E. S., Ferguson, T. S., and Parthasarathy, T. (eds.), *Stochastic Games and Related Topics: In Honor of Professor LS Shapley*, Theory and Decision Library, Series C, Game Theory, Mathematical Programming and Operations Research, vol.7, Kluwer, pp. 45–57.

Catoni, O., Oliu-Barton, M., and Ziliotto, B. (2021, November) Constant Payoff in Zero-Sum Stochastic Games, *Annales de l'Institut Henri Poincaré* (*Probabilités et Statistiques*), **57**(4), 1888–1900.

Chakrabarti, S. K. (2003) Pure Strategy Markov Equilibrium in Stochastic Games with a Continuum of Players, *Journal of Mathematical Economics*, **39**(7), 693–724.

Chatterjee, K., Alfaro, L. D., and Henzinger, T. A. (2008) Termination Criteria for Solving Concurrent Safety and Reachability Games. In Matheiu, C. (ed.), *Proceedings of the Twentieth Annual ACM-SIAM Symposium on Discrete Algorithms*, Association for Computer Machinery and Society for Industrial and Applied Mathematics, pp. 197–206.

Chatterjee, K., Doyen, L., and Henzinger, T. A. (2009) A Survey of Stochastic Games with Limsup and Liminf Objectives. In Albers, S., Marchetti-Spaccamela, A., Matias, Y., Nikoletseas, S., and Thomas, W. (eds.), *International Colloquium on Automata, Languages, and Programming*, Springer, pp. 1–15.

Chatterjee, K., Doyen, L., and Henzinger, T. A. (2013) A Survey of Partial-Observation Stochastic Parity Games, *Formal Methods in System Design*, **43**(2), 268–284.

Chatterjee, K. and Henzinger, T. A. (2012) A Survey of Stochastic ω-Regular Games, *Journal of Computer and System Sciences*, **78**, 394–413.

Chatterjee, K., Majumdar, R., and Henzinger, T. A. (2008) Stochastic Limit-Average Games Are in EXPTIME, *International Journal of Game Theory*, **37**, 219–234.

Condon, A. (1992) The Complexity of Stochastic Games, *Information and Computation*, **96**(2), 203–224.

Cottle, R. W. and Dantzig, G. B. (1968) Complementary Pivot Theory of Mathematical Programming, *Linear Algebra and Its Applications*, **1**, 103–125.

Cottle, R. W., Pang, J. S., and Stone, R. E. (1992) *The Linear Complementarity Problem*, SIAM.

Coulomb, J. M. (2003) Stochastic Games without Perfect Monitoring, *International Journal of Game Theory*, **32**, 73–96.

Couwenbergh, H. A. M. (1980) Stochastic Games with Metric State Spaces, *International Journal of Game Theory*, **9**, 25–36.

Duffie, D., Geanakoplos, J., Mas-Colell, A., and McLennan, A. (1994) Stationary Markov Equilibria, *Econometrica*, **62**(4), 745–781.

Eibelshäuser, S. and Poensgen, D. (2019) *Markov Quantal Response Equilibrium and a Homotopy Method for Computing and Selecting Markov Perfect Equilibria of Dynamic Stochastic Games*, Preprint.

Eilenberg, S. and Montgomery, D. (1946) Fixed Point Theorems for Multi-Valued Transformations, *American Journal of Mathematics*, **68**(2), 214–222.

Etessami, K., Wojtczak, D., and Yannakakis, M. (2019) Recursive Stochastic Games with Positive Rewards, *Theoretical Computer Science*, **777**, 308–328.

Fan, K. (1952) Fixed-Point and Minimax Theorems in Locally Convex Topological Linear Spaces, *Proceedings of the National Academy of Sciences of the United States of America*, **38**(2), 121–126.

Filar, J. A. and Raghavan, T. E. (1984) A Matrix Game Solution of the Single-Controller Stochastic Game, *Mathematics of Operations Research*, **9**(3), 356–362.

Filar, J. A. and Tolwinski, B. (1991) On the Algorithm of Pollatschek and Avi-Itzhak. In Raghavan, T. E. S., Ferguson, T. S., Parthasarathy, T., and Vrieze, O. J. (eds.), *Stochastic Games and Related Topics*, Kluwer, pp. 59–70.

Filar, J. and Vrieze, K. (1997) *Competitive Markov Decision Processes*, Springer Science and Business Media.

Fink, A. M. (1964) Equilibrium in a Stochastic *n*-Person Game, *Journal of Science of the Hiroshima University*, **28**, 89–93.

Flesch, J., Schoenmakers, G., and Vrieze, K. (2008) Stochastic Games on a Product State Space, *Mathematics of Operations Research*, **33**, 403–420.

Flesch, J., Schoenmakers, G., and Vrieze, K. (2009) Stochastic Games on a Product State Space: The Periodic Case, *International Journal of Game Theory*, **38**, 263–289.

Flesch, J., Thuijsman, F., and Vrieze, K. (1996) Recursive Repeated Games with Absorbing States, *Mathematics of Operations Research*, **21**, 1016–1022.

Flesch, J., Thuijsman, F., and Vrieze, K. (1997) Cyclic Markov Equilibria in Stochastic Games, *International Journal of Game Theory*, **26**, 303–314.

Fortnow, L. and Kimmel, P. (1998) Beating a Finite Automaton in the Big Match. In Moss, L. S. (ed.), *Proceedings of the Seventh Conference on Theoretical Aspects of Rationality and Knowledge*, Morgan Kaufmann, pp. 225–234.

Friedlin, M. and Wentzell, A. (1984) *Random Perturbations of Dynamical Systems*, Springer.

Gensbittel, F. and Renault, J. (2015) The Value of Markov Chain Games with Incomplete Information on Both Sides, *Mathematics of Operations Research*, **40**(4), 820–841.

Gillette, D. (1957) Stochastic Games with Zero Stop Probabilities. In Kuhn, H. W. and Tucker, A. W. (eds.), *Contributions to the Theory of Games*, vol.3, Princeton University Press, pp. 179–187.

Gimbert, H., Renault, J., Sorin, S., Venel, X., and Zielonka, W. (2016) On Values of Repeated Games with Signals, *The Annals of Applied Probability*, **26**(1), 402–424.

Glicksberg, I. L. (1952) A Further Generalization of the Kakutani Fixed Point Theorem, with Application to Nash Equilibrium Points, *Proceedings of the American Mathematical Society*, **3**(1), 170–174.

Govindan, S. and Wilson, R. (2010) A Global Newton Method to Compute Nash Equilibria, *Journal of Economic Theory*, **110**(1), 65–86.

Guo, X. and Hernández-Lerma, O. (2005a) Nonzero-Sum Games for Continuous-Time Markov Chains with Unbounded Discounted Payoffs, *Journal of Applied Probability*, **42**, 303–320.

Guo, X. and Hernández-Lerma, O. (2005b) Zero-Sum Continuous-Time Markov Games with Unbounded Transition and Discounted Payoff Rates, *Bernoulli*, **11**(6), 1009–1029.

Hansen, K. A., Ibsen-Jensen, R., and Miltersen, P. B. (2011) The Complexity of Solving Reachability Games Using Value and Strategy Iteration. In Kulikov, A. and Vereshchagin, N. (eds.), *International Computer Science Symposium in Russia*, Springer, pp. 77–90.

Hansen, K. A., Ibsen-Jensen, R., and Neyman, A. (2018) The Big Match with a Clock and a Bit of Memory. In Tardos, E., Elkind, E., and Vohra, R. (eds.), *Proceedings of the 2018 ACM Conference on Economics and Computation*, Association for Computer Machinery, pp. 149–150.

Hansen, K. A., Ibsen-Jensen, R., and Neyman, A. (2021) Absorbing Games with a Clock and Two Bits of Memory, *Games and Economic Behavior*, **128**, 213–230.

Hansen, K. A., Koucky, M., Lauritzen, N., Miltersen, P. B. and Tsigaridas, E. P. (2011) Exact Algorithms for Solving Stochastic Games. In Fortnow, L. and Vadhan, S. (eds.), *Proceedings of the Forty-Third Annual ACM Symposium on Theory of Computing*, Association for Computer Machinery, pp. 205–214.

Harris, C., Reny, P., and Robson, A. (1995) The Existence of Subgame-Perfect Equilibrium in Continuous Games with Almost Perfect Information: A Case for Public Randomization, *Econometrica*, **63**(3), 507–544.

He, W. and Sun, Y. (2017) Stationary Markov Perfect Equilibria in Discounted Stochastic Games, *Journal of Economic Theory*, **169**, 35–61.

Heller, Y. (2012) Sequential Correlated Equilibria in Stopping Games, *Operations Research*, **60**(1), 209–224.

Herings, P. J. J. and Peeters, R. J. A. P. (2004) Stationary Equilibria in Stochastic Games: Structure, Selection, and Computation, *Journal of Economic Theory*, **118**, 32–60.

Hörner, J., Rosenberg, D., Solan, E., and Vieille, N. (2010) On a Markov Game with One-Sided Information, *Operations Research*, **58**, 1107–1115.

Horst, U. (2005) Stationary Equilibria in Discounted Stochastic Games with Weakly Interacting Players, *Games and Economic Behavior*, **51**, 83–108.

Jaśkiewicz, A. and Nowak, A. S. (2005) Nonzero-Sum Semi-Markov Games with the Expected Average Payoffs, *Mathematical Methods of Operations Research*, **62**, 23–40.

Jaśkiewicz, A. and Nowak, A. S. (2006) Zero-Sum Ergodic Stochastic Games with Feller Transition Probabilities, *SIAM Journal on Control and Optimization*, **45**, 773–789.

Jaśkiewicz, A. and Nowak, A. S. (2011) Stochastic Games with Unbounded Payoffs: Applications to Robust Control in Economics, *Dynamic Games and Their Applications*, **1**, 253–279.

Jaśkiewicz, A. and Nowak, A. S. (2018a) Zero-Sum Stochastic Games. In Başar, T. and Zaccour, G. (eds.), *Handbook of Dynamic Game Theory*, vol. 1, Springer, pp. 1–65.

Jaśkiewicz, A. and Nowak, A. S. (2018b) Non-Zero-Sum Stochastic Games. In Başar, T. and Zaccour, G. (eds.) *Handbook of Dynamic Game Theory*, vol. 1, Springer, pp. 281–344.

Jasso-Fuentes, H. (2005) Noncooperative Continuous-Time Markov Games, *Morfismos*, **9**, 39–54.

Jovanovic, B. and Rosenthal, R. W. (1988) Anonymous Sequential Games, *Journal of Mathematical Economics*, **17**(1), 77–87.

Jurdziński, M., Paterson, M., and Zwick, U. (2008) A Deterministic Subexponential Algorithm for Solving Parity Games, *SIAM Journal on Computing*, **38**(4), 1519–1532.

Kakutani, S. (1941) A Generalization of Brouwer's Fixed Point Theorem, *Duke Mathematical Journal*, **8**(3), 457–459.

Khan, M. A. and Sun, Y. (2002) Non-Cooperative Games with Many Players. In Aumann, R. and Hart, S. (eds.), *Handbook of Game Theory with Economic Applications*, vol. 3, North Holland, pp. 1761–1808.

Kocel-Cynk, B., Pawłucki, W., and Valette, A. (2014) A Short Geometric Proof that Hausdorff Limits Are Definable in any O-minimal Structure, *Advances in Geometry*, **14**(1), 49–58.

Kohlberg, E. (1974) Repeated Games with Absorbing States, *The Annals of Statistics*, **2**(4), 724–738.

Korevaar, J. (2004) *Tauberian Theory: A Century of Developments*, Springer.

Kumar, P. R. and Shiau, T. H. (1981) Existence of Value and Randomized Strategies in Zero-Sum Discrete Time Stochastic Dynamic Games, *SIAM Journal on Control and Optimization*, **19**, 617–634.

Laraki, R. and Sorin, S. (2015) Advances in Zero-Sum Dynamic Games. In Aumann R. J. and Hart S. (eds.), *Handbook of Game Theory with Economic Applications*, vol. 4, Elsevier, pp. 27–94.

Lehrer, E. and Monderer, D. (1994) Discounting versus Averaging in Dynamic Programming, *Games and Economic Behavior*, **6**, 97–113.

Lehrer, E., Solan, E., and Solan, O. N. (2016) The Value Functions of Markov Decision Processes, *Operations Research Letters*, **44**, 587–591.

Lehrer, E. and Sorin, S. (1992) A Uniform Tauberian Theorem in Dynamic Programming, *Mathematics of Operations Research*, **17**, 303–307.

Levy, Y. (2013a) Discounted Stochastic Games with No Stationary Nash Equilibrium: Two Examples, *Econometrica*, **81**(5), 1973–2007.

Levy, Y. (2013b) Continuous-Time Stochastic Games of Fixed Duration, *Dynamic Games and Applications*, **3**(2), 279–312.

Levy, Y. J. and McLennan, A. (2015) Corrigendum to "Discounted Stochastic Games with No Stationary Nash Equilibrium: Two Examples," *Econometrica*, **83**(3), 1237–1252.

Liggett, T. M. and Lippman, S. A. (1969) Stochastic Games with Perfect Information and Time Average Payoff, *SIAM Review*, **11**, 604–607.

Maitra, A. and Parthasarathy, T. (1970) On Stochastic Games, *Journal of Optimization Theory and Applications*, **5**, 289–300.

Maitra, A. and Sudderth, W. (1993) Borel Stochastic Games with Lim sup Payoff, *The Annals of Probability*, **21**, 861–885.

Maitra, A. and Sudderth, W. (1998) Finitely Additive Stochastic Games with Borel Measurable Payoffs, *International Journal of Game Theory*, **27**(2), 257–267.

Maitra, A. and Sudderth, W. (2012) *Discrete Gambling and Stochastic Games*, Springer Science and Business Media.

Maschler, M. (1967) The Inspector's Non-Constant-Sum Game: Its Dependence on a System of Detectors, *Naval Research Logistics Quarterly*, **14**(3), 275–290.

Maschler, M., Solan, E., and Zamir, S. (2020) *Game Theory*, Cambridge University Press.

Mashiah-Yaakovi, A. (2014) Subgame Perfect Equilibria in Stopping Games, *International Journal of Game Theory*, **43**(1), 89–135.

Mertens, J. F. (2002) Stochastic Games. In Aumann, R. J. and Hart, S. (eds.), *Handbook of Game Theory with Economic Applications*, vol. 3, Elsevier, pp. 1809–1832.

Mertens, J. F. and Neyman, A. (1981) Stochastic Games, *International Journal of Game Theory*, **10**, 53–66.

Mertens, J. F. and Parthasarathy, T. (1987) Equilibria for Discounted Stochastic Games, CORE Discussion Paper No. 8750. Also published in Neyman, A. and Sorin, S. (eds.), *Stochastic Games and Applications,* NATO Science Series, Kluwer, pp. 131–172.

Mertens, J. F., Sorin, S., and Zamir, S. (2015) *Repeated Games*, Cambridge University Press.

Monderer, D. and Sorin, S. (1993) Asymptotic Properties in Dynamic Programming, *International Journal of Game Theory*, **22**, 1–11.

Nash, J. F. (1950) Equilibrium points in n-person games, *Proceedings of the National Academy of Sciences of the United States of America*, **36**(1), 48–49.

Neumann, J. V. (1928) Zur theorie der gesellschaftsspiele. *Mathematische Annalen*, **100**(1), 295–320.

Neyman, A. (2013) Stochastic Games with Short-Stage Duration, *Dynamic Games and Applications*, **3**, 236–278.

Neyman, A. (2017) Continuous-Time Stochastic Games, *Games and Economic Behavior*, **104**, 92–130.

Neyman, A. and Sorin, S. (eds.) (2003) *Stochastic Games and Applications*, Springer Science and Business Media.

Nowak, A. S. (1985a) Existence of Equilibrium Stationary Strategies in Discounted Noncooperative Stochastic Games with Uncountable State Space, *Journal of Optimization Theory and Applications*, **45**, 591–620.

Nowak, A. S. (1985b) Universally Measurable Strategies in Zero-Sum Stochastic Games, *Annals of Probability*, **13**(1), 269–287.

Nowak, A. S. (1986) Semicontinuous Nonstationary Stochastic Games, *Journal of Mathematical Analysis and Applications*, **117**, 84–99.

Nowak, A. S. (2003a) Zero-Sum Stochastic Games with Borel State Space. In Neyman, A. and Sorin, S. (eds.), *Stochastic Games and Applications*, NATO Science Series, Kluwer, pp. 77–91.

Nowak, A. S. (2003b) *N*-Person Stochastic Games: Extensions of the Finite State Space Case and Correlation. In Neyman, A. and Sorin, S. (eds.), *Stochastic Games and Applications*, NATO Science Series, Kluwer, pp. 93–106.

Nowak, A. S. (2003c) On a New Class of Nonzero-Sum Discounted Stochastic Games Having Stationary Nash Equilibrium Points, *International Journal of Game Theory*, **32**, 121–132.

Nowak, A. S., and Raghavan, T. E. S. (1992) Existence of Stationary Correlated Equilibria with Symmetric Information for Discounted Stochastic Games, *Mathematics of Operations Research*, **17**(3), 519–526.

Nowak, A. S. and Raghavan, T. E. S. (1993) A Finite Step Algorithm via a Bimatrix Game to a Single Controller Non-Zero Sum Stochastic Game, *Mathematical Programming*, **59**(1–3), 249–259.

Oliu-Barton, M. (2014) The Asymptotic Value in Finite Stochastic Games, *Mathematics of Operations Research*, **39**(3), 712–721.

Oliu-Barton, M. (2020) New Algorithms for Solving Zero-Sum Stochastic Games, *Mathematics of Operations Research*, **46**(1), 255–267.

Parthasarathy, T. and Raghavan, T. E. S. (1981) An Orderfield Property for Stochastic Games when One Player Controls Transition Probabilities, *Journal of Optimization Theory and Applications*, **33**, 375–392.

Parthasarathy, T. and Sinha, S. (1989) Existence of Stationary Equilibrium Strategies in Non-Zero Sum Discounted Stochastic Games with Uncountable State Space and State-Independent Transitions, *International Journal of Game Theory*, **18**(2), 189–194.

Peleg, B. (1969) Equilibrium Points for Games with Infinitely Many Players, *Journal of the London Mathematical Society*, **1–44**, 292–294.

Puterman, M. L. (1994) *Markov Decision Processes: Discrete Stochastic Dynamic Programming*, Wiley.

Raghavan, T. E. S. (2003) Finite-Step Algorithms for Single-Controller and Perfect Information Stochastic Games. In Neyman, A. and Sorin, S. (eds.), *Stochastic Games and Applications*, Kluwer, pp. 227–251.

Raghavan, T. E. S., Ferguson, T. S., and Parthasarathy, T. (1991) *Stochastic Games and Related Topics: In Honor of Professor LS Shapley*, Theory and Decision Library, Series, C, Game Theory. Mathematical Programming and Operations Research, vol. 7, Kluwer.

Raghavan, T. E. S. and Filar, J. A. (1991) Algorithms for Stochastic Games – A Survey, *ZOR – Methods and Models of Operations Research*, **35**, 437–472.

Raghavan, T. E. S. and Syed, Z. (2002) Computing Stationary Nash Equilibria of Undiscounted Single-Controller Stochastic Games, *Mathematics of Operations Research*, **27**(2), 384–400.

Raghavan, T. E. S. and Syed, Z. (2003) A Policy-Improvement Type Algorithm for Solving Zero-Sum Two-Person Stochastic Games of Perfect Information, *Mathematical Programming, Series. A*, **95**, 513–532.

Ramsey, F. P. (1930) On a Problem of Formal Logic, *Proceedings of the London Mathematical Society*, **30**, 264–286.

Renault, J. (2006) The Value of Markov Chain Games with Lack of Information on One Side, *Mathematics of Operations Research*, **31**, 490–512.

Renault, J. (2011) Uniform Value in Dynamic Programming, *Journal of the European Mathematical Society*, **13**, 309–330.

Renault, J. (2012) The Value of Repeated Games with an Informed Controller, *Mathematics of Operations Research*, **37**(1), 154–179.

Renault, J. (2014) General Limit Value in Dynamic Programming, *Journal of Dynamics and Games*, **1**(3), 471–484.

Renault, J. (2019) A Tutorial on Zero-Sum Stochastic Games, arXiv:1905.06577.

Renault, J. and Venel, X. (2017) Long-Term Values in Markov Decision Processes and Repeated Games, and a New Distance for Probability Spaces, *Mathematics of Operations Research*, **42**(2), 349–376.

Renault, J. and Ziliotto, B. (2020a) Limit Equilibrium Payoffs in Stochastic Games, *Mathematics of Operations Research*, **45**(3), 889–895.

Renault, J. and Ziliotto, B. (2020b) Hidden Stochastic Games and Limit Equilibrium Payoffs, *Games and Economic Behavior*, **124**, 122–139.

Rosenberg, D., Solan, E., and Vieille, N. (2002) Blackwell Optimality in Markov Decision Processes with Partial Observation, *The Annals of Statistics*, **30**, 1178–1193.

Rosenberg, D., Solan, E., and Vieille, N. (2003) The MaxMin of Stochastic Games with Imperfect Monitoring, *International Journal of Game Theory*, **32**, 133–150.

Rosenberg, D., Solan, E., and Vieille, N. (2004) Stochastic Games with a Single Controller and Incomplete Information, *SIAM Journal on Control and Optimization*, **43**, 86–110.

Rosenberg, D., Solan, E., and Vieille, N. (2009) Protocol with No Acknowledgement, *Operations Research*, **57**, 905–915.

Ross, S. M. (1982) *Introduction to Stochastic Dynamic Programming*, Academic Press.

Shapley, L. S. (1953) Stochastic Games, *Proceedings of the National Academy of Sciences of the United States of America*, **39**, 1095–1100.

Shiryaev, A. N. (1995) *Probability*, Springer.

Shmaya, E. and Solan, E. (2004) Two-Player NonZero-Sum Stopping Games in Discrete Time, *The Annals of Probability*, **32**(3B), 2733–2764.

Simon, R. S. (2006) Value and Perfection in Stochastic Games, *Israel Journal of Mathematics*, **156**(1), 285–309.

Simon, R. S. (2007) The Structure of Non-Zero-Sum Stochastic Games, *Advances in Applied Mathematics*, **38**, 1–26.

Simon, R. S. (2012) A Topological Approach to Quitting Games, *Mathematics of Operations Research*, **37**, 180–195.

Simon, R. S. (2016) The Challenge of Non-Zero-Sum Stochastic Games, *International Journal of Game Theory*, **45**(1–2), 191–204.

Solan, E. (1998) Discounted Stochastic Games, *Mathematics of Operations Research*, **23**, 1010–1021.

Solan, E. (1999) Three-Player Absorbing Games, *Mathematics of Operations Research*, **24**(3), 669–698.

Solan, E. (2003) Continuity of the Value in Competitive Markov Decision Processes, *Journal of Theoretical Probability*, **16**, 831–845.

Solan, E. (2008) Stochastic Games. In Liu, L. and Tamer Özsu, M. (eds.), *Encyclopedia of Database Systems*, Springer.

Solan, E. and Solan, O. N. (2020) Quitting Games and Linear Complementarity Problems, *Mathematics of Operations Research*, **45**(2), 434–454.

Solan, E. and Solan, O. N. (2021) Sunspot Equilibrium in Positive Recursive General Quitting Games. *International Journal of Game Theory*, **50**, 1–19.

Solan, E., Solan, O. N., and Solan, R. (2020) Jointly Controlled Lotteries with Biased Coins, *Games and Economic Behavior*, **119**(2020), 383–391.

Solan, E. and Vieille, N. (2001) Quitting Games, *Mathematics of Operations Research*, **26**, 265–285.

Solan, E. and Vieille, N. (2002) Correlated Equilibrium in Stochastic Games, *Games and Economic Behavior*, **38**, 362–399.

Solan, E. and Vieille, N. (2010) Computing Uniform Optimal Strategies in Two-Player Stochastic Games, *Economic Theory*, **42**, 237–253.

Solan, E. and Vieille, N. (2015) Stochastic Games: A Perspective, *Proceedings of the National Academy of Sciences of the United States of America*, **112**(45), 13743–13746.

Solan, E. and Vohra, R. (2001) Correlated Equilibrium in Quitting Games, *Mathematics of Operations Research*, **26**, 601–610.

Solan, E. and Vohra, R. (2002) Correlated Equilibrium Payoffs and Public Signalling in Absorbing Games, *International Journal of Game Theory*, **31**, 91–121.

Solan, E. and Ziliotto, B. (2016) Stochastic Games with Signals. In Thuijsman, F. and Wagener, F. (eds.), *Advances in Dynamic and Evolutionary Games*, Birkhäuser, pp. 77–94.

Sorin, S. (1986) Asymptotic Properties of a Non-Zerosum Stochastic Games, *International Journal of Game Theory*, **15**, 101–107.

Sorin, S. (2002) *A First Course on Zero-Sum Repeated Games, Mathématiques and Applications*, vol. 37, Springer-Verlag.

Sorin, S., Venel, X., and Vigeral, G. (2010) Asymptotic Properties of Optimal Trajectories in Dynamic Programming, *Sankhya: The Indian Journal of Statistics A*, **72**(1), 237–245.

Sorin, S. and Vigeral, G. (2015) Reversibility and Oscillations in Zero-Sum Discounted Stochastic Games, *Journal of Dynamic Games*, **2**(1), 103–115.

Szczechla, W. W., Connell, S. A., Filar, J. A., and Vrieze, O. J. (1997) On the Puiseux Series Expansion of the Limit Discount Equation of Stochastic Games, *SIAM Journal on Control and Optimization*, **35**(3), 860–875.

Takahashi, M. (1964) Stochastic Games with Infinitely Many Strategies, *Journal of Science of the Hiroshima University Series A-I*, **26**, 123–134.

Thuijsman, F. and Raghavan, T. E. S. (1997) Perfect Information Stochastic Games and Related Classes, *International Journal of Game Theory*, **26**(3), 403–408.

Thuijsman, F. and Vrieze, O. J. (1991) Easy Initial States in Stochastic Games. In Raghavan T. E. S., Ferguson, T. S., Parthasarathy, T., and Vrieze, O. J. (eds.), *Stochastic Games and Related Topics*, Kluwer, pp. 85–100.

Venel, X. and Ziliotto, B. (2016) Strong Uniform Value in Gambling Houses and Partially Observable Markov Decision Processes, *SIAM Journal on Control and Optimization*, **54**(4), 1983–2008.

Vieille, N. (2000a) Equilibrium in 2-Person Stochastic Games I: A Reduction, *Israel Journal of Mathematics*, **119**, 55–91.

Vieille, N. (2000b) Equilibrium in 2-Person Stochastic Games I: The Case of Recursive Games, *Israel Journal of Mathematics*, **119**, 93–126.

Vieille, N. (2000c) Large Deviations and Stochastic Games, *Israel Journal of Mathematics*, **119**, 127–144.

Vieille, N. (2000d) Solvable States in n-Player Stochastic Games, *SIAM Journal on Control and Optimization*, **38**(6), 1794–1804.

Vieille, N. (2002) Stochastic Games: Recent Results. In Aumann, R. J. and Hart, S. (eds.), *Handbook of Game Theory with Economic Applications*, vol. 3, Elsevier, pp. 1833–1850.

Vigeral, G. (2013) A Zero-Sum Stochastic Game with Compact Action Sets and No Asymptotic Value, *Dynamic Games and Their Applications*, **3**, 172–186.

Vrieze, O. J. and Thuijsman, F. (1989) On Equilibria in Repeated Games with Absorbing States, *International Journal of Game Theory*, **18**, 293–310.

Vrieze, O. J., Tijs, S. H., Raghavan, T. E. S., and Filar, J. A. (1983) A Finite Algorithm for the Switching Control Stochastic Game, *Operations-Research-Spektrum*, **5**(1), 15–24.

Wei, Q. and Chen, X. (2016) Stochastic Games for Continuous-Time Jump Processes under Finite-Horizon Payoff Criterion, *Applied Mathematics and Optimization*, **74**, 273–301.

Zachrisson, L. E. (1964) Markov Games. In Auslander, L. A. and Aumann, R. J. (eds.), *Advances in Game Theory*, Princeton University Press, pp. 211–253.

Zhang, W. (2018) Continuous-Time Constrained Stochastic Games under the Discounted Cost Criteria, *Applied Mathematics and Optimization*, **77**, 275–296.

Ziliotto, B. (2016a) A Tauberian Theorem for Nonexpansive Operators and Applications to Zero-Sum Stochastic Games, *Mathematics of Operations Research*, **41**(4), 1522–1534.

Ziliotto, B. (2016b) General Limit Value in Zero-Sum Stochastic Games, *International Journal of Game Theory*, **45**(1–2), 353–374.

Ziliotto, B. (2016c) Zero-Sum Repeated Games: Counterexamples to the Existence of the Asymptotic Value and the Conjecture maxmin $= \lim v_n$, *The Annals of Probability*, **44**(2), 1107–1133.

Ziliotto, B. (2018) Tauberian Theorems for General Iterations of Operators: Applications to Zero-Sum Stochastic Games, *Games and Economic Behavior*, **108**, 486–503.

Zwick, U. and Paterson, M. (1996) The Complexity of Mean Payoff Games on Graphs, *Theoretical Computer Science*, **158**(1–2), 343–359.

Index

Printed in the United States
by Baker & Taylor Publisher Services